TURBULENT MIXING
IN NONREACTIVE
AND REACTIVE FLOWS

A PROJECT SQUID WORKSHOP

TURBULENT MIXING IN NONREACTIVE AND REACTIVE FLOWS

Edited by

S. N. B. Murthy

Project SQUID Headquarters
Thermal Sciences and Propulsion Center
School of Mechanical Engineering
Purdue University
West Lafayette, Indiana

PLENUM PRESS • NEW YORK AND LONDON

Library of Congress Cataloging in Publication Data

Project Squid Workshop on Turbulent Mixing in Nonreactive and Reactive Flows,
  Purdue University, 1974.
  Turbulent mixing in nonreactive and reactive flows.

  At head of title: A Project Squid workshop.
  Workshop sponsored by Project Squid and the Air Force Office of Scientific Research.
  Includes bibliographies.
  .1. Fluid dynamics—Congresses. 2. Mixing—Congresses. 3. Turbulence—Congresses. I.
Murthy, S. N. B. II. Project Squid. III. United States. Air Office of Scientific Re-
search. IV. Title.
QA929.P7   532'.0527   75-22329

Proceedings of the Project SQUID WORKSHOP ON TURBULENT
MIXING IN NONREACTIVE AND REACTIVE FLOWS, held at
Purdue University, West Lafayette, Indiana, May 20-21, 1974,
and sponsored by the Office of Naval Research (Project SQUID),
and the Air Force Office of Scientific Research.

©1975 Plenum Press. New York
Softcover reprint of the hardcover 1st edition 1975
A Division of Plenum Publishing Corporation
227 West 17th Street, New York, N.Y. 10011
United Kingdom edition published by Plenum Press, London
A Division of Plenum Publishing Company, Ltd.
Davis House (4th Floor), 8 Scrubs Lane, Harlesden, London, NW10 6SE, England

This work relates to the Department of the Navy Contract
No. N00014-67-A-0226-0005 issued by the Office of Naval Research under
contract authority No. NR 098-038. However, the content does not necessarily
reflect the position or policy of the Department of the Navy or the Government,
and no official endorsement should be inferred. The United States Government
has a royalty-free, nonexclusive and irrevocable license throughout the world
for Government purposes to publish, translate, reproduce, deliver, perform,
dispose of, and to authorize others to do so, all or any portion of this work.

ISBN 978-1-4615-8740-8   ISBN 978-1-4615-8738-5 (eBook)
DOI 10.1007/978-1-4615-8738-5

Preface

Turbulence, mixing and the mutual interaction of turbulence and chemistry continue to remain perplexing and impregnable in the frontiers of fluid mechanics. The past ten years have brought enormous advances in computers and computational techniques on the one hand and in measurements and data processing on the other. The impact of such capabilities has led to a revolution both in the understanding of the structure of turbulence as well as in the predictive methods for application in technology.

The early ideas on turbulence being an array of complicated phenomena and having some form of reasonably strong coherent structure have become well substantiated in recent experimental work. We are still at the very beginning of understanding all of the aspects of such coherence and of the possibilities of incorporating such structure into the analytical models for even those cases where the thin shear layer approximation may be valid. Nevertheless a distinguished body of "eddy chasers" has come into existence.

The structure of mixing layers which has been studied for some years in terms of correlations and spectral analysis is also getting better understood. Both probability concepts such as intermittency and conditional sampling as well as the concept of large scale structure and the associated strain seem to indicate possibilities of distinguishing and synthesizing 'engulfment' and molecular mixing. At the same time, sufficient data are not available at high Reynolds and Mach numbers and thus the effects of compressibility (arising from different causes), for example, can be understood only when the effects of dilatation on strain become established on firm grounds.

The modelling of turbulent mixing has progressed beyond the possibilities at the time of the 1968 Stanford Conference. Many problems of practical importance can now be modelled for numerical solution. This effort of describing practical situations with the minimum amount of unambiguous input should be watched with great interest.

The mutual interaction of turbulence and chemistry is an extremely difficult problem. Progress in measurements is the key to advances in this field.

This volume presents the current status of several facets of the problem of turbulent mixing as viewed by a distinguished body of investigators. The papers are the outcome of a workshop type conference on turbulent mixing organized by Project SQUID for the Office of Naval Research and co-sponsored by the Office of Scientific Research, U. S. Air Force. In addition to the papers presented at the workshop and the discussion thereon, a paper (also invited to the workshop) outlining some aspects of the 1974 Southampton Colloquium on Coherent Structures is included. In order to provide a background to the discussions at the meeting, a review was written to examine in some depth the status of the subject and it is presented here as a position paper on turbulent mixing.

There are reasonable grounds to believe that exciting developments are being made in the understanding and calculation of turbulent mixing processes with at least a modest accounting of molecular level chemical processes. It is hoped that the volume will display some of the excitement which the investigators are finding continuously and will provide also a number of entry points into the developments in this field of great scientific and technological interest.

Contents

Introduction

The attainment of homogeneity at the molecular level is obviously the ultimate measure of the effectiveness of mixing. When turbulence is present in the fluid flows involved in mixing, the entrainment or "engulfment" of the fluid and the final process of molecular mixing should be understood individually. Chemical action, which is essentially a molecular level process, can only occur where the fluids are molecularly mixed.

The progress in the modelling and computation of turbulent mixing processes deserves continuous attention. While the primary concern in modelling is obviously to introduce the minimum number of assumptions and to use the most unambiguous input information, it is clear that a satisfactory synthesis of models and flows (jets, wakes and shear layers) can only be based upon a proper understanding of the structure of turbulence in turbulent mixing.

Interesting developments are occurring in the measurement, identification and analysis of turbulence structure. Coherence in turbulence and the presence of large scale and strong eddy structure in mixing layers are beginning to be well accepted. Some hope is being expressed that it may in fact be possible to utilize the large scale eddy structure information in modelling, although there are still basic questions in regard to the manner in which such structure arises and sustains itself. There are also many aspects of the presence of coherent structure that are not understood at high Reynolds and Mach numbers.

In many practical flows, the Reynolds, Mach and Schmidt numbers are of fundamental significance. In problems involving energy transfer, the Prandtl number needs to be properly defined. Where chemical action occurs, a method of incorporating the turbulent stress-strain information into the calculation of chemical activity (reaction rate, enthalpy generation, product formation) is of vital importance. In modelling flows with chemical action, it is also useful to establish the relation between incompressible, inert mixing process and compressible, chemically active processes.

In view of the great importance of the problem of turbulent mixing in many aspects of chemical propulsion and chemical lasers, a Workshop was organized in May 1974 on Turbulent Mixing in Non-Reactive and Reactive Flows. The primary objective of the conference was to bring together those who are actively engaged in researches in this field and to generate an understanding of the various points of view regarding further developments both in regard to phenomenology and calculations.

This volume is devoted to the proceedings of the Workshop. It also includes a paper which was to have been presented at the Workshop on coherent structures in turbulence based to some extent on the 1974 Southampton Colloquium on Coherent Structures in Turbulence. A position paper on turbulent mixing which is intended to review the researches in this field is also presented here.

* * *

One way of introducing this volume is to outline the opening session of the Workshop.

S. N. B. MURTHY

We have very great pleasure--Dr. Goulard and I--in welcoming you to this Workshop on Turbulent Mixing.

The Workshop is sponsored by the Office of Naval Research and the Air Force Office of Scientific Research through Project SQUID at Purdue University. Purdue University and the School of Mechanical Engineering, in which Project SQUID is located, have always been very generous in their support of Project SQUID activities.

I wish to say a few words about the Workshop program.

The Workshop is concerned with mixing in free turbulent shear flows.

Broadly, we discuss today various approaches to the <u>calculation</u> of turbulent mixing flows--the associated experiments, models and computer studies--all of them characterized by relatively high Reynolds numbers.

It is clear, there is no obviously satisfactory way of modeling the mixing of even simple shear flows and obtaining, for example, a parameter such as the spreading rate taking account of the transverse pressure gradient. If we consider multiple-equation models or those involving directly the Reynolds stresses, it is necessary to establish that they are not only rational but without inconsistencies in regard to accepted fluid mechanical laws. A general theory for mixing all types of mixing layers, jets and wakes, with appreciable free stream turbulence, is obviously some distance away, since quite small variations in the flow conditions--initial or boundary conditions--can lead to large complexities.

In many practical mixing studies, transport and chemical action become quite important. Scalar quantity behavior is therefore of special interest in mixing studies. Scalar quantities are also of significance in that they can be used as tracers or indicators in mixing experiments. There have been advances in establishing the conditions under which the chemical or at least the inert scalar field can be superimposed on the turbulent field but many basic questions remain.

High speed, variable density flows with enthalpy generation and multiple species present compressibility effects which require to be distinguished from one another in various flows. The usual methods of undertaking this based on similarity or transformation obviously require more detailed information regarding the relations between the production terms, the turbulence diffusion terms and the pressure diffusion terms.

Tomorrow, we discuss the structure of the mixing process. There is increasing evidence for the existence of some form of coherence in regard to the large scale structure of turbulence during mixing. It appears that in principle this should lead to a unified point of view in mixing problems. However, such observations have also raised some questions, for example:

(i)   What is the interconnecting mechanism between the large and small scale processes in mixing and such conventional parameters as Reynolds stress, entrainment, Prandtl and Schmidt numbers and enthalpy release in a turbulent flame.

(ii)   What is the nature of measurements--space-time correlations, spectral analysis, pdf and intermittency data--which are required to understand the structure of turbulence in mixing.

In very general terms, the three aspects of experiments which require further clarification are experimental configurations, measurement techniques and their accuracy, and data processing, for example through conditional sampling and pattern recognition. When considering experimental configurations, we should emphasize not only critical experiments but also those which are clearly definable and therefore become part of a universal set which can be studied from differing points of view by different investigators.

In analysis, the great ideas of instability and wave-like structure, vortical structure and the associated strain rate and statistical continuum theories based on functional representation have each helped some advances in understanding the strong interactions between different scales of motion in turbulence. The manner in which such ideas can be applied to nonhomogeneous turbulent shear flows and mixing problems continues to be an open problem.

It is the expectation in all these workshops that several things can be achieved--centered 'round a free exchange of ideas. I would merely express the hope that the problem of turbulent mixing will not remain exactly where it is now after the two days of discussion.

I have now the greatest pleasure to present to you Dr. W. B. Cottingham, Head of the School of Mechanical Engineering, who is here to speak to us on behalf of the University, and Dr. R. Roberts, Director of the Power Program, Office of Naval Research.

W. B. COTTINGHAM

The program indicates that Dr. Arthur Hansen, the President of the University, is going to give the introduction this morning. I am not Dr. Arthur Hansen--nor do I want to be. Dr. Hansen sends his regrets that he could not be with you today to welcome you to Purdue University and to indicate to you the important role that he thinks this workshop plays in the technology of this country. He called me, actually, about the middle of the week--this past week-- and said, "I cannot be there, would you fill in?", and since he pays my check, I said, "Yes, sir."

We do welcome you to Purdue, though, and those of you for whom this represents your first visit to this campus, I think you will probably agree with me that it is somewhat of a phenomena in the middle of the Indiana cornfields. We are very proud of this institution, the engineering, science, technology and agriculture that it represents.

In looking at the program today, I am very impressed by the list of guest speakers that are on that program. Professor Murthy should be congratulated on putting together this workshop. I cannot recall in my twenty years' association with this university, any meeting or workshop in which such an impressive collection of individuals has been at one meeting. In particular, I want to welcome our guests from overseas. Thank you for being here. I am sure you found the trip from Chicago to Lafayette much more hazardous than the trip from Paris or London, or wherever, to Chicago.

I know absolutely nothing about turbulent mixing except what I see in my coffee cup when I add the cream to it through bleary eyes in the morning. I hope to be elucidated during the course of the day and perhaps on leaving today and tomorrow, we will all know a great deal more about turbulent mixing.

On behalf of the School of Mechanical Engineering, Project SQUID, Office of Naval Research, Air Force Office of Scientific Research, we welcome you to the conference and hope it brings a great deal of benefit to you.

R. ROBERTS

I wish to welcome you here on behalf of the Office of Naval Research and the Office of Scientific Research of the U. S. Air Force. This is one of a series of workshops which we have been sponsoring--sometimes with the cooperation of the Office of Scientific Research and other agencies or by ourselves--on selected aspects of various subjects related to problems in the field of chemical propulsion. It is not necessary to attempt to explain to this group the importance of the subject of this workshop, turbulent mixing with and without chemical reaction, to this and other areas of technology.

The objective of these workshops for the sponsors is to learn about the current state of knowledge, the major problem areas and the possibilities--experimental and theoretical--for advancing our understanding of the phenomena involved. We also hope that it will assist the research and development community in its assessment of current knowledge and influence their future research and engineering activities. Thus, we are looking forward to the results of this discussion as guidance to the future research programs of the sponsors and to the activities of the research and development community carrying out these and related efforts.

The sponsors wish to thank the organizer of the symposium, Professor S. N. B. Murthy, and the Director of Project SQUID, Professor R. Goulard, for bringing together this outstanding group of participants, and Purdue University for the excellent facilities and outstanding hospitality. We also wish to thank the participants, especially the speakers and the members of the panel for contributing their knowledge and time, essential to the success of the Workshop.

*   *   *

It is a great pleasure to acknowledge here the continuous assistance and encouragement received from Dr. Ralph Roberts and Mr. James R. Patton, Jr., Power Program, Office of Naval Research, in the organization of the Workshop. Their interest in the creation of Forums for assessing the direction of scientific and technological progress is well known. Dr. B. T. Wolfson co-ordinated the AFOSR participation in the Workshop and we are very appreciative of his support and guidance.

Lastly, the manuscript copy was typed by Mrs. Barbara Pound and the illustrations were prepared by Mr. Wayne Anderson. This book should be testimony to their elegant workmanship. We wish to thank them for their efforts.

S. N. B. Murthy

TURBULENT MIXING IN NON-REACTIVE AND REACTIVE FLOWS:

A REVIEW

S. N. B. Murthy

School of Mechanical Engineering

Purdue University, West Lafayette, Indiana

ABSTRACT

A short review is presented on turbulent mixing of shear flows involving ordinary and chemically reactive fluids. The two aspects of advances in this area are (i) modelling of mixing layers (including statistical continuum theories) and (ii) elucidation of the structure of mixing processes (especially the strong, coherent large scale structure). The occurrence of chemical reaction presents many complications in regard to both of those and it appears correlations and spectral analysis are insufficient for a detailed understanding of the processes involved. Nevertheless progress is being made both in experimentation and theory. Attention is focussed on some of the current ambiguities and controversies.

CONTENTS

1. OBJECTIVES OF THE REVIEW

The objectives of this brief review on turbulent mixing are:

1. to provide a summary statement on the current status
    of knowledge in simple free turbulent mixing;

2. to discuss some of the recent approaches to the problem
    of nonhomogeneous mixing involving multiple reactive
    species and heat; and

3. to set the background for a discussion of the manner in
    which some of the more fruitful approaches to the
    problem of turbulent mixing may be pursued.

What will be said in this review will undoubtedly not cover all
aspects of this problem; we shall generally restrict attention to
high Reynolds number flows away from physical boundaries. The
ultimate emphasis here will be on reactive mixing of turbulent
shear flows.

1.1 General Observations

The problem of reactive mixing of turbulent shear flows is of some technological importance, it is well known, in jet propulsion, combustion, chemical lasers and environmental studies. The direct contribution of the technological importance of the subject to the development of turbulent mixing theory has been in the evolution of a number of methods of making ad hoc calculations of the mean flow properties. The improvements in predictive methods, based on heuristic reasoning, have, of course, been encouraged by advances in computer technology and computational mathematics. One by-product of the extensive use of computers is the realization that a general-ized model is not as effective as a specialized model in many situa-tions.

One may have wished that the technological importance of the subject may have also led to extensive measurements of the relevant flow fields. Unfortunately, there is no agreement in sight on any of the following questions:

what parameters are of interest in mixing;

what should be measured to aid in the determination
    of those parameters;

what is the accuracy and applicability of the possible
    measurement and sampling techniques; and, finally

what flow configurations best permit a unified picture
    of turbulent mixing to evolve.

The current situation is such that even for the more simple types of engineering calculations, the turbulent information has quite often been generated by trial and error. We are still in need of reliable data and theory, for example, for the Prandtl and Schmidt numbers in turbulent flows and for the enthalpy pro-duction in turbulent flames.

It may be disputable that the evolution of theoretical models has progressed slightly ahead in this field compared to measure-ments, but it is difficult to counteract the argument that the postulated phenomenology has invariably tended to dictate the measurements. On the other hand, there can be no rational experi-ments without a point of view. One may therefore hope that two or three definitive flow configurations will be selected for intensive study from several points of view and adopting different measurement techniques. This is the first question in measurements--for example, the possibility of setting up a mixing flow configuration with definable non-equilibrium characteristics and a flow configuration

with reactants suitable for dilution and reaction control, in each
case with properly specified initial conditions. The measurement
techniques have improved steadily over the years, as well as the
ability to extract complicated statistical properties and conditional probabilities through analogue and digital processing of
data. The method of combining hot-wire anemometry, laser-Doppler
velocimetry or holography and Raman-laser spectroscopy in the
measurement of turbulence and associated scalar properties is the
second question. For example, a fundamental limitation of laser-
Doppler velocimetry is the Doppler ambiguity introduced by the
finiteness of the scattering volume. It is possible therefore
that it will supplement rather than displace the use of standard
anemometry. The Raman scattering technique has to be compared
with the applicability of the Rayleigh scattering technique in
given flows with respect to the scattered intensity and the difficulties of separating the stray reflections from the scattered
light. The inclusion of the necessary electronics for conditional
sampling is a third question, especially since conditional sampling
techniques may also be used to determine the influence of upstream
history on the motion of selected parts of the turbulence. Finally,
some of the physical experiments have to be supplemented by computer
experiments, for example in regard to pressure fluctuations. A
fourth question therefore is the integration of computer experiments
into certain aspects of physical experiments.

Advances in computer and computational sciences have also
raised the question of whether the flow development can be computed
directly for given initial conditions without including specific
physical models for the flow processes. In connection with the
development of mixing layers, a computer experiment has been
reported (Kodomtsev and Kostomorov, 1972) wherein the Euler equations are solved without the viscous terms for a splitter plate
(two-dimensional) flow with vortices prescribed as initial conditions. Suggestions for such calculations have been made from time
to time. An interesting question in this connection pertains to
the type of initial conditions to be prescribed.

1.2 Outline

First, we shall summarize the experimental results on turbulent mixing of shear flows. Next, we shall outline briefly
modelling of turbulent flows including scalar transport and chemical reaction. That will be followed by a discussion of some
representative approaches in statistical continuum theory.

In a later part of the review, we shall examine the question
of superposing chemical reaction on turbulent mixing. In view of
the importance of turbulent flames in this context, a short discussion on the advances in turbulent diffusion flame theory is included.

It will no doubt be felt that four aspects of the problem of turbulent mixing and transport are dealt with only indirectly, namely (i) the influence of high Mach number induced compressibility effects in relation to other causes of compressibility, (ii) the presence of large free stream turbulence and its interaction with the shear flow turbulence leading to non-self preserving flows, (iii) the interaction between combustion and turbulence, and (iv) the synthesis of jets, wakes and mixing layers. Nevertheless, it is hoped that the gradual evolution of a point of view concerning the structure of the mixing layer and the manner in which such structure can be incorporated into the analytical models will appear in perspective.

2. MIXING OF SHEAR LAYERS

It has been engineering practice to discuss the development of mixing layers in terms of the following parameters.

(a) spreading rate

(b) entrainment rate and

(c) shear stress distribution.

The influence of Reynolds number, Mach number, velocity ratio and density ratio on the spreading parameter of a mixing layer has been discussed in detail by Birch and Eggers (1972). The thickness of the mixing layer can be defined in several ways. A common definition is "the velocity profile maximum-slope thickness" which can also be interpreted as a "vorticity thickness."

$$\delta_\omega = \frac{U_1 - U_2}{(\partial U / \partial y)_{max}}$$

$$= \frac{1}{|\omega|_m} \int_{-\infty}^{\infty} |\omega| \, dy$$

since $\omega = - \partial U / \partial y$, U being the mean velocity.

The entrainment rate of fluid into a mixing layer has also been defined in various ways. One should perhaps mention the recent attempts (Brown and Roshko, 1974) at considering the non-turbulent fluid injested into the turbulent mixing layer in terms of the fluid drawn into the large scale eddies present in the mixing layer; see Section 2.3.1 for further details. These ideas are still in a very tentative stage of development. Entrainment is looked upon here as "engulfment" and molecular processes are taken into account during the mixing.

When chemical reaction is involved, it is usual to consider the following additional parameters.

(i)   reaction zone thickness and spread,

(ii) volumetric heat release, and

(iii) reactedness distribution of the mixture.

The simplest flow configuration involving mixing is the two-dimensional mixing of two perfect fluids. One can then consider a hierarchy of problems which can be included in such a flow configuration. To start with, let there be two flows, built up on either side of a flat plate of negligible thickness, which come into contact downstream of the trailing edge of the plate. In general, the problem is to describe the mixing taking place between the two streams under various conditions as follows:

(i)   Each of the flows consists of the same single-component inert-fluid and is identical to the other in all respects except that one of the streams is moving faster than the other.

(ii) In (i), the turbulence in one of the streams is slightly altered.

(iii) In (i), (a) the density or (b) the temperature is slightly disturbed.

(iv) In (i), a finite difference in turbulence is introduced between the two flows.

(v)   In (i), a finite difference in (a) the density or (b) the temperature is introduced between the two flows. The density difference can arise on account of temperature, Mach number, or molecular weight.

(vi) In (i), the two flows are replaced by chemically reactive fluids. The variables in this modified configuration are:

1. the number of species, including the inert diluents,

2. the rate of reaction,

3. the order of the reaction,

4.   the reversibility of the reaction,

5.   the proportion of the species in relation
     to stoichiometry,

6.   the thermic nature of the reaction, and

7.   the molecular diffusion characteristics
     of the different species.

In each of the flows, the nonlinear nature of the stochastic
variables and the governing equations and the coupling between the
velocity and the scalar fields where they are present calls for an
evaluation of many different approaches both in theory and in
experiment.

One can visualize more complicated flows in practice involving,
for example, axisymmetry (and, more generally, three-dimensionality),
rotation, shock and expansion waves, transonic conditions, changes
in turbulence structure and transition. In chemically reactive
systems other complexities can arise through the coupling between
molecular interactions and turbulence.

In order to crystallize ideas, we shall describe some basic
experiments conducted on mixing layers. All of the experiments
pertain to a simple mixing layer formed between two parallel
streams or between one two-dimensional stream and a stationary
body of fluid. No chemical reaction is involved in any of the
experiments. For a given velocity ratio between the two streams,
the major interest is in the following:

(a)   the spreading parameter of the mixing layer and the
      mean velocity profile;

(b)   the structure of the mixing layer turbulence;

(c)   the turbulent shear stress distribution; and

(d)   the intermittency profiles.

2.1 Some Experimental Results

The three important aspects of measurements in turbulent
mixing and transport are the following:

1. Determination of experimental configurations which
will permit systematic variation of initial conditions and
the near field in terms of parameters that can be included
on a rational basis.

2. Determination of measurements that are of significance on the basis of commonality for different flows and
that distinguish the different flows.

3. Evaluation of measurement techniques and data
processing techniques.

It is clear that detailed flow measurements are not available
in high speed, high temperature or high concentration flows.
Measurements are urgently required at least to establish predictive
methods based upon parametric integration in such cases.

In optimizing the flow configurations, it is necessary to take
into account (a) the influence of geometry, (b) the controllability
of initial conditions, (c) the need for identification of successive
regimes in a developing flow and (d) the effect of boundary conditions.

The instrumentation that is available currently may be classified as follows:

1. Optical observation
    a. schlieren technique
    b. interferometric technique
        i. photographic
        ii. holographic
2. Pitot tubes
    a. differential pitot tube
3. Thermocouples
4. Sampling probe with quenching
    a. probe for concentration fluctuations
5. Hot-wire anemometry
    a. heat-pulse anemometer
    b. velocity-concentration probes
6. Optical anemometry
    a. particle-timing
    b. particle-tracing
    c. crossed-beam correlation
7. Scattered light intensity
    a. Raman scattering
    b. Rayleigh molecular scattering
    c. infrared scanner

The development of instrumentation appears to be of the
greatest significance in regard to the following:

(a) Flow parameters which can be introduced into the
     Reynolds equations without time-averaging.

(b) Flow parameters that can distinguish compressibility
effects due to various causes.

(c) Flows with large simultaneous differences in density
and temperature.

(d) Pressure fluctuations in moving fluids.

(e) Large scale eddy structure and motion.   And,

(f) Intermittency.

An analysis of the optimum measuring techniques for different
stochastically varying quantities in turbulent flows has not been
carried out but adequate information concerning the various measuring techniques, data-acquisition and data processing is now available for such an analysis to be attempted.

2.1.1 The first detailed measurements (Liepmann and Laufer,
1947) were made in the mixing layer developed between a jet of air
(emerging from the contraction of a two-dimensional wind tunnel)
and the still air on one of its boundaries. The flow field was
incompressible. Hot wire anemometers were the principal tools of
measurement. Both the double correlation coefficient as well as
the microscale turbulence were processed. The laminar boundary
layer at the jet mouth was 1 mm. thick and the transition to turbulence was estimated to occur in the mixing layer in about 6 cms.
The fully developed turbulent velocity profiles were obtained at
about 30 cms. distance from the jet mouth.

The measurements of u and v (the fluctuating components of
velocity) indicate that their lateral distributions remain the same
at different distances from the jet mouth; that u/U_0 is considerably
higher across the mixing zone than v/U_0; and v reaches a maximum
value somewhat closer to the free stream than u does.

The measurements of correlation between the longitudinal
fluctuations at two different points at various stations along the
mixing layer show that the microscale of turbulence is constant
across the mixing region. From the measured distributions, the
turbulence scale could also be estimated. At a given distance from
the jet mouth, the turbulence scale appears to decrease from the
free stream side to the outer edge. If we compare the stream-wise
distribution of scale and microscale, the turbulence scale varies
linearly while the microscale probably varies parabolically.

In the calculation of shear stress from the measured velocity
profile it was assumed that the shear stress must be a maximum
where $\partial^2 U/\partial y^2 = 0$. It was felt that the flow at the outer edge of

the mixing layer could not be adequately described. Considering
the energy balance of fluctuating motion, the dissipation (viscous)
was related to a single microscale of turbulence even though it was
clear that the turbulence in the mixing layer was not isotropic.
Then it was observed that the production, dissipation and diffusion
of energy reached their maximum values in the same region where the
velocity distribution had an inflection point. It was also observed
that while the energy transport through a unit surface area was
small in the center of the mixing region, no such statement could
be made at the edges of the mixing region in view of the difficulty
of estimating triple correlation terms.

   2.1.2 The next detailed measurements made on a similar flow
configuration (Wygnanski and Fiedler, 1970) in incompressible flow
incorporated more refined hot-wire anemometers and associated
electronics. Higher order products of velocity fluctuations and
their spatial derivatives and space-time correlations as well as
intermittency were measured. The basic signal that was employed
to distinguish between the turbulent and non-turbulent regions was
$(\partial^2 u/\partial t^2)^2 + (\partial u/\partial t)^2$ and it was processed according to Heskestad
(1965). In order to obtain the details of the instantaneous flow,
velocity profiles were obtained conditioned to specific location
of the interface. Also, zone averages were measured from which
through appropriate electronics one could separate the turbulent
fluctuations inside the mixing region from the potential fluctuations outside the turbulence interface.

   One feature introduced in these experiments was a trip wire
which was placed just upstream of the start of the mixing region
in order to obtain a transition to turbulence in a comparatively
short distance.

   In general, it is argued that the mixing layer in these experiments may be looked upon as consisting of a wake-like flow on the
high velocity side and a jet-like flow on the quiescent side. The
two interfaces on either side of the mixing zone are shown to behave
independently of each other. Thus, to quote, "the mixing region as
a whole does not flap like a flag about some average location." It
will also become evident later that the energy transport mechanisms
at the two interfaces are also different from each other.

   From the spatial variation of the intermittency factor it
appears that bodies of irrotational flow exist throughout the mixing region. Considering the integral scales, the integral scale
on the high velocity side is probably lower than that on the low
velocity side. It is also shown that the turbulent front is more
flat on the high velocity side.

   From the mean velocity data, the turbulent zone appears to
spread more rapidly on the quiescent side compared to the high

speed side. Considering the instantaneous picture of the flow, the fluid on the high speed side between turbulent bulges is decelerated while the fluid on the quiescent side is accelerated, the latter to a much greater extent, with large changes in flow direction.

The axial velocity profiles conditioned to a specific location of the turbulent interface were obtained both on the high velocity and the quiescent side; the conventional mean velocity profile could be obtained from those using the average frequency of the pulses at the detector. Within the mixing zone, $d<U>/d\eta$ appears to be constant but it depends upon the location of the interface. Outside the mixing zone, the behavior of $d<U>/d\eta$ on the high speed and the quiescent sides are significantly different. On the high speed side $<U>/U_0$ falls linearly after a rather sudden change from its uniform value and on the low speed side $<U>/U_0$ falls nonlinearly and at a much lower rate to zero velocity. Thus, the turbulence in the mixing region is not governed by a constant mean shear.

The average distribution of shear stress shows that the maximum value arises where $d^2(\overline{U}/U_0)/d\eta^2$ is zero; the measured and calculated shear stress distributions agree well. From the zone average distribution of $\overline{uv}$, the profile of $\overline{uv}$ in the turbulent region shows that it is almost constant in the quiescent side and drops very quickly on the high velocity side; it is thus asymmetrical with respect to the average distribution.

The conventional average distributions of the three velocity fluctuations show no unusual feature except that the lateral and transverse velocity fluctuations have a maximum displaced towards the high speed side compared to the axial velocity fluctuations whose maximum value seems to occur at the center of the mixing zone. From the zone-average measurements of the three velocity fluctuations, two conclusions may be drawn: (a) the intensities in the turbulent zone are nonhomogeneous and non-isotropic and (b) the intensities in the potential region are not small even when weighted by the intermittency factor.

In regard to energy balance, within the framework of boundary layer approximation for the mean velocity, the production, convection and diffusion terms were calculated from the measured quantities. In calculating the dissipation term, isotropy was not assumed. The conventional averages as well as the (turbulent) zone averages were examined for each of the important terms in the energy balance equation. The two distributions are similar only in the center of the flow. The diffusion term, which is unsymmetrical with respect to the center of the flow and has a maximum slightly towards the high velocity side, varies in a fashion which indicates transport of energy from the center towards the quiescent side. On that side the energy so transported is lost to viscous dissipation. On the other hand, on the high velocity side, the energy balance is achieved by the convective loss being made up by the diffusion gain and pressure transport, rather as in a wake flow. One can deduce similar conclusions about the differences between the two boundary surfaces on the basis of the difference in $\bar{V}$ values (turbulent zonal average), although point averages without time delay do not provide much information about the lateral energy transport.

When one examines the spread of the mixing zone in these experiments, it is found that the mixing zone spreads to a greater extent into the quiescent side. Considering $\overline{uq^2}$ and $\overline{vq^2}$ in the turbulent zone, q^2 representing the turbulence energy, it was found that on the high speed side most of the energy was transported in the axial direction while on the quiescent side the energy was transported in the lateral direction.

Two point space-time correlations were also measured, the R_{11} correlation by keeping one probe stationary and moving the other probe in the U direction and the R_{12} correlation by moving both the probes laterally. One conclusion from those is that self-preservation is attained as far as the various scales are concerned although the distance to attain self-similarity for intensity is not the same as that for the correlation coefficients. When one examined the lateral distribution of R_{11} and R_{12} it was found that the variation of scales was not continuous with respect to the lateral distance but it was such as to indicate a clear division of flow into two zones. In each part of the flow the integral scale was approximately constant and the transition from one scale to another seems to be connected with the intermittency factor. These results again seem to indicate that the processes in the outer and inner fields in the mixing layer can be looked upon as being different from one another.

2.1.3 The next set of experiments which we shall discuss here as background investigations is one pertaining to the mixing layer formed between two parallel streams (Spencer and Jones, 1971, 1972). Two velocity ratios, 0.3 and 0.6, the latter with weaker shear, were examined. A specially designed pressure transducer was employed to determine the pressure fluctuations. It is based on measuring the velocity fluctuations of a fluid bled into the stream through a capillary tube. The pressure transducer along with the velocity probe provided the pressure-velocity correlations.

In order to understand the coupling between the fluctuations in the surrounding fluid and in the adjacent turbulent fluid, the intermittency profiles were examined and it is clear that the turbulence structure includes significant elements with scales of the same order as the mixing layer width and that entrainment extends well into the central region. The surrounding fluid maintains on entrainment its mean velocity well into the mixing layer, but the velocity fluctuation levels become quite large for such entrained fluid. The turbulent fluid reaching the mixing zone boundaries also has large intensity. The fluctuations diminish in intensity in the outer irrotational flow (Stewart, 1956) according to a power law, the decrease being faster on the low velocity side and at large axial distances. It was also observed that while the three velocity fluctuations were similar in their distributions across the mixing layer, the peak of the v-component intensity appeared slightly towards the low velocity side.

The distributions of skewness and flatness factors for the three velocity fluctuations were determined in the turbulent and non-turbulent elements and also as a total signal. The latter indicate a highly non-Gaussian behavior but the separate flow elements show an essentially Gaussian behavior except for the non-turbulent fluid in the high velocity side. The effect of the convected turbulent elements is different on the low and high velocity sides and this determines the positive or negative skewnesses in those regions.

Comparing the velocity and pressure fluctuations, the experiments seemed to indicate that the pressure fluctuations attained self-preservation considerably later in the flow. This fact also presents certain ambiguities about the location of the fully developed region. The fluctuating pressure shows a Gaussian character in the central region and a skewed behavior at the edges. The spectral distribution of static pressure fluctuations is very similar to that of the u-fluctuations. The velocity-pressure correlations, $\overline{up}$ and $\overline{vp}$, were also measured, as stated earlier. In order to understand the significance of those correlations, one has to take into account (a) the large contribution of the local velocity field, (b) the manner in which the local characteristics of the mean pressure distribution are convected by large scale fluctuations, and (c) the pressure fluctuations coming into existence on account of fluid motions in the surrounding fluid. This is quite a complicated interaction process to take into account. Considering the energy balance equation and the pressure transport term, it turned out that the pressure term contribution does not integrate to zero across the mixing layer even though it is a lateral transport term. It was also found that simple adjustment of the dissipation term at the edges of the mixing layer would not correct this.

The one-dimensional energy spectra for the three velocity components and the fluctuating pressure were evaluated in the spectra-frequency domain. The spectrum of u-component fluctuations showed

that the large and small scale motions were widely separated.
Hence the presence of an initial subrange is suggested, especially
since there is no overlap in the ranges. The u-component fluctuations have a large percent of their energy in the low frequency
domain while the v and w components have most of their energy in
an intermediate spectral region.

From a comparison of the center line spectra at a certain
section in the flow with the spectra of the external flow (high as
well as low velocity sides), it was found that the irrotational
fluctuations were concentrated in a narrow spectral interval with
the peaks occurring at approximately the same frequency. This
behavior was the same whether the velocity ratio was 0.6 or 0.3.
In both cases the peaks occurred at Raichlen's characteristic
energy-containing frequency. The integral length scale L_{ux}
evaluated at that frequency increases linearly with distance and
it is also related directly to the mixing layer width. Such considerations have been noted as indicative of the presence of large
scale structure in the flow.

Two point space-time correlations were also measured for the
axial component of fluctuating velocity using hot-wire anemometry
to obtain axial, transverse and lateral separations (Jones et al.,
1973). The measured data were processed to obtain the space
correlations ($\tau = 0$). The data were analyzed as narrow and broad
band signals.

The space-time correlations for axial separations were obtained
at a cross-section 22.0 in. away from the end of the splitter plate
(the boundary layer displacement thickness on each side being 0.025
in.) at five transverse locations. Assuming that correlated volumes
of turbulence convect in the axial direction, the envelope of the
$R_{uu}(y, \Delta x, \tau)$ correlation distributions was taken to be the turbulence auto-correlation in the convected frame. At each transverse
location, the convected frame auto-correlations ($R_L(\tau)$), as well as
the axial space-time correlations for a narrow band of frequencies
($R_L(\tau, f)$), were found to vary exponentially with respect to time
non-dimensionalized with the convected frame integral time scale.
The ratio of the convected frame integral time scale to that for a
narrow band of frequencies, $\tau/\tau_L(f)$, varied linearly with frequency
for low frequencies. The difficulties of relating the convected
frame integral time scale to a flow field parameter such as the
mean strain rate seem to be unavoidable.

The various aspects of the existence of a coherent structure
in the mixing zone and the implications of the distribution of
scales and convection velocities were also examined.

2.1.4 Another set of experiments (Patel, 1973) have been conducted at the exit of a blower cascade wind tunnel utilizing the same basic experimental configuration as the experiments of Liepmann and Laufer and Wygnanski and Fiedler. The test condition Reynolds number was 5.28×10^5 per meter and the freestream turbulence intensity at the origin of mixing was 0.56%.

The growth of the plane mixing layer was linear with distance. The initial growth presented some difficulty on account of the uncertainties associated with pitot and hot-wire measurements on the one hand and the geometry of the experimental configuration on the other.

In a similar manner, there is considerable ambiguity and discrepancy in the distribution of shear stress. The principal difficulty is in locating the edges of the mixing layer, especially on the streaming side. There are several ways of establishing the shear stress distribution and each of them is subject to some criticism: (a) using the integrated momentum equation with some definition of the edges of the mixing layer; (b) using the mean velocity profiles with some assumption pertaining to the maximum of the shear stress in relation to the maximum of the gradient in mean velocity; (c) using a method similar to (b) but with an assumption about some other point in the velocity profile and (d) using a method similar to (b) and with the velocity specified at one point in the profile based on the Townsend large eddy equilibrium hypothesis. It is important to recognize that small changes in boundary conditions can introduce wide differences in the manner in which the transverse velocity becomes zero at the edges.

The measurements again confirmed the self-preserving nature of the normal Reynolds stresses. The turbulence intensity distribution with respect to the longitudinal and the transverse components showed the relative influence of the production, diffusion and dissipation terms.

2.2 Experiments on the Structure of Mixing Layers

Recent experimental results are beginning to yield some understanding of the structure of mixing layers. Broadly the appeal is either to the coherence of the large scale structure of turbulence or to the outer edge intermittency in mixing layers. The basic ideas in either case are not new but the prospects for developing mixing layer models based on further experimental results appear important to pursue.

2.2.1 Consider the mixing layer developed between two parallel streams coming off a splitter plate, the type of flow configuration employed by Brown and Roshko (1972). One of the striking optical observations made in this flow configuration is the coherence and organization of the large scale structure in the mixing layer. No external disturbances were introduced and the organized structure was observed both with velocity differences alone and with density differences alone between the two mixing streams. Similar organized, large-scale structure has been reported elsewhere (Winant and Browand, 1973). It must be remembered that in all cases, the Reynolds number has been low and the flows on either side of the splitter plate have not been turbulent. It is essential to test the appearance of large scale, organized structure at higher Reynolds numbers. Meanwhile, the presence of large scale structure at the lower Reynolds numbers should be considered as compatible with the turbulent nature of the mixing layer. In the two-dimensional configurations under discussion, it appears that except for small scale lateral disturbances, the large scale convoluted surface is basically two-dimensional. No such visual observations have been reported for axisymmetric flows. The large scale structure in the two-dimensional configuration has been reported as persisting for an appreciable distance and to yield a linearly growing mixing layer. The scales and spacings of the eddies, of course, cannot increase continuously. A process of amalgamation of the eddies or a process of "vortex pairing" (Winant and Browand, 1974) must eventually set in, possibly due to the small differences in the eddies. The precise process of the amalgamation of two eddies is not clear (for example, if an instability sets in) but the amalgamation itself seems to occur by the rolling of one of the eddies over the other and the feeding of vorticity from one to the other.

In order to define the large scale structure, one can examine the spacing of the eddies and the life of an eddy (in terms of length traversed) before its possible amalgamation. It appears that the spacing of the eddies can be related (Brown and Roshko, 1974; Winant and Browand, 1974; Spencer and Jones, 1971) to the mean thickness of the mixing layer in a universal fashion irrespective of the density or velocity ratios of the two streams engaged in mixing. No such rule has yet been deduced for the distance an eddy will travel before it may become "paired" with another.

One point of view, that requires further clarification, in regard to the large scale structure is that the vortices may also be looked upon as waves, the latter stages of the laminar instability of a mixing layer. The finite amplitude waves may be the result of the growth and distortion of infinitesimal waves in relation to the vortex. However, there may simply be an irregularity in the occurrence of vortices and a kind of lateral motion. This point of view may in fact lead to a method of explaining the pairing of vortices as well. Nevertheless, at this stage one can only wait for more experimental results.

It is too early to draw any quantitative results from the experiments to set up a model for the turbulence of the mixing layer in terms of the behavior of the large scale eddies. While several methods of calculating the development of the mixing layer have been discussed earlier, a method of incorporating the large eddy behavior into the phenomenology of mixing layer turbulence, in place of intermittency for instance, is not yet clear (see Coles, p. 663 in NASA SP 321, 1972). However, a basic point of view is developing in regard to the structure of the mixing layer and the presence of large scale eddies on an orderly basis may be central to many aspects of the growth of the mixing layer.

A beginning has been made (Winant and Browand, 1974) on a theory for the formation of the large scale structure and the "pairing" of the eddies. One considers first the vorticity origi-nally distributed in the shear layer and the initial nonlinear growth of a row of vortical entities containing most of the vortic-ity in the steady shear layer; the part played by phase-locking has been known for some time. Then the mixing layer is considered as a double row of vortices, each row of vortices consisting of an infinite number of vortices of the same sign and strength; this is a solution of the inviscid equations (Stuart, 1967). Then the individual vortices begin to pair and rotate around each other as a consequence of instability arising from small disturbances. Con-sider now that the rows of vortices are governed by two parameters, one related to the small differences in horizontal distances between the vortices in a row and another related to the concentrated vor-ticity in the vortices. Then the growth rate of the mixing layer can be related to a function of the foregoing two parameters only. Furthermore, by constructing the appropriate stream function and the vorticity distribution, one can calculate the paths described by a pair of amalgamating vortices as well as the velocity fluctua-tions for given values of vorticity concentration and total phase shift across the mixing layer. At present there seems to be ade-quate corroboration between the measured large scale fluctuations and the calculated fluctuations.

Another question is whether the large scale eddies extend across the entire flow (Wygnanski and Fiedler, 1970). In the experiments on the mixing layer development between a two-dimen-sional stream and a stationary fluid body, the convection velocity of the energy containing eddies was obtained from the cross-correlations of the U-signal and in view of the strong variation of the convection velocity in the lateral direction, it is at no point in the cross-section larger than the local mean velocity. The presence of the large gradient of convection velocity has been attributed to the large scale eddies not extending across the entire flow. Another factor which has been cited in support of this conclusion is that, when the wave-number dependence of the convection velocity is examined at several locations across the mixing zone, the convection velocity is the same at low frequency, that is for larger eddies, only at some lateral locations (high velocity side). In this experiment there is no dependence between the motions of the (turbulent, non-turbulent) interfaces enclosing the mixing region. Perhaps, one should take into account the asymmetry in this flow configuration, which also affects the energy balance.

In another experiment on the mixing of two parallel streams, two point space-time correlations of the axial component of fluctuating velocity have been obtained (Jones et al., 1973) to determine the spatial variations of turbulence structure. The space correlations which indicate the spatial extent of correlated regions have been obtained at several locations along the flow in the axial, lateral and transverse directions. From the axial and lateral correlations, one can observe that, in the outer intermittent regions, there is a periodic structure in the axial direction and a highly correlated region for large distances in the lateral direction. In the outer intermittent region, it has been stated that the turbulent energy was concentrated near 100 Hz. This is taken to be indicative of the mixing layer displaying the large orderly structure.

Some experiments conducted on the development of the mixing zone between parallel streams of hydrogen and fluorine (Shackleford et al., 1973) have revealed further details regarding the large scale structure in the turbulent part of the mixing layer. The experiments were conducted to gain an understanding of the mixing process in chemical laser configurations; the velocities in the two streams are therefore supersonic ($M \simeq 4$). The nozzle contours corrected for boundary layer thickness were designed to yield uniform parallel flow. The temperatures of the two streams were controlled independently by electric arc heating. Continuous visual and spectroscopic observations were the principal means of studying the mixing layer. A pitot probe was also employed. An IR scanner was used to obtain the infrared emission from the reaction zone, which is a function of the vibrationally excited HF along the line of sight. In addition to the HF concentrations in vibrational quantum states, rotational temperatures of HF were also deduced from the variation in intensity due to the rotational energy of the upper level. The concentration of HF in the vibrational ground state was determined from the gain of individual lines using a source capable of emitting lines with known spectral contours.

Two different injectors, one smaller than the other by a factor of $\sqrt{1/2}$, were employed in the investigations; the larger injector could not be operated at sufficiently high pressure (the nozzle exit pressure, 1 to 13 torr, matching the discharge cavity pressure) to obtain turbulent mixing.

In regard to the structure of the mixing layer, it appears that one could identify three regimes along the direction of flow: a laminar region giving rise to a transitional region which would then produce a fully turbulent region. The axial distance for the beginning of the transitional regime was estimated to be about 300 boundary layer momentum thicknesses. It has been suggested that the orderly, large-scale structure is essentially characteristic of the transition region and that such a structural coherence needs to be established at high Reynolds numbers and in the fully turbulent region.

Another observation is that the lateral spreading of the mixing layer is independent of both the temperature and velocity ratios of the two streams. Regarding the temperature of the mixing zone in the transition zone, there appears to be some doubt concerning the spectroscopic measurements since such measurements may be biased towards eddies containing reaction products while the time-averaged temperatures at the edges of the mixing layer are considerably smaller.

It is significant to point out that the presence of orderly structure has also been observed in axisymmetric configurations (Grant et al., 1973; Graham et al., 1973). Axisymmetric jets of propane diluted with nitrogen or methane discharging into air at about atmospheric pressure and yielding a lifted diffusion flame were examined using stroboscopic shadowgraphy, hot wire anemometry, the gated mass-spectrometric technique and Rayleigh scattering. The region of interest was from the exit plane of the jet to the flame location. The formation of ring vortex cores and the presence of unmixed oxygen entrainment were studied when a small velocity fluctuation was impressed on the jet (by means of a loud speaker located in a settling chamber upstream of the jet). Following the observation of an orderly large scale structure in stroboscopic shadowgraphs (see also Davies, 1966 in this connection), the growth of the interfacial waves at preferred wave lengths and the generation of vortex rings by the folding back of the waves was analyzed using hot wire anemometry. For a stationary observer, the waves grow in space but not in time. Over a range of flow rates (or Strouhal numbers based on the diameter of the jet) the waves of the preferred wave length are amplified and one has a nearly single valued spectrum of vortex size. Next, the mass-spectrometric data show on the axis of the jet that the air is entrained periodically into the core of the jet and that the distinct masses of air and fuel will persist without mixing almost upto the location of the flame. Mass-spectrometric data, of course, can be time-averaged and

one will observe a great amount of mixing in the time-averaged
sense. Finally, the structure of the vortex rings in the shear
region of the jet flow has been observed using Rayleigh scattering.
That the vortex rings are stratified is evident from the double
peaks in the concentration fluctuation signal. The signal from
the centerline of the jet shows clearly the unmixedness of oxygen
and the fuel.

The implication of the large scale structure for the stabiliza-
tion of the flame at its location is a separate subject. The
important things to observe here are that the interfacial waves and
the large scale structure are connected directly and that vortex
rings formed from the waves are transported over appreciable distances causing entrainment of reasonably distinct entities of the
surrounding fluid.

2.2.2 The most recent experiments which illustrate the significance of the intermittency at the outer edge of a mixing layer
have been conducted in flows involving scalar quantities (Stanford
and Libby, 1973 and La Rue and Libby, 1973). One set of experiments deals with the mixing between helium injected through the wall
of a duct conveying air and the other set with the wake of a heated
cylinder. In both cases therefore the excursions of the scalar
quantities at the interface are of considerable significance. The
principal measuring tool consists of a hot-wire anemometer probe
(an extension of the idea introduced by Way and Libby, 1970) which
can measure two velocity components and the concentration (c) at
the probe location. The probe utilizes a normal film and wire to
measure the u and c components and an independent swept film to
measure the v component of velocity. Extensive investigations have
been undertaken in regard to probe calibration and data reduction.

In the pipe flow experiments with helium injection, one first
of all obtains time series for u, v and c variables and from these,
the probability density functions. The pdf can then be employed to
obtain the desired moments and cross-correlations.

Another, and very important, result that can be obtained from
the measurements is the possibility of establishing by conditional
sampling (Kovasznay, 1970) the occurrence and non-occurrence of a
chosen event at a particular location of the sensor. Thus, considering a point at the interface between two fluids undergoing
mixing, it is possible to discriminate the probe signals (on the
basis of two values, zero or one) to establish the percentage of
time a certain event occurs, for example turbulence or the presence
of a scalar quantity and also the rms value and the various moments
which can be associated with one or the other fluid and also with
the "backs" ("upstream") and "fronts" ("downstream") of the interfaces. On this basis, even though it may seem somewhat arbitrary,

it has been suggested that one can utilize discriminating techniques by which a type of random telegraph signal can be generated.

We should perhaps note two aspects of the measurements: one, the measured data are supposed to yield the dynamic variables and not, for instance, the density-weighted Favre quantities and two, the density fluctuations have to be deduced from the concentration fluctuations assuming the pressure is constant.

Another point about the data processing in the experiments is that the zero-one-signal is generated through a gate which does not take into account the "interfaces" where the property under consideration changes value. It is implicit that the passage of the "interfaces" at a given space-point contribute little to the time-averaged value at that space-point.

One significant result to be obtained from the measurements is that the probability density distributions can be very highly skewed in regions where the mixing is occurring. In other words, a near-Gaussian and symmetric distribution is evident only where intense mixing has occurred.

Another outcome of the experiments is the attempt at generating the so-called conditioned equations which will apply to the fluctuating interface between a turbulent shear layer and the surrounding irrotational fluid. This is based on an idea that the behavior of the interface may be a significant indicator.

In discussing the development of the conditioned equations, it is useful to consider a scalar quantity "tacked-on" to the fluid (a suggestion made by Bradshaw, 1972). In order to obtain the conditioned equations, Libby considers a scalar quantity whose value at a space-time point can be assigned to be zero or one. Suppose then we consider a dynamic variable and its correlation with the scalar quantity; we can define a conditioned mean value for the dynamic variable.

The zero-one scalar quantity function now needs to be modelled. Libby adopts a diffusion equation (see Saffman, 1970) which includes a creation term for the scalar quantity. It is then necessary to relate the creation term to the intermittency and this is done by postulating that the volumetric rate of creation is due to changes in the value of the scalar quantity; in other words, when there are no interface crossings, there is no net rate of creation.

The next step is to establish a phenomenological basis for the creation of scalar quantity. This matter is again in considerable doubt since no model exists for a satisfactory accounting of the asymptotic behavior of the creation term towards the edges and no

logical connection exists between such a creation term and the
stress distribution, for instance.

It is also necessary to relate the conditioning function to
the turbulence. This is sought to be done utilizing the Reynolds
stress equation in irrotational flow, as deduced by Corrsin and
Kistler (1955). Several new terms arise because of the inclusion
of the conditioning function. This in turn means that one has to
relate the variation of the scalar quantity in the flow direction
with the geometry of the interface. In view of the zero-one nature
of the conditioning function, this becomes the problem of relating
the point statistics associated with the scalar quantity to the
instantaneous geometry of the interface.

Finally, one can incorporate these ideas into the usual turbulent flow equations under high Reynolds number flow (boundary layer
type) assumptions. Two sets of equations result, one for the conditioned variables and the other for the unconditioned variables.

An attempt has been made by Libby to apply this method to the
problem of a two-dimensional mixing layer by introducing a mixing
layer spreading rate and two interface surfaces on either side of
the mixing layer. Assuming furthermore similarity, the governing
equations are deduced for conditioned velocities and intermittency.
While solutions have been generated and compared with experimental
results, essentially on a parametric basis, two factors which
require further clarification are: one, the type of conditional
sampling required in mixing layer studies and two, the inherent
difference between the present modelling procedure and that involved,
say, in a mixing length equation. Nevertheless, Libby's approach
may have important implications in the mixing of scalar quantities.

2.2.3 When chemical reaction occurs in the mixing layer, some
of the questions discussed in the last section on the structure of
mixing layers have a direct bearing on the reaction rate and the
heat release. Since chemical reaction occurs on the molecular scale,
the manner in which mixing occurs on that scale is related to the
details of the entrainment process; the latter in turn is sought to
be related to the eddy structure or to the intermittency in the
mixing layer. However, this is obviously difficult to accomplish
at this stage. In the case of diffusion flames (Williams, 1974),
one can obtain the heat release per unit area of a flame sheet by
making it proportional to $D_c \equiv \rho D |\nabla c|$, which is evaluated at the
stoichiometric value; ρ and D are the local density and diffusion
coefficient respectively. The distribution function for the quantity D_c can be obtained from the joint distribution function for c
and $|\nabla c|$ which itself can be replaced by the conditioned density at
c equal to the stoichiometric value. In experimentation this calls
for advances both in sensing and in data processing. However, the

problem is still to relate the energy release per unit area to the volumetric energy release. In order to do that, it is essential to take account of the non-Gaussian character of the distribution function which is evident in measurements.

2.3. Effect of Density Difference

An important problem in turbulent mixing is one involving a large density difference between the two mixing streams (Abramovich, 1963) and we shall discuss here (see also Section 5.0) some aspects of that problem. One question in regard to such mixing is the spreading of the mixing layer for different density ratios. The thickness of the mixing layer can be understood in terms of mean fluid flow properties but one has to take into account the inter-mittent nature of the interfaces enclosing the mixing region. The growth of the mixing layer cannot yet be related to the compres-sibility effects because one needs to be able to distinguish between density differences arising on account of concentration, velocity or temperature differences between the two streams.

Consider, for example, two parallel streams on either side of a splitter plate coming into contact at the end of the plate and forming a mixing layer; the velocity and density ratios of the two streams may be denoted U_1/U_2 and ρ_1/ρ_2. At high Reynolds numbers, the mean flow may be considered to be independent of molecular diffusion and to obey similarity rules.

It may be recalled here that a convenient measure for the spreading rate of the mixing layer can be based on the velocity-profile maximum-slope thickness or a vorticity thickness. In either case one is assigning a characteristic length scale for the mixing layer. The spreading of the mixing layer is then a function of the velocity ratio (r) and the density ratio (s). From an appropriate origin of the mixing layer, a constant rate of growth of the mixing layer can be considered as an indication of the attainment of similarity.

Some experiments with $rs^2 = 1$ have been reported (Brown and Roshko, 1971, 1974; Rebello, 1973) in a two-dimensional flow con-figuration. These splitter plate experiments, conducted at low Mach numbers, employed a density ratio of seven and therefore the usual ideas of relating the velocity and concentration profiles will probably not apply, the Schmidt numbers being considerably less than one. The gases employed were nitrogen and helium and the Reynolds number for nitrogen flow was of the order of 10^5. The boundary layers were laminar in those experiments and there are at the moment no other results in other cases. Some questions have been raised regarding the developed or self-similar nature of the flow. However,

the measured (Brown and Rebello, 1972) root mean square concentration fluctuation does display similarity. Such self-similarity, of
course, may require very large, of the order of 10^2-10^3 momentum
thicknesses, distances (Bradshaw, 1966) from the initiation of the
mixing layer. Some questions have also been raised regarding the
influence of the tunnel walls but there seems to be adequate demonstration that equilibrium is obtained in an adverse pressure gradient.

Another important observation in the splitter plate experiments
(Brown and Roshko, 1971) comes from the density data obtained using
an aspirating density probe (Brown and Rebello, 1972). (Some questions exist regarding the functioning of the density probe itself.)
When the density probe was traversed across the mixing layer, it
was observed that the fluctuations of density were almost equal to
the density difference. It appears that while each gas penetrates
into the region of the other gas, there is little molecular mixing
and the interface separates the two gases even though it must not
be forgotten that the interface is highly convoluted and viscosity
and diffusivity will tend to smear the interface. This observation
has obvious implications for reactive systems and, in any case, for
the entrainment process. At this stage, it appears difficult to
show a direct connection between the large coherent structure and
the entrainment process as opposed to the influence of the behavior
of the turbulent-nonturbulent interface.

3. MODELLING OF MIXING LAYERS

In a broad sense, the problem on nonhomogeneous turbulent mixing can be formulated in terms of two questions:

   (a)   given a body of fluid in contact with another in
         each of which motions can be expected to display
         some form of coherence and to conform to certain
         equations and given that at some initial instant
         the velocity of the fluid is a random function
         of position described by certain probability laws,
         to determine the probability laws that govern the
         motion of the fluid in the contact region at
         subsequent times; and

   (b)   given a statistical description of each of the
         species present in the two bodies of fluids and
         given the instantaneous rates of reaction in
         terms of kinetic equations, to determine the
         probability laws that describe the distribution
         of each of the species at subsequent times.

Such broad questions may appear essentially useless even though entirely valid. They have to be kept in mind, however, because it is yet difficult to point to what one wishes to know in general in mixing without emphasis on the governing conditions of specific problems. In fact one is at a loss to obtain a general measure for mixing. The attainment of homogeneity at the sub-microscopic level may be defined as the state of complete mixing but the presence of scalar quantities and chemical action, in addition to simple momentum transfer, requires examination of other scales and intensities. Thus, in addition to the time scales of turbulent convection and diffusion, one has to take in account the reactive time scale. The effect of concentration intensity can only be assessed in relation to the number of species involved (including the diluent), the order of the reaction and the reaction rate. In general, reactive flows involve both concentration and temperature, even though the isothermal approximation is often introduced in theory as an effective means of decoupling the momentum balance and the mass conservation equations.

We should recognize the following hierarchy of problems:

   (i)   determination of momentum transfer in turbulent mixing;

   (ii)   determination of the concentration and the thermal fields in the absence of convective motion;

   (iii) modifications introduced by the convective motion; and

   (iv)   modification introduced by the chemical reaction dynamics.

Thus, in the final problem, reaction is superimposed on convection and diffusion. The fine structure of scalar fields naturally depends upon the effects of the interaction between the three processes.

3.1. Fine Structure of Turbulence

The physical mechanisms by which the fine structure is established in the absence of reaction have been studied by a number of people (Townsend, 1951; Batchelor, 1959; Batchelor et al., 1959; Saffman, 1963; Corrsin, 1964; Gibson, 1968). Weakly diffusive scalars are distinguished from strongly diffusive scalars. In all cases it is assumed that the time scales for the changes in velocity and scalar fine structure are small compared with time scales for the large structure that universal equilibrium statistical

distributions will develop. It has then been shown that the largest
wave number perturbations of the scalar field in all cases are
influenced by the local rate of strain. In the case of strongly
diffusive scalars, with molecular diffusivity large compared with
the kinematic viscosity of the fluid, that is the Prandtl number
being very much smaller than one, the direct interaction of velocity
fluctuations of a wave number--sufficiently separated from the
inverse of the Kolmogoroff length scale ($L_K \equiv (\nu^3/\varepsilon)^{\frac{1}{4}}$, ε being the
mean rate of viscous energy dissipation per unit mass) and the
Obukhov-Corrsin length scale ($L_c \equiv (D^3/\varepsilon)^{\frac{1}{4}}$) --will determine the
perturbations of the scalar distribution for scales smaller than
the Obukhov-Corrsin length scale. These models of course apply to
regions of uniform gradient. On the other hand, suppose we write
the conservation equation for a scalar property θ in terms of the
scalar flux q (normal to fluid flux) as follows.

$$\frac{\partial \theta}{\partial t} + \underline{u} \cdot \underline{\nabla} \theta = D \nabla^2 \theta \qquad\qquad\qquad (3.1.1)$$

$$\frac{\partial \theta}{\partial t} = - \underline{\nabla} \cdot \underline{q} \qquad\qquad\qquad (3.1.2)$$

$$\underline{q} \equiv \underline{u}\, \theta - D \underline{\nabla} \theta \qquad\qquad\qquad (3.1.3)$$

It can be seen that the production of scalar fine structure must be
important where scalar gradients are small since the action of fluid
motion is to produce small scale fluctuations in the θ-distribution.
In view of this, it has been concluded that the perturbations in the
scalar distribution are governed by the number and distribution of
points in the fluid at which the scalar gradient vector is zero. It
must be emphasized that this postulate is based on physical and
geometrical considerations.

     In a density-sensitive heating mechanism due to chemical action,
it can also be shown (Eschenroeder, 1965) that a spectrally-selective
internal power source arises in addition to the external source that
is feeding energy into the turbulence. Using a relationship expressing the balance of energy transfers in wave number space and a particular form of the homogeneous source term (taken as a function of
the energy addition rate), it is shown that the external source
exerts the most intense influence on the energy-containing region
of the spectrum and hence introduces the greatest distortion in a
region surrounding the spectral peak. As the source effectiveness
increases, non-equilibrium effects become dominant.

     Suppose we consider a dynamically passive reactant consisting
of a single specie undergoing a second order reaction. From the
exact equations for a statistically homogeneous scalar field, it
can be shown (Corrsin, 1958; O'Brien, 1968) that, <u>in the absence of</u>

<u>diffusion</u>, all single point functions of the concentration field, such as mean concentration, mean square concentration, etc., decay at a rate independent of turbulence. The important thing to note is that the reaction rate has no dependence on scale and therefore turbulence can play no direct role in decreasing scalar intensity but only alter the scale of the concentration field. Turbulent convection, on the other hand, can interact with reaction by an increase in the mean square gradients of concentration through turbulence and enhance diffusion. In any case, based on an exact result that can be obtained in first order reactions and the assumption that the reactions are not strongly selective spectrally, attempts have been made (O'Brien, 1969) to show that the spectral insensitivity of single-species-reactions may be quite general. The limitations of such theories are obvious.

Another thing to note in the reactive case is that one has to assume a non-zero mean concentration since the reaction rate depends directly on the local concentration. Thus, three nonlinear interactions become important: those among the velocity and concentration fluctuations induced by convection, those among the concentration fluctuations induced by the second-order reaction and those between the mean concentration and the fluctuations, the latter being especially important in shear flows and totally absent in isotropic turbulent mixing, although the linear interaction is present under all conditions.

It may be observed in regard to the fluctuations that, when the typical amplitude of fluctuations becomes large, a skewed fluctuation distribution becomes unavoidable (assuming the local concentration is always positive) and therefore, a Gaussian pdf is not admissible as it leads to negative values.

We may also note here Corrsin's early decay postulates for the mean concentration and rms fluctuations, that three asymptotic ranges exist characterized by the ratio of a time scale of reaction to a time scale for turbulent mixing. These asymptotic ranges have been explored by others in detail. In contrast to the two limiting situations of slow reaction and the very rapid reaction, the case in which the ratio of the two times is near unity is especially difficult. The rapid reaction case is characterized by the fact that the homogeneous kinetic rate becomes quite different from a two-species diffusion-controlled reaction. The rate of reaction in this case is governed by mixing. The depletion rate of the mean concentration has to be considered in relation to that of the fluctuations and one should expect intermittency due to lack of stoichiometry locally.

3.2. Modelling of Shear Flows

In general no great advances have been made in the possibility
of employing a non-Newtonian constitutive equation for the so-called
"turbular" fluids. A transport rate with memory (Lumley, 1967) and
a non-local eddy viscosity (Nee and Kovasznay, 1969) have been
investigated. But otherwise Newtonian stress-strain relations have
remained universal. However, phenomenological theories, it is well
known, have assumed a turbulent fluid to have its own characteristic
properties as long as boundaries and interfaces are not involved.

There is no reason to discard the Navier-Stokes (N-S) equations,
with the mean and the fluctuating components of fluid and flow properties introduced by Reynolds, as being inadequate for turbulent
mixing problems, although they are insufficient to clarify the structure of a mixing layer. By taking moments of the local space
averaged N-S equations, one can obtain transport equations for the
Reynolds stress or for the turbulent energy. The equations for the
transport of Reynolds stress and energy can then be shown to represent a balance between:

A. 1. generation by interaction with mean flow;

2. transport or diffusion by

        velocity fluctuations and

        pressure fluctuations;

3. transport and destruction or dissipation by
    viscous forces; and

4. redistribution by the interaction between pressure
    fluctuations and the fluctuating rate of strain;
    and

B. 5. convection or advection.

A particular case of interest is the state of so-called "local
equilibrium" wherein all of the transport terms are zero. That
state is related to special conditions on the velocity and length
scales.

In general, the equations of motion for the velocity moments
of all orders are coupled on account of the nonlinearity of the
N-S equations. Several mathematical difficulties arise in handling
the coupled equations especially at high Reynolds numbers on account
of the strong dissipation in such flows. On the other hand, at high
Reynolds numbers the mean Stokes stresses are negligible and also
most of the viscous terms.

Regarding item (3) in the foregoing, it may be observed that the rate of energy supply and thus the rate of energy dissipation are controlled by the larger eddies which do not depend upon viscosity. The small eddies receive energy from the large eddies by a process of random vortex stretching, and therefore remain isotropic. The small eddies contribute little to the Reynolds stresses and the Reynolds stresses become independent of viscosity.

If the total temperature is constant, one can relate, within boundary layer type approximations, the normal gradient of shear stress to the total pressure gradient along a streamline.

In the energy balance equation, it is clear that the overall redistribution by the pressure fluctuations will become equal to zero since the pressure term depends upon the whole flow. However, the anisotropy of turbulence or the difference between the different components of the Reynolds stress tensor is due to the pressure-strain interaction terms, though we have no measurements to illustrate the connection.

Now, turbulence is by definition rotational. Turbulence may in fact be thought of as an "entanglement" of rolled vortex sheets, stretched in the preferred direction of mean motion. One may therefore setup a vorticity balance equation for turbulent motion and examine its implications. It may be noted here that while vortex stretching dominates both the Reynolds stress transport equations as well as the equations for the turbulence length scales, it is not possible to write the latter equations entirely in terms of the vorticity. One point to note about the vorticity balance equation is that it will not have the pressure terms in incompressible flow since pressure fluctuations do not directly affect the vorticity therein.

However, in the case of compressible flows, one has to reckon with pressure fluctuations which may be of the same order as the fluctuation in vorticity, temperature and density. The establishment of the fluctuation modes (vorticity, pressure and entropy, say) which interact with each other in various flows is an extremely difficult task. The work on a homogeneous flow field (Chu and Kovasznay, 1958) with small $\Delta \overline{M}$ and $\Delta \overline{T}$ (M and T being the Mach number and temperature, respectively) is well known in this connection. That is second order theory and it has been shown (Kovasznay, 1962) how the second order theory is still inadequate for compressible shear flows since the presence of a mean gradient must be regarded as a separate mode.

In examining the mean compressible turbulent flows, the turbulent stress tensor can be shown (Laufer, 1969) to take the same form as the incompressible stress tensor and furthermore, the addition of mean turbulent stresses acting on fluid elements along mean streamlines can be shown to be consistent both with respect to momentum balance and mechanical energy balance. However, we do not have adequate information concerning the importance of turbulent dissipation and pressure diffusion terms in relation to the production terms.

In accounting for compressibility effects, it is generally agreed (Markovin, 1962) that the rms fluctuating Mach numbers are small for free stream Mach numbers under five and that compressibility effects may be neglected so long as the resulting pressure fluctuations are small compared to the mean pressure. However, it is possible that the latter criterion may set a different (lower?) limit to Mach number in mixing layers for the neglect of compressibility.

Suppose we now consider a process involving streamwise density gradients, it appears that one can still accept Markovin's hypothesis. However, there is experimental evidence (see Bradshaw, 1972, for various references to experimental studies) that mean dilatation does affect the structure of turbulence. It has also been shown that extra rates of strain applied to thin shear layers have given rise to Reynolds stresses an order of magnitude larger than those that can be calculated. Therefore, it has been argued that the effect of dilatation can be taken into account by increasing the dissipation (destruction) term in a semi-empirical manner. The rationale here is that the contribution of the pressure gradient to the mean acceleration can be calculated using Markovin's "strong Reynolds analogy" (instantaneous total temperature constant and pressure fluctuations small) and that this will provide the needed correction to the dissipation term provided one assumes that the full effect of dilatation is felt gradually as the turbulence structure changes. The question then is how rapid a density gradient can be accommodated within this approximation.

At the same time, if we consider a mixing problem involving density differences, it seems quite important to identify the cause for the density difference between the mixing streams. The Howarth-Dorodnitsyn transformation for density does not specify a cause for the density changes. However, in turbulent mixing involving such quantities as $\overline{p'v}$, $\overline{\rho'v}$ and $\overline{vT'}$, it is essential to take into account the difference between Mach number-, temperature-, and molecular weight-induced density changes. In mixing problems one is interested in the mean velocity, temperature and concentration profiles and not merely in the stress distribution. Accordingly, the usual methods of truncated equations or transformations will only yield results of doubtful validity. The uncertainties may only be removed by systematic experiments although the measurements will be extremely complicated (Liepmann and Laufer, 1949; Brown and Roshko, 1971; Birch and Eggers, 1972).

A provisional analysis has been attempted (Libby, 1972) on the spreading characteristics of a two-dimensional mixing layer as observed in experiments in the two limiting cases, (i) the low speed isothermal mixing of two gases of different molecular weights and (ii) the compressible adiabatic mixing of a high speed flow with the same gas in a quiescent state when the stagnation temperature is the same everywhere. The experimental results (Brown and Roshko, 1971) seem to indicate little alteration of the spreading rate with large density differences in the first case and a significant reduction in the spreading rate with Mach number in the second case, though the latter is controversial. The analysis is based on an eddy viscosity model related to a length associated with the scale of the mixing layer and to the square root of the turbulent kinetic energy. Five parameters are examined which are related to the pressure rate of strain, the diffusion and the dissipation of turbulent kinetic energy, the ratio of specific heats and a characteristic Reynolds number based on a scale length of the order of the mixing layer thickness. It is shown through an analysis of the experimental results that the Mach number effect is perhaps best reproduced through an adjustment of the characteristic Reynolds number and the other parameters are not of equal significance. There is no understanding at this time as to why the characteristic Reynolds number should be of significance only up to a certain value of Mach number. The low speed isothermal mixing case results seem to indicate independence of the spreading rate with respect to density difference but the analysis shows a connection between the density and the velocity ratios of the mixing streams on the one hand and the spreading rate on the other.

In the foregoing analysis, the pressure effects do not explain the compressibility effect on mixing when the flows are supersonic. On the basis of an analysis of the governing equations (Brown and Roshko, 1974), it has however been shown that pressure-velocity correlations can account for the differences between incompressible and supersonic flows. At supersonic velocities, the Reynolds equations of course involve Mach number dependent terms containing pressure-velocity correlations. However, this should not be regarded as a settled question, especially since no measurements are available of the pressure-velocity correlations. Some order of magnitude estimates have also been made for the shear stress and the rms fluctuation level of the transverse component of velocity in terms of the time-averaged velocity of the center of mass of the particles at any point, the constant of proportionality being a measure of the mixing layer spreading rate. At high Mach numbers, it seems to be implied that there is a decrease in intermittency and that the spreading rate decrease with increase in Mach number is related to decrease in v.

3.3. Chemically Reactive Flows

We shall next discuss the hierarchy of equations describing turbulent reacting flows (O'Brien, 1971; Donaldson and Hilst, 1972; Libby, 1972). The simplest but least satisfactory way of calculating such flows is to calculate the time rate of production of each species using the instantaneous rate with time mean values of specie concentration, density and temperature. The assumption is incorrect when the scale of turbulence is large. Thus, considering a reaction involving two species A and B, if the equations governing the reactions are given by

$$\frac{DC_A}{Dt} = - K_1\, C_A C_B$$
$$\frac{DC_B}{Dt} = - K_2\, C_A C_B$$

$$(3.3.1)$$

where C_i are the mass fractions of the species and K_1 and K_2 are the reaction rates, the equations for the time rates of change of the mean values of C_A and C_B are _not_

$$\frac{D\bar{C}_A}{Dt} = - \bar{K}_1\, \bar{C}_A \bar{C}_B$$
$$\frac{D\bar{C}_B}{Dt} = - \bar{K}_2\, \bar{C}_A \bar{C}_B$$

$$(3.3.2)$$

but are given by the following so long as K_1' and K_2' are equal to zero.

$$\frac{D\bar{C}_A}{Dt} = - \bar{K}_1\, (\bar{C}_A\, \bar{C}_B + \overline{C_A'\, C_B'})$$
$$\frac{D\bar{C}_B}{Dt} = - \bar{K}_2\, (\bar{C}_A\, \bar{C}_B + \overline{C_A'\, C_B'})$$

$$(3.3.3)$$

The foregoing equations make abundantly clear that only when the fluctuations do not exist or when they are uncorrelated is the average reaction rate governed by the average concentrations.

In order to deal with Eqn. 3.3.3, it is necessary to have a prediction equation for $\overline{C_A' C_B'}$ and this has been shown to be

$$\frac{D \ \overline{C_A' C_B'}}{Dt}\bigg|_{chem} = - \ K_1 \left(\overline{C}_A \ \overline{C_B'^2} + \overline{C}_B \ \overline{C_A' C_B'} + \overline{C_A' C_B'^2} \right)$$

$$- \ K_2 \left(\overline{C}_B \ \overline{C_A'^2} + \overline{C}_A \ \overline{C_A' C_B'} + \overline{C_A'^2 C_B'} \right) \qquad (3.3.4)$$

which introduces several new terms. While the prediction equations for the variances, $C_A'^2$ and $C_B'^2$, do not introduce any further new terms, one does need prediction equations for the third order correlation terms in Eqn. 3.3.4.

One can at once ask two questions (Hilst, 1973): when can the concentration fluctuations be neglected and when can the third order correlations be neglected. Regarding the first, the ratio of the reaction rates with and without including the fluctuations is given by

$$1 + \frac{\overline{C_A' C_B'}}{\overline{C}_A \overline{C}_B} \ .$$

Accordingly, the potential for order of magnitude changes in reaction rates exists when the joint distributions of the reactant concentrations are skewed towards large values of these concentrations, that is when the variance to mean squared value ratios can be large compared to one. On the other hand, the concentration fluctuations play little part when C_A and C_B are randomly distributed. Whether the chemical reaction rate is accelerated or depressed by the inclusion of fluctuations depends upon whether the fluctuations are correlated positively or negatively, respectively. In the latter case, when the correlation is negative, one should examine the interaction between chemical and diffusion equations.

Regarding the influence of the third order correlation terms in Eqn. 3.3.4, they control the rate of change of $\overline{C_A' C_B'}$ and hence have a time-integrated effect on the $\overline{C_A' C_B'}$. In order to determine the effect of the higher order correlations, it is necessary to examine the distribution functions from which the mean values and the moments are derived. The neglect of the higher order terms can lead to highly erroneous results.

One may turn attention again to Eqn. 3.3.3 and ask under what conditions one may neglect $\overline{C_A' C_B'}$ in comparison with $C_A C_B$. In a manner similar to that of the equation for the Reynolds stress tensor, one can write an equation for the substantive derivative of $\overline{C_A' C_B'}$ as follows, provided no large gradients are present, and assuming incompressible flow for simplicity (Donaldson, 1972).

$$\frac{D}{Dt} \overline{C'_A C'_B} = \frac{D}{Dt} \overline{C'_A C'_B} \bigg)_{chem} - 2D\, g^{mn} \left(\overline{C'_A, m\, C'_B, n} \right) \qquad (3.3.5)$$

where g^{mn} is the contravariant form of the metric tensor g_{mn}. The second term on the right hand side of Eqn. 3.3.5 is nothing but the diffusion term and one can therefore introduce a diffusion length or time as follows.

$$L_D^2 = 2D\, \tau_D \qquad (3.3.6)$$

In a similar manner, one can also introduce a chemical time with respect to the first term on the right hand side of Eqn. 3.3.5 by writing

$$\frac{D\, \overline{C}_A\, \overline{C}_B}{Dt} \bigg)_{chem} = -\, \tau_C \left(\overline{C}_A\, \overline{C}_B + \overline{C'_A\, C'_B} \right) \qquad (3.3.7)$$

where

$$\tau_C = \frac{1}{K_1\, \overline{C}_B + K_2\, \overline{C}_A}$$

The interpretation of Eqn. 3.3.6 is that chemical action alone will drive $\overline{C}_A \overline{C}_B$ to the negative of $\overline{C'_A C'_B}$ with a characteristic time τ_C and that the reaction between A and B will stop before the exhaustion of A or B unless A and B are perfectly mixed. The important point to remember is that we are considering a situation in the absence of diffusion.

It is clear that whether the removal of species A and B from the flow is governed by reaction rates or is limited by molecular diffusion is determined by whether the ratio of τ_D to τ_C is much smaller than one or much larger than one respectively. If the dissipative or diffusive scale of turbulence is very small and τ_D/τ_C is small, $\overline{C'_A C'_B}$ may be neglected and molecular diffusion can be expected to keep the two species well mixed. In assessing a given experiment, of course, one has to take into account the integral scale of turbulence as well as its relation to the diffusive scale of turbulence.

Separating the diffusion-chemistry model from the turbulence model, one can write (Donaldson, 1973) the equation for the rate of change of $\overline{u_i C'_A}$ and $\overline{C'_A T'}$ as follows assuming incompressible flows.

$$\frac{D}{Dt} \overline{u_i C_A'} = - \overline{u_i u_j} \frac{\partial C_A}{\partial x_j} - \overline{u_j C_A'} \cdot \frac{\partial u_i}{\partial x_j}$$

$$- \frac{\partial}{\partial x_j} \left(\overline{u_i u_j A_A'} \right) + \frac{g}{T_0} \overline{C_A' T'}$$

$$+ \nu_0 \left\{ \frac{\partial^2 \overline{u_i C_A'}}{\partial x_j^2} - 2 \overline{\frac{\partial u_i}{\partial x_k} \frac{\partial C_A'}{\partial x_k}} \right\}$$

$$- \frac{\overline{\partial p' C_A'}}{\partial x_i} + \overline{p' \frac{\partial C_A'}{\partial x_i}} - K_A \left\{ \overline{C}_B \overline{u_i C_A'} + \overline{C}_A \overline{u_i C_B'} \right.$$

$$\left. + \overline{u_i C_A' C_B'} \right\} \tag{3.3.8}$$

$$\frac{D \overline{C_A' T'}}{Dt} = - \overline{u_j T'} \frac{\partial \overline{C}_A}{\partial x_j} - \overline{C_A' u_j} \frac{\partial \overline{T}}{\partial x_j}$$

$$- \frac{\partial}{\partial x_j} \left(\overline{u_j C_A' T'} \right)$$

$$+ \nu_0 \left\{ \frac{\partial^2 \overline{C_A' T'}}{\partial x_j^2} - 2 \overline{\frac{\partial C_A'}{\partial x_k} \frac{\partial T'}{\partial x_k}} \right\}$$

$$- K_A \left\{ \overline{C}_B \overline{C_A' T'} + \overline{C}_A \overline{C_B' T'} + \overline{T' C_A' C_B'} \right\} \tag{3.3.9}$$

We need in addition equations for the specie concentration fluctuation C_A', the $\overline{C_A' C_B'}$ and the variance $\overline{C_A'^2}$.

$$\frac{D C_A'}{Dt} = - U_j \frac{\partial \overline{C}_A}{\partial x_j} - u_j \frac{\partial C_A'}{\partial x_j} + \frac{\partial}{\partial x_j} \left(\overline{u_j C_A'} \right)$$

$$+ \nu_0 \frac{\partial^2 C_A'}{\partial x_j^2} - K_1 \left(C_A' \overline{C}_B + \overline{C}_A C_B' + C_A' C_B' - \overline{C_A' C_B'} \right) \tag{3.3.10}$$

$$\frac{D\ \overline{C'_A C'_B}}{Dt} = -\ \overline{U_j C_B}\frac{\partial \overline{C}_A}{\partial x_j} -\ \overline{U_j C_A}\frac{\partial \overline{C}_B}{\partial x_j} -\ \frac{\partial}{\partial x_j}\left(\overline{u_j C'_A C'_B}\right)$$

$$+\ \nu_0 \left\{\frac{\partial^2 \overline{C'_A C'_B}}{\partial x_j^2} - 2\ \overline{\frac{\partial C'_A}{\partial x_k}\cdot\frac{C'_B}{\partial x_k}}\right\}$$

$$-\ \left.\frac{D\ \overline{C'_A\ C'_B}}{Dt}\right|_{chem} \tag{3.3.11}$$

$$\frac{D\ \overline{C'^2_A}}{Dt} = -\ 2\ \overline{C'_A u_j}\frac{\partial \overline{C}_A}{\partial x_j} -\ \frac{\overline{\partial u_j C'^2_A}}{\partial x_j}$$

$$+\ \nu_0 \left\{\frac{\partial^2 \overline{C'^2_A}}{\partial x_j} - 2\ \overline{\frac{\partial C'_A}{\partial x_k}\frac{\partial C'_A}{\partial x_k}}\right\}$$

$$-\ 2\ K_A \left\{\overline{C}_B\ \overline{C'^2_A} + \overline{C}_A\ \overline{C'_A C'_B} + \overline{C'^2_A C'_B}\right\} \tag{3.3.12}$$

It is now necessary to model the third-order correlations, pressure correlations and dissipation terms in addition to the third-order chemistry correlations and the temperature-chemistry correlation. One can then set up the set of simultaneous equations and solve them for given initial distributions of the chemical and flow parameters as well as the length scale distributions in the turbulence.

At present this is rather in the nature of a proposal for further investigation.

3.3.1 Another hierarchy of equations has been set up (Libby, 1973) for an infinitely fast chemical reaction involving two species with one product and under highly diluted conditions. Denoting the mass fractions of the species by Y_A and Y_B, the fast reaction is denoted by the relation

$$\overline{Y_A\ Y_B} \cong 0 \tag{3.3.13}$$

with the instantaneous rate of production of each species being given by the combined influence of convection, diffusion and fast chemistry. The equations are of course unclosed in regard to various order correlations and also the dissipation terms. When closure approximations are introduced, the resulting set of equations, involving $\overline{Y}_1$, $\overline{Y}_2$ and fluctuating quantities $\overline{\rho\, Y_j'^2}$ in conjunction with the mean element concentrations, can be solved for the mean concentration.

The hierarchy of equations so obtained contains cross-diffusion terms of the type $\rho D(\partial Y_A'/\partial x_k)(\partial Y_B'/\partial x_k)$. Such a term involves two scalar quantities neither of which is conserved in general. For passive scalar fields (arbitrary Prandtl number) it is known that the scalar dissipation rate is related to the turbulent convection and molecular diffusion at the interfaces with small scalar gradients. It is conjectured that similar may be the case with cross-diffusion terms arising at the reaction interfaces.

This leads one to the question of the nature of the interfaces. The reaction itself must be confined to the part of the interface where the reactants are contiguous. However, apart from the fact that the reactants cannot coexist together, the volumes containing one or the other reactant will also contain the product. The spreading of the product will arise through the straining and molecular diffusion processes in the original volumes giving rise to the formation of new product and its dissemination into other volumes.

The infinite reaction rate limit also permits an estimate of a characteristic reaction zone width (Pearson, 1963). If t is the elapsed time of the reaction, the characteristic width of the reaction zone becomes

$$L_R \sim \frac{D^{1/2}}{K^{1/3}} t^{1/6} \tag{3.3.14}$$

provided one assumes that the dominant processes within the reaction zone are diffusion and chemical reaction rather than the convective field, the latter merely serving to establish a time scale. On the other hand, if one proceeded according to the uniform strain theory (Batchelor, 1959) and assumed that the characteristic time scale of the reaction zone is determined by γ^{-1}, where γ is the magnitude of the rate of strain (associated with the Kolmogoroff scale eddies), the characteristic width of the reaction zone becomes (Gibson and Libby, 1973)

$$L_R \sim \frac{D^{1/2}}{K^{1/3}} \gamma^{-1/6} \tag{3.3.15}$$

It has been argued (O'Brien, 1973) that the characteristic width should not be related to the uniform rate of strain since it would imply (a) a near statistical equilibrium to exist at the reaction zone and (b) a characteristic velocity of propagation of the

reaction zone which does not permit (being too large) the reaction
zone to align itself perpendicular to the least principal rate of
strain axis. Gibson and Libby seem to indicate that the reaction
surface will align itself perpendicular to the least principal rate
of strain axis so long as the reaction surface is different from
the fluid motion velocity only by a diffusional velocity, very
small compared to the diffusional velocity $(DY)^{\frac{1}{2}}$ which is characteristic of a region of thickness $(D/Y)^{\frac{1}{2}}$, the Batchelor length scale,
L_B.

     In addition to L_R and L_B, there is the viscous zone,
$L_V = (\nu/D)^{\frac{1}{2}} L_B$, with the Schmidt number $N_{SC} = \nu/D$. An analysis of
the reaction zone formed during the conductimetric titration of a
weak base (ammonium hydroxide) and a weak acid (acetic acid) has
displayed the relation $L_R < L_B < L_V$. A number of questions remain
in regard to such experiments including the resolution of the probe,
what specie is in fact being sensed by the probe (the Gibson-Libby
experiments sensing the product) and so on. One interesting question is whether it is possible to distinguish between the characteristic length, L_{NR}, obtained from the time scale for non-reactive
mixing and L_R. In a given elapsed time t, the rate of loss of a
scalar due to simple mixing is $O[tu/\lambda_\theta]$ where u is the rms turbulence level and λ_θ is the Taylor microscale of the scalar field.
Comparing this with the rate of diffusion of the scalar into the
reaction zone, one can show under certain approximations that

$$L_{NR} \sim L_B N_{SC}^{-\frac{1}{4}}$$ (3.3.16)

and this should be compared with the expression for L_R provided in
Eqn. 3.3.15 or 3.3.14.

3.4. Modelling of Non-Reactive Flows

     The modelling of turbulence is unavoidable as will have become
evident from the discussion in the preceding. In particular, one
does not have intrinsic inter-relationships between the averaged
quantities in various moment equations. Thus, referring to the
Reynolds stress balance discussed in the beginning of this Section,
it is necessary to represent the transport and redistribution terms
as functions of the Reynolds stresses and the mean velocity field.
In addition, we need associated equations for the eddy length scales
and the velocity scales. The eddy length scales will in general
have to be represented by transport equations. The Reynolds stresses
may be expressed as functions of the mean velocity field as in the
mixing length or eddy viscosity hypotheses. Higher order closures
will involve differential equations for turbulence quantities, for
example, second-order or Reynolds stress closure employed for the

closure of the Reynolds stress transport equation. It is important in setting up such approximate closure schemes that each of the processes involved in obtaining a balance of the Reynolds stresses is taken into account. In a similar manner, it will also have been observed in Sections 3.1 and 3.2 that several closure approximations will be required in regard to the scalar quantities and the scalar-velocity correlations.

One method of classifying the various models of turbulence is as follows (Launder and Spalding, 1972), wherein the complexity of the model increases down the classification, although one cannot always claim greater satisfaction from the more complex model.

1. The mixing length hypothesis for the transfer of momentum.

    1.1 van Driest's hypothesis and its modifications.

2. The mixing length hypothesis for the transfer of scalar properties based on the diffusional approximation.

    2.1 Differential equations for the concentration fields along with length scales for transport.

3. A differential transport equation with the time-averaged turbulence kinetic energy as the variable and with the diffusion, production and dissipation terms modeled appropriately, and the length-scale distribution prescribed.

    3.1 Bradshaw's modelling of the energy equation with the kinematic shear stress directly proportional to the turbulence energy and the diffusive flux of energy proportional to the product of the energy and a diffusion velocity of the large eddies of turbulence.

    3.2 Nee and Kovasznay transport equation for the kinematic turbulent viscosity.

4. Transport equations for <u>both</u> the turbulence energy <u>and</u> the length scale.

5. Multi-equation "stress" models with various closure schemes.

    5.1 Equations for scalar fluctuations.

In each case, two general questions can be raised: (i) does the approach lead to a specific advance in understanding or computing the flow fields and (ii) is the method capable of systematic improvement through experimental (physical or computational) verification and subsequent changes. The latter question is especially important since the generalization of a model becomes effective only insofar as it does not demand basic changes in our understanding of the exact equations governing the process. Of course, one can dispute even the exact Reynolds stress (local) equations as not being truly representative of the turbulence terms. However, transport equations for length scales, for example, differ widely from each other and lead to different kinds of ambiguities if a transport equation for Reynolds stress is deduced from a transport equation for length scales. There is, unfortunately, no independent exact equation for length scales or eddy viscosity.

Considering the various closure schemes, the first order closures involve various approximations for the eddy viscosity and mixing length. The next level of sophistication, the second-order closure schemes have shown considerable promise and point to both specific measurements as well as to systematic analysis of the applicability of various models (Fendell, 1971; Bush and Fendell, 1972, 1973). Some of the length scale equations can also be used for second-order closure. It has been suggested (Bradshaw, 1972) that the next level of sophistication is a type of third-order closure wherein a transport equation is written for the dissipation rate, ε. At high Reynolds numbers, the rate of dissipation is equal to the rate of transfer of energy from the energy-containing large eddies to the small eddies and the energy transfer depends upon triple correlations.

Thus, in the absence of scalar quantities, one can use one Reynolds stress transport equation and an algebraic relation for the length scale provided one is dealing with thin shear layers. In more complex flows, it may be necessary both to include transport equations for all non-zero Reynolds stresses as well as a transport equation for length scales. The question of whether two or more than two (in each case the length scale equation being included as one) equations are required cannot be answered in a general fashion, though multiple equations seem to be more appropriate when convection and diffusion transport are appreciably significant (Henjalic and Launder, 1970).

Regarding chemically active flows involving scalar quantities, it is necessary as pointed out earlier to obtain closure approximations for the triple correlation terms. The direct approach here is to use the lower-order moments to generate at least part of the information regarding the third-order correlations. A more complicated approach is to set up prediction equations at the third-order and then seek closure by examining the appropriate higher order moments.

We shall conclude here with some general remarks regarding the transport equations.

(1) It will be recalled that the pressure fluctuations are divided into a part caused by interactions of velocity fluctuations and a second part produced by the interaction of the mean velocity gradients with velocity fluctuations. It has now become accepted that the pressure-strain part, the first part, should be related, not to the mean velocity gradients (Rotta, 1951), but to the turbulence terms (Rotta, 1962). Thus, the pressure-strain term is related to the anisotropy as follows

$$\overline{p' \left(\frac{\partial U_i}{\partial x_j} + \frac{\partial U_j}{\partial x_i} \right)} = -2 C_1 \varepsilon \frac{\overline{u_i u_j} - \overline{q^2}\, \delta_{ij}/3}{\overline{q^2}}$$

$$= - C C_1 \frac{\overline{q^2}}{2} \frac{\overline{uv}}{L} \tag{3.4.1}$$

since the dissipation

$$\varepsilon = C \left(\frac{\overline{q^2}}{2} \right)^{3/2} \cdot \frac{1}{L} \tag{3.4.2}$$

In the foregoing, C and C_1 are constants of proportionality.

However, mean velocity gradients do arise in various closure models (Bradshaw, 1971; Daly and Harlow, 1970) and more specific connection remains to be established.

Now, the pressure-strain correlation can be expressed in terms of the interaction due to fluctuating quantities and that due to the mean rate of strain. The proposal for the first of these (Rotta, 1951) is to make it proportional to the local level of anisotropy. It has been proposed recently (Launder et al., 1973) that the second interaction may also be expressed similarly through its proportionality to the mean strain. However, it has turned out that such modelling of the pressure-strain terms does not in fact provide a satisfactory Reynolds stress closure. Whether this is due to the influence of secondary strain terms (Bradshaw, 1973) or some other cause is not clear at the moment.

(2) In regard to the turbulent transport processes the bulk convection hypothesis (Townsend, 1956; Bradshaw, 1971) does have more physical justification. This approximation for the diffusion process renders the differential turbulence energy equation in hyperbolic form. However, the gradient-diffusion hypothesis is the more commonly employed. Progress in modelling this term has not been considered important since the diffusion terms are small.

Chou (1945) and Davidov (1961) have provided elaborate transport equations for the diffusion terms. The modelling of the triple-correlation term for velocity is at present in some doubt (Mellor and Herring, 1973; Hanjalic and Launder, 1972).

(3)   The length scales which have been examined so far are (i) the characteristic length scales of turbulent motion, (ii) the correlation integral scales (Rotta, 1951), (iii) the dissipation length parameter, $(\overline{U_i^2})^{3/2}/\varepsilon$ (Daly and Harlow, 1970) and (iv) the length scale based on the mean square vorticity or frequency of the energy containing eddies, $(\overline{U_i^2})^{3/2}/\omega$ (Saffman, 1970). Exact transport equations can be derived only for (ii) and (iii). The transport equation (Rotta, 1971) is similar to the Reynolds stress transport equation except that (a) the length scale may be assumed to have no preferred direction in a given case, and (b) the transfer of energy to smaller scales by vortex stretching will cause a reduction in the length scale.

The length scale equation may be written as follows (Rotta, 1951; Wolfshtein, 1970).

$$\frac{D}{Dt}\left(\frac{q^2\lambda}{2}\right) = \frac{\partial}{\partial x_k}\left[(\nu + D_1\, q\, \lambda)\,\frac{\partial}{\partial x_k}\left(\frac{q^2\lambda}{2}\right)\right.$$

$$\left. + D_2 q^3\lambda\,\frac{\partial\lambda}{\partial x_k}\right] - D_3\lambda\,\overline{U_i U_j}\,\frac{\partial U_i}{\partial x_j}$$

$$- D_4 q^3 \tag{3.4.3}$$

where λ is proportional to all length scales and $D_1, \ldots D_4$ are four constants which must be determined from data.

An analysis of the length scale equation shows that the generation or production term is extremely important and at the same time the diffusion term is not dominating in its influence (Rotta, 1971; Rodi and Spalding, 1970).

In view of the fact that the length scale does not diffuse at a rate proportional to its local gradient, it has been more common to employ a variable combining the kinetic energy of turbulence and the length scale. The general form of the equation for such a variable, z, can be written

$$\rho\,\frac{Dz}{Dt} = \frac{\partial}{\partial y}\left[\frac{\mu_t}{\sigma_z}\,\frac{\partial z}{\partial y}\right] + z\left[c_1\,\frac{\mu_t}{k}\left(\frac{\partial u}{\partial y}\right)^2 - c_2\,\frac{\rho\frac{z}{k}}{\mu_t}\right] + S_z \tag{3.4.4}$$

which can be compared with the equation for k, namely

$$\rho \frac{Dk}{Dt} = \frac{\partial}{\partial y} \left[\frac{\mu_t}{\sigma_k} \frac{\partial k}{\partial y} \right] + k \left[\frac{\mu_t}{k} \left(\frac{\partial u}{\partial y} \right)^2 - c_D \frac{\rho_k^2}{\mu_t} \right]$$

(3.4.5)

In Eqn. 3.4.3, μ_t denotes turbulent viscosity, c_1 and c_2 are constants, σ_z is the Prandtl/Schmidt number for z and S_z is a secondary source or sink term. Both Eqns. 3.4.3 and 3.4.4 are two-dimensional boundary layer type equations for high Reynolds number flows.

3.5. Transport Equations

In the development of transport equations for scalar quantities, the turbulent diffusion approximation (considering for example a two-flux model (Launder and Spalding, 1972) and distinguishing between the mean value of the scalar variable and the average of the mass-average values traveling in the positive and negative lateral directions) is of course restricted in application unless the length scale for the scalar quantity as well as the turbulence energy field variation are both small.

An equation for the transport of the mean square fluctuation of a scalar quantity, g, has been obtained (Spalding, 1971) by analogy with the stress equation (see Eqns. 3.4.4 and 3.4.5) as follows

$$\rho \frac{Dg}{Dt} = \frac{\partial}{\partial y} \left[\frac{\mu_t}{\sigma_g} \frac{\partial g}{\partial y} \right] + c_{g1} \mu_t \left(\frac{\partial \overline{f}}{\partial y} \right)^2 - c_{g2} \frac{\rho k^{\frac{1}{2}}}{\rho} g$$

(3.5.1)

where c_{g1} and c_{g2} are constants, $\overline{f}$ is the time-averaged value of the scalar quantity. Models based on Eqns. 3.4.3, 3.4.4 and 3.5.1 cannot account explicitly for such phenomena as unmixedness and intermittency.

3.5.1 In the limit of homogeneous turbulence and isothermal flows, a closure scheme for second order chemical reactions has been proposed (Lin and O'Brien, 1972) based upon a procedure developed (Orzag, 1967) for turbulence dynamics. The closure scheme has been called the Inequality Preserving Closure Scheme which permits an ordered, though not unique, way of developing closures which satisfy prescribed inequality constraints. The inequalities that must be satisfied are associated with non-negative random variables and the requirement that the closure should asymptotically predict the first and second moments in the limit of no molecular diffusion.

The mass conservation equations for the local random concentration of species, after suitable non-dimensionalizing, become

$$\frac{\partial \Gamma_A}{\partial t} + N_D \, N_{Pe} \, \nabla \cdot \left(\underline{u} \Gamma_A \right) = N_D \, \nabla^2 \, \Gamma_A - \beta \, \frac{\Gamma_B(0)}{n \Gamma_A(0)} \, \Gamma_A \, \Gamma_B \qquad (3.5.2)$$

$$\frac{\partial \Gamma_B}{\partial t} + N_D \, N_{Pe} \rho \nabla \cdot \left(\underline{u} \Gamma_B \right) = N_D \, D_{BA} \, \nabla^2 \, \Gamma_B - \Gamma_A \, \Gamma_B \qquad (3.5.3)$$

where N_{Pe} is the Peclet number, β is the stoichiometric ratio and D_{BA} is the ratio of diffusivities of B and A species.

The closure which has been proposed is as follows

$$\overline{\gamma_A \gamma_A' \gamma_B} = \theta \left\{ \overline{\Gamma}_A + \frac{\overline{\gamma_A^2}}{\overline{\Gamma}_A} + \frac{\overline{\gamma_A \gamma_B}}{\overline{\Gamma}_B} + \frac{\overline{\gamma_A \gamma_B}}{\overline{\Gamma}_A^2} \frac{\overline{\gamma_A^2}}{\overline{\Gamma}_B} - \overline{\Gamma}_A \right\} \overline{\gamma_A' \gamma_B}$$

$$+ \frac{\overline{\gamma_A \gamma_A'} \cdot \overline{\gamma_A \gamma_B}}{\overline{\Gamma}_A} \qquad (3.5.4)$$

where γ and γ' are the fluctuation in concentration about the mean at two spatially separated points and θ is a constant determined by the initial statistical conditions; for example $\theta = 1$ if Γ_A and Γ_B are log normally distributed. The closed set of moment equations can then be solved numerically for given initial conditions and N_D, D_{AB} and β.

4. STATISTICAL CONTINUUM THEORIES

It is the general objective of statistical continuum theories to obtain a set of determinate equations for turbulent flow systems in contrast to the Reynolds transport equations which contain undetermined quantities. However, in view of the dynamic non-linearities (with or without a chemical reaction) in turbulent flow, the closure difficulties of the moment equations are formally the same as in the moment equations deduced from the Navier-Stokes equations for the turbulence problem. We shall not say anything here about the moment closure approaches. Instead, we shall discuss three models, two of them based upon a distribution function approach and the third based on a model equation for the distribution function of the velocities in turbulent flow.

4.1. Edwards' Model

The first approach to be discussed here is based on the premise that the problem of turbulence can be cast in the form of differential equations in function space using an appropriate probability density functional (Hopf, 1952; Chandrasekhar, 1956; Kraichnan, 1959; Edwards, 1964). In particular, one can consider the problem of a randomly excited turbulence (the forcing function being related in shear flow to the mean rates of strain); the input energy cascades from low wave numbers to high wave numbers through the action of the inertial terms in the equations of motion. Starting with a Liouville-type equation, itself deduced from the Navier-Stokes equations in function space, it has been shown that

(a) the resulting nonlinear integral equation for the energy spectrum of the velocity field is analogous to the Peierls-Boltzmann equation for phonon scattering;

(b) an expansion parameter exists in turbulence problems such that the Boltzmann-type equation may be replaced, with a prior knowledge of the spectral form in the inertial range, by a local differential equation, namely, the Fokker-Planck equation; and

(c) an analogue of the Fokker-Planck equation can be derived for the energy occupation of wave-number space, noting that the energy cascade is local in wave number. However, it should be pointed out that no criterion for the accuracy of the expansion is established.

It is then found (Edwards and McComb, 1971) that the notions of a generalized diffusivity and a generalized viscosity are fruitful and mathematically tractable in the steady state; the introduction of a generalized viscosity is the principal artifice in obtaining closure of the equations. It relates the response of the turbulent system to an infinitesimal perturbation and ensures positive definitions of spectra. However, this formulation still leaves completely open the question of a connection between the generalized diffusivity and viscosity. It has been argued that the generalized viscosity may be selected on the basis of maximizing the entropy of turbulence, though this is open to question in some respects.

Finally, it has been shown (Edwards and McComb, 1972) how differential equations (in the centroid variables) can be obtained for the mean velocity, kinetic energy and shear stress by the expansions of the nonlinear integral transport equations in a series of homogeneous kernels. Those kernels have a structure similar to the basic transport term in homogeneous turbulence. It appears that further developments are required before one can show, for example in a pressure-driven channel flow, that the transport and dissipation terms determine the velocity correlations and the high frequency components of the energy spectrum in effect determine the form of the mean velocity profile.

4.1.1 An extension of this approach to chemically reactive systems has been attempted (Hill, 1969). It is assumed that the molecular diffusivity of the species and the reaction rate are constant. However, the invariance properties and the inequalities have to be satisfied; in particular, all statistical functions of the concentration have to be independent of the velocity field in homogeneous turbulence in the limit of zero diffusivity (O'Brien, 1966). Using the Lagrangian equation for the concentration field and the conditional Eulerian pdf's, it has been shown by Hill that the zero-diffusivity invariance is more general than indicated by O'Brien and in fact applies to arbitrary isochoric motions provided the concentration is initially homogeneous, and also for multipoint functions.

The mean decay rates and the two point correlation functions of the concentration field can be obtained from the probability density functions. In order to determine the one and two-point pdf's of the concentration field, one can set up the Liouville equation, as discussed earlier, for the conditional probability functional. The single point pdf is then obtained by multiplying the Liouville equation by the full pdf of the velocity field for all time. The resulting equation includes the convective and the diffusive terms. The convective term has been approximated by expansion as stated earlier. The diffusion term is more difficult to handle. In practice, it is very difficult to carry out even this approximate procedure without additional assumptions. In any case, once a hierarchy of multipoint pdf equations is obtained in x-space, any closure of the hierarchy can be shown to preserve the zero diffisivity invariance and to yield the exact result in the absence of diffusion. Further, in the case of a second order reaction, if the independent variable namely concentration is replaced by the specific volume of reactant, the moment equations satisfy the invariance, are closed at each level and give the exact result when diffusion is neglected.

4.2. O'Brien's Model

The second approach (Dopazo and O'Brien, 1973) in recent statistical continuum theory of turbulent mixing is also based upon a functional equation formulation. A one-step, second-order, irreversible, exothermic chemical reaction (dynamically passive reactants) is examined using the pdf of the temperature and the concentration of species. The very rapid reaction examined is for the ignition period of the mixture. The velocity field is assumed to be with zero mean and unaffected by mass production or chemical heat. The Arrhenius rate constant is taken to be temperature dependent. The molecular diffusivities are assumed equal and the Lewis number is taken as unity. An equation for the single point pdf is then generated from the conservation equations, in a way rather similar to the approach of Edwards, under the assumptions of (a) negligible reactant consumption, (b) homogeneity and (c) the smallness of the correlation between the temperature and the concentration fields initially and during the initiation of the fast reaction. The equation for the pdf becomes the following

$$\frac{\partial}{\partial t} P(T, t) + C_1 \frac{\partial}{\partial t} \left[e^T P(T, t) \right]$$

$$= - \lim_{r \to 0} \nabla^2_r \frac{\partial}{\partial t} \int T' P(T, T'; r, t) \, dT' \qquad (4.2.1)$$

where $P(T, t)$ is the pdf of the temperature field at time t and $P(T, T'; r, t)$ is the joint pdf at time t separated by distance $\underline{r}$. In the foregoing equation, C_1 is assumed to be a constant, independent of time. It depends upon the concentrations and the correlation. There is, of course, no experimental confirmation of the assumptions made here.

The equation for the pdf is again not closed on account of the presence of the two joint pdf in the diffusion term. It may be pointed out that if convection had been included, a second term (the nonlinear convective term) would have arisen in the equation and so there would have arisen a second cause for the equation to remain unclosed.

The closure for the equation with only diffusion is obtained by assuming a Gaussian conditional expected value for the temperature at a point given the temperature at a neighboring point. This method of obtaining closure is similar to the one employed by Lundgren (1969) who introduced the BGK relaxation term. Suppose $E(T'|T; \underline{r}, t)$ is the conditional expected value of temperature fluctuation at the specific location and time. Then, Eqn. 4.2.1 becomes

$$\frac{\partial}{\partial t} P(T; t) + C_1 \frac{\partial}{\partial t} \left[e^T P(T; t) \right]$$

$$= \frac{\partial}{\partial T} \left\{ P(T; t) \left[-\lim_{r \to 0} \nabla^2 \underline{r} \, E(T'|T; \underline{r}, t) \right] \right\} \qquad (4.2.2)$$

If $P(T'|T; \underline{r}, t)$ is conditionally Gaussian,

$$E(T'|T; \underline{r}, t) = \overline{T}(t) + \rho(\underline{r}, t) [T - \overline{T}(t)] \qquad (4.2.3)$$

where $\overline{T}(t)$ is the mean of the random variable at time t and $\rho(\underline{r}, t)$ is the auto-correlation. If the turbulence is isotropic, $\rho(\underline{r}, t)$ becomes $\rho(r, t)$.

Finally, the probability density function equation may be written as follows:

$$\frac{\partial}{\partial t} P(T; t) + C_1 \frac{\partial}{\partial T} \left[e^T P(T; t) \right]$$

$$= C_2 \frac{\partial}{\partial T} \{[T - \overline{T}(t)] P(T; t)\} \qquad (4.2.4)$$

where C_2 is the inverse of the second Damkohler number based on the dimensional Taylor's microscale, the latter taken to be a constant.

The introduction of the dimensional microscale into the second Damkohler number is based on the suggestions of Corrsin (1958) and Kovasznay (1958). First, the microscales of the scalar and the velocity fields are related through the Schmidt number. Second, it is suggested that the intensity of turbulence in itself is insufficient to characterize the details of the turbulence and that the velocity gradient or the vorticity based on the microscale is also significant. However, there is no critical test to distinguish the influence of the turbulence scale and the microscale.

The first-order hyperbolic differential equation, 4.2.4, with variable coefficients can then be investigated to determine the evolution of $P(T, t)$ with time. However, the actual computation has presented several difficulties.

The application of the foregoing approximate method to turbulent shear flows is yet to be demonstrated.

4.3. Chung's Model

The third approach of interest here is one (Chung, 1970, 1972, 1973) wherein a model equation is developed for the distribution function of the velocities in turbulent flows.

The fundamental bases of this approach are the following: (a) At high Reynolds numbers, a statistical separation exists between higher equilibrium wave numbers and lower non-equilibrium wave numbers and the generalized Brownian motion is an adequate

description of the result (in the spectral plane) that the non-
equilibrium degree of freedom decays through its interaction with
all other degrees of freedom. It is an acceptable description once
the clear separation in the characteristic times of the large and
small eddies is recognized. (b) The transport of the various
properties in turbulence is due to the larger eddies and one can
assume that the statistical property of a fluid element is substantially due to the larger eddies. It is therefore possible to
consider a single significant dynamical scale associated with the
larger eddies. If one postulated more than one dynamical scale,
it is necessary to establish some connection between the two scales
(Bywater and Chung, 1973). There is at present no systematic way
of relating such scales. (c) In view of (a) and (b), one can
(see Chandrasekhar, 1943) set up Langevin's linear stochastic equation to represent the decay of the (fluid element) momentum associated with the lower nonequilibrium wave numbers. The important
boundary condition for this equation is that as the characteristic
decay rate of lower wave numbers tends to infinity and the molecular
dissipation tends to zero, the distribution function for a transferable property must become Maxwellian. It will be noted that the
distribution function is introduced as a weighting function and in
an average sense. In the final analysis we are concerned only with
certain moments of the distribution function. (d) The distribution
function can then be described by a modified form of Fokker-Planck
equation which is equivalent to the Langevin's equation. That
integral equation is converted into a differential equation by
Taylor series expansion. (e) The various order moment equations
deduced from the generalized moment equation corresponding to the
differential equation can be shown to be not totally inconsistent
with the corresponding order moment equations deduced from the
Navier-Stokes equations. In fact, the characteristic times for the
equilibrium and nonequilibrium degrees of freedom, the molecular
dissipation and the molecular diffusion are obtained from such
consistency considerations and assumptions pertaining to the
characteristic lengths of the larger eddies and the applicability
of the universal equilibrium theory.

One of the important steps in the foregoing is the introduction
of the Fokker-Planck equation for the distribution function. One
argument for the introduction of that equation is the assumption
that pressure fluctuations have a randomizing effect. In other
words, the Fokker-Planck equation is a means of modelling eddy
scrambling.

In applying this theory to chemically reactive flows, the
distribution function of the transferable quantity is written as
follows

$$F^{(\alpha)}\left[X_i, U_i, t\right] = n^{\alpha}\left[X_i, U_i, t\right] f\left[X_i, U_i, t\right] \qquad (4.3.1)$$

where f dx du is the occupancy probability of the fluid element, n, the concentration of a transferable property and α is a constant equal to or greater than zero. The probability average of n^{α} is then

$$< n^{\alpha} > = \int F^{(\alpha)} \, d\underline{u}. \qquad (4.3.2)$$

Now, one can deduce the generalized moment equation including n and compare the different order moment equations obtained therefrom with the moment equations obtained from the classical equations.

In the case of a single chemical specie diluted in an inert medium, the modified Fokker-Planck equation for f is complete in itself for the so-called homologous case--stationary, homogeneous field with uniform velocity and concentration gradients. The determination of the concentration field is then straightforward in problem formulation.

The next order of complication arises when we consider, for example, two uniform, isotropic, parallel streams in contact which yield a homologous region both with respect to the velocity and the concentration fields. The two parallel streams could have different, but at the same time constant, velocities and concentrations. If the concentration in the two streams is totally independent of the velocities therein, no inhomogeneities will arise in the mixing layer.

Finally, one has the case where chemical reaction can occur between the two species. Any formulation of that problem within this framework will require assumptions pertaining to (a) the Damkohler number, (b) the molecular Schmidt number, (c) the mean gradients of velocity and concentration in the mixing layer, and (d) the coupling between the concentration fluctuations and the structure of turbulence in the total flow field of interest. Hence the turbulence transport and the mean square fluctuation of concentration will become a function of those parameters.

4.3.1 We shall discuss here the application of the foregoing theory to the study of a diffusion flame. Within the framework of this method of problem formulation, the reaction zone has been assumed to be embedded within the homologous zone. Reactants A and B, under highly diluted conditions and moving in parallel streams, come into contact at a thick mixing zone. Under the one-dimensional approximation, without convection, the boundaries of the thick zone are fixed. The fluid within this zone is non-homogeneous, while outside the distribution of fluid properties is isotropic and Maxwellian in both of the streams. The thick zone satisfies the homologous conditions. Inside this thick zone, there is a reaction zone which also has a finite thickness. The reactants and the product are distributed throughout the thick mixing zone but with the following restraints: (a) on the side of reactant A in the mixing zone outside the reaction zone, there is only reactant A and product; (b) on the side of reactant B in the mixing zone outside the reaction zone, there is only reactant B and product; and (c) in the reaction zone, there is product as well as the unreacted parts of the reactants. The latter requires further explanation. It is assumed that while chunks of reactants A and B meet within the reaction zone, the reaction itself is controlled entirely by molecular diffusion of one reactant into the other across the interface separating the chunks of the two reactants. It is then assumed that (a) instantaneously some of the chunks of reactants find themselves in orientations which facilitate molecular diffusion and hence reaction, while others are in orientations where they slip past each other without any reaction; and (b) the reaction zone thickness is determined by the rate of chemical action and the diffusion. Within the one-dimensional no-convection approximation, the boundaries of the reaction zone thickness are fixed. If infinitely fast chemistry is assumed, the reaction zone thickness becomes nothing but that of the order of the integral scale of turbulence. It is useful to recall here that the flow has been assumed to be governed by two characteristic lengths, L, the characteristic length of the flow field, and λ, the characteristic length of the large scale eddies, treated as uniform in this approximation. Thus, the thickness of the reaction zone becomes directly proportional to λ and inversely (and nonlinearly) proportional to the normal gradient of the mean velocity. We may also note here that the boundaries of the reaction zone will remain sharp with discontinuous gradients as long as the Damkohler number is assumed to be infinitely large but the assumption of large but finite chemical reaction rates will smoothen the profiles at the edges of the reaction zone. It should be clear from the foregoing that an extremely idealized situation with a specific model for the occurrence of chemical reaction is being considered here. Whether this model can be improved systematically is entirely dependent upon establishing experimental results in support of this physical picture of a reaction front.

We have described earlier the analytical formulation of the problem on the basis the generalized moment equation deduced from the Fokker-Planck equation and the particular moment equations deduced from the generalized moment equation. Such particular moment equations have to be generated for the reactants, the diluent, the product and the enthalpy. Each of these involves the distribution function and the next major question is an approximation for the distribution function in the two streams. Here an approximation is introduced based on the assumption that a general distribution function applicable in the mixing zone, with emphasis on the possible skewness of such a distribution function, can be approximated by the sum of two half-Gaussian distributions (Liu and Lees, 1961). The only sense in which such equivalence can be considered is that the moments generated from such distributions can be arbitrarily and on an integral basis be equated to the moments generated from any actual non-Gaussian distribution. However, since the reaction zone thickness and structure have already been assumed to be fixed, the approximation concerning the distribution function is consistent. On the other hand, this still leaves open the question of differences between the distribution functions for the various quantities in the mixing zone and again it is assumed that in each case the distribution function can be written as the sum of the two half-Gaussian distributions.

If we concentrate on the thickness of the reaction zone, the distribution of the product and the distribution of the enthalpy, according to this model, assuming the Arrhenius rate of instantaneous reaction, the governing parameters become (a) λ/L, (b) the ratio h_1/h_2 corresponding to the half-Gaussian distributions of temperatures assumed and (c) the mean velocity gradient. The gross features of the present stage of this model should be obvious. One can in fact summarize the implications of the model as follows: if a flow field is homologous with respect to all its properties in the absence of reaction and if a reaction is permitted in a finite but initially unknown part of that flow field, what is the departure from the homologous character of the flow field that will arise taking account of the product and enthalpy generation? The answer to this question has been sought in terms of arbitrarily postulated distribution functions with properties related to the integral properties of the distribution functions that may be obtained in reality.

4.4. Spectral Structure of Shear Flows

In view of the importance of the behavior of wave number spectra in the basic studies on turbulence, we shall touch briefly upon the spectral structure of turbulent scalar and reactive fields. It is well known (Kolmogorov, 1941) that at high Reynolds numbers, the small scale components of turbulence (small eddies, high wave numbers) can be considered to be steady, isotropic and independent of the large scale components (large eddies, low wave numbers) and that energy is transferred from the larger eddies to the smaller eddies. By dimensional analysis, the length and velocity scales for the dissipating eddies can be shown to be $\eta = (\nu^3/\varepsilon)^{\frac{1}{4}}$ and $v = (\nu\varepsilon)^{\frac{1}{4}}$. If L is the characteristic length of the energy containing eddies, then $k \gg k_0$ ($= 1/L$) represents the universal equilibrium range of wave numbers, $k_0 \ll k \ll k_K$ ($= 1/\eta$) represents the inertial subrange and $k \geq k_K$ represents the dissipation subrange. The behavior of the small scale eddies can be obtained explicitly in the inertial subrange where dissipation is negligible and the transfer of energy is dominantly by inertia forces. The dissipation range is much more difficult to handle. The theories (for example, Obukhov, 1941; Heisenberg, 1948) developed for the universal range are unlikely to be applicable in the dissipative range. Various phenomenological (Townsend, 1951) and approximate theories (Kraichnan, 1959, 1962; Edwards, 1964) are available.

In respect of scalar quantities such as concentration and temperature the inertial subrange has been investigated (Obukhov, 1949; Corrsin, 1951) to determine the small scale structure of the fluctuating field. If χ is the rate at which the scalar quantity is fed into the small scale components, the spectrum function for the scalar field $\theta(\underline{x}, t)$ is given by dimensional analysis as follows for the convective subrange provided the Reynolds and the Peclet numbers are sufficiently large.

$$F(k) = A_1 \chi \, \varepsilon^{-1/3} \, k^{-5/3} \tag{4.4.1}$$

where A_1 is a dimensionless constant. Batchelor (1959) has suggested for large Schmidt numbers that one should consider a viscous convective subrange, $k_K \ll k \ll (\varepsilon/\nu D^2)^{1/4}$, and a viscous convective subrange. Based on a uniform straining model, he has obtained

$$F(k) = - \chi \, \alpha^{-1} \, k^{-1} \, \exp\left[\frac{D}{\alpha} k^2\right] \tag{4.4.2}$$

where $\alpha = -\frac{1}{2} (\varepsilon/\nu)^{\frac{1}{2}}$. Furthermore, an inertial diffusive subrange has been postulated (Batchelor et al., 1959) for small Schmidt numbers, $k_c = (\varepsilon/D^3)^{1/4} \ll k \ll k_K$ for which

$$F(k) \cong 1/3 \, \chi \varepsilon^{2/3} \, D^{-3} \, k^{-17/3}. \tag{4.4.3}$$

The spectra for scalar quantities has been studied (Corrsin, 1964) for turbulent mixing with second-order chemical reaction in the universal range.

Another approach to the study of the transfer of turbulent kinetic energy and scalar quantities at large wave numbers is based upon a simple continuous spectral cascading process (Pao, 1965). The entire universal equilibrium range of wave numbers can be covered by this. Considering that the cascading is mainly due to turbulent convection which in turn is dependent upon ε, the cascading rates are assumed to be dependent upon ε and k.

For a second-order chemical reaction involving reactants and product, one can then obtain the spectrum function as follows.

(a) In the universal equilibrium range, one considers $k > k_\theta$ where k_θ is based on the smaller of the integral scales associated with the scalars. One finds then

$$F_{\overline{\theta^2}} (k) \sim k^{-7/3} \exp\left[-3/2 \, A_1 \, k^{4/3} + 3 \, A_2 \, k^{-2/3}\right] \qquad (4.4.4)$$

(b) In the inertial convective range, with diffusivity and viscosity negligible,

$$F_{\overline{\theta^2}} (k) \sim k^{-5/3} \exp\left[3A_3 \, k^{-2/3}\right] \qquad (4.4.5)$$

and (c) in the viscous-convective and viscous-diffusive ranges, if we assume that the cascading rate depends upon the local least turbulent straining rate,

$$F_{\overline{\theta^2}} (k) \sim k^{n_1} \exp(-n_2 \, k^2) \qquad (4.4.6)$$

where n_1 is related to the diffusivity and the strain rate and n_2 is inversely proportional to the strain rate.

5. SIMPLIFIED APPROACHES TO REACTIVE FLOWS

We will now ask a specific question: What are the conditions under which turbulent mixing theory can be combined with the theory for stochastically distributed reactive systems to yield a method of analyzing reactive turbulent flows of the type we have referred to earlier (Lin and O'Brien, 1972; Bush and Fendell, 1973)? Some of the difficulties associated with this postulate should be clear from the earlier discussion on the non-Gaussian distribution of reactant concentration and the persistence of the influence of high order fluctuation moments.

5.1. An Independence Principle

First, we shall discuss an independence principle (Corrsin, 1964; O'Brien, 1969) which can be stated as follows: consider a second-order reaction involving a single specie in a turbulent convection and molecular diffusion field. The mass conservation equation can be written as

$$\frac{\partial \Gamma}{\partial t} + \underline{\nabla} \cdot \left(\underline{U}_\Gamma\right) = D \, \nabla^2 \, \Gamma - K \, \Gamma^2 \qquad (5.1.1)$$

where Γ is the random concentration of the single specie, $\underline{U} \, (\underline{x}, \, t)$

the velocity field, D the diffusion constant, and K the reaction rate. Consider a homogeneous, incompressible flow with zero mean velocity. Writing

$$\gamma\ (\underline{x},\ t) = \Gamma\ (\underline{x},\ t) - <\Gamma(t)>$$

$$= \text{fluctuation about the mean}$$

and

$$\phi\ (\underline{k},\ t) = \text{transform of} <\Gamma\ (\underline{n},\ t)\ \Gamma\ (\underline{n} + \underline{r},\ t) >,$$

the question that is being asked is as follows. If $\phi_C(\underline{k},\ t)$ is the solution of the first moment of the conservation equation and the spectrum function equation in Fourier space when D = 0, $\underline{U}$ = 0, and if $\phi_C(\underline{k},\ 0)\ Q(\underline{k},\ t)$ is the solution of the same equations when K = 0, $Q(\underline{k},\ 0)$ = 1, then an approximation to the solution, $\phi(\underline{k},\ t)$, of the equations for arbitrary D and $\underline{U}$ is given by

$$\phi_I = \phi_C\ (\underline{k},\ t)\ Q(\underline{k},\ t).$$

This has been referred to as the independence hypothesis. Physically, the hypothesis implies that the amplitude of Fourier modes of the concentration in the given flow-reaction configuration is governed by the product of its time history due to mixing and diffusion and its time history due to reaction.

A point to observe here is that since the first moment equation and the spectrum function equation are not closed, some form of closure (at a higher moment level) is needed. This has been given in the form of a consistency condition whereby only the spectral decay due to reaction exhibits wave length similarity with a constant length scale. Though some useful information can be extracted on mean and rms concentration based on the independence hypothesis, a number of questions remain, especially in regard to the role of turbulence and the detailed spectral transfer due to reaction.

Now, the consistency condition may be written as

$$\phi_C\ (\underline{k},\ t) = F\ (t)\ \phi_C\ (\underline{k},\ 0),$$

$$F(0) = 1. \tag{5.1.2}$$

It follows from this that

$$\frac{<\gamma^2(t)>_I}{<\gamma^2(0)>} = \frac{<\gamma^2(t)>_K}{<\gamma^2(0)>} \cdot \frac{<\gamma^2(t)>_M}{<\gamma^2(0)>} \tag{5.1.3}$$

where the subscripts K and M denote conditions with reaction alone and mixing alone, respectively. It is unfortunate that there are not enough experimental data to determine the applicability of the hypothesis. The hypothesis seems to have been most successful in the final period turbulence problem wherein the role of turbulent convection on the scalar spectrum for most wave numbers is less important than the role of molecular diffusion.

While the discussion on the independence hypothesis was begun by considering a second-order reaction involving a single specie, it is not possible to demonstrate the same kind of hypothesis in the case of a reaction (assumed stoichiometric) involving two reactants and a product (O'Brien, 1971). However, in the case of a slow reaction regular closure schemes seem unavoidable. On the other hand, in the case of a fast reaction, one can proceed by distinguishing an initial stage of reaction from a final stage. It has been shown that during the initial stage, reaction is the dominant factor in determining the decay of random fields. The initial stage period of course depends upon the shortest time scale of turbulent convection and diffusion and the reaction rate. Exact stochastic solutions can be shown to exist during the initial stage for appropriate initial conditions.

Now, using the solutions of the initial stage as the initial conditions, one can proceed to examine the final stage of the reaction where diffusion is the governing factor. If the diffusivities of the two species are unequal, an approximate solution is all that is possible for mean square concentrations. The time scale of decay of each species becomes related to the time scale associated with mixing of a single species, that is, the solutions of the reacting species can be constructed if the probability laws of the nonreacting turbulent mixing problem are known.

Before leaving this subject we may refer to two special features of the solutions for rapidly reacting mixtures. First, during the initial stage of the reaction the relative intensity of concentration fluctuations may become quite large simply depending upon the log-normal distribution parameter of the initial state and also it has a lower bound. This fact combined with the large positive lower bounds for the asymptotic skewness and kurtosis of the concentration fields seems to indicate that the probability distribution of the concentration is not even approximately normal. Second, when the diffusivities of the two reactants are widely different, the smaller of the diffusivities determines the decay rate of stoichiometric reactions. Thus, single specie experiments may not become of value in examining two-specie reactions in this regime.

5.2. Toor's Approach

We will next turn to another approach for very rapid reactions (Toor, 1962). It will have been realized by now that the basic difficulty in turbulent reactive systems is not the determination of the exchange flux of the species involved but the establishment of the details of the concentration profiles and the reaction distribution. On a time-averaged basis, the turbulent systems differ from the non-turbulent systems only in the fact that the interface region in the turbulent system cannot be a thin surface even with rapid reaction. When two turbulent streams come into contact, if the boundary concentrations are not fluctuating and the diffusivities of the two streams are equal, one can calculate the mass exchange flux in the same manner as in non-turbulent flow situations for rapid reactions. However, one cannot say that given the time-averaged concentration profiles in a non-reacting turbulent system one can establish in a corresponding reactive system the time-average rate of reaction even for a rapid reaction. The second approach to this problem--the first being the independence hypothesis--consists in making approximations by means of which a knowledge of the time-average behavior in the equivalent non-reactive system will permit the time-average behavior of the reactive system provided the rms concentration fluctuations are also known in the equivalent non-reactive system. This statement is in a way equivalent to the consistency condition that was employed in the deduction of the independence hypothesis. Incidentally, we may note here that there are very few data for temperature and mass fraction fluctuations especially in reactive systems.

The nature of the approximation involved is very simple (Bush and Fendell, 1973). There are several steps in setting up the approximation. Let us consider an irreversible homogeneous reaction between two reactants (highly diluted) A and B as follows.

$$aA + bB \xrightarrow{\overline{K}} product$$

where a and b are stoichiometric coefficients and $\overline{K}$ is the time-averaged reaction rate. If C denotes the concentration and

$$C^* = bC_A - aC_B,$$

the mass transfer equation for equal diffusivities (D_1) of the two species, including reaction, may be written

$$\frac{D}{Dt} C^* = D_1 \nabla^2 C^* \tag{5.2.1}$$

Next, the molar flux at the boundary separating the species can be written as

$$\underline{N}^* = bN_A - aN_B$$

$$= D_1 \nabla C^* + \underline{U} \ C^*$$

where $\underline{U}$ is the velocity vector.

On introducing the time-averaged and fluctuating quantities into Eqn. 5.2.1, it follows that

$$\frac{\partial}{\partial t} C^* + \nabla \cdot (\underline{U} \ \overline{C}^* + \underline{u} \ C^{*\prime}) = D_1 \nabla^2 \overline{C}^* \tag{5.2.2}$$

Now, we assume for equivalence of the reactive and the non-reactive systems that the geometry, the turbulent fields and the boundary conditions are the same in the two systems.

The first step then is that certain transformations can be shown to exist by means of which Eqns. 5.2.1 and 5.2.2 can be shown to be the same as simple mass transfer equations by replacing C^* with the no-reaction value.

Suppose now that the concentration of A at boundary 1 is C_{A1} and the concentration of B at boundary 2 is C_{B2}. The second step then is to relate $\overline{C}^*$ to the time-average behavior of non-reactive systems by means of the relation

$$\frac{\overline{C}^* - C_2^*}{C_1^* - C_2^*} = \frac{b \ \overline{C}_A - a \ \overline{C}_B + a \ C_{B2}}{b \ C_{A1} + a \ C_{B2}} = f(Z_1, \ \ldots \ Z_n) \tag{5.2.3}$$

where Z is the space or time-coordinate.

Having established the time-averaged value, the next question is the determination of the concentration profiles. Introducing the instantaneous concentration field and the no-reaction case variables in Eqn. 5.2.3, it follows that

$$\frac{\overline{C}_0 + C_0^\prime - C_{20}}{C_{10} - C_{20}} = f_1 \ (Z_1, \ \ldots \ Z_n). \tag{5.2.4}$$

where the subscript 0 indicates the no-reaction case. If we now examine the time-averaged concentration profile, it becomes a function of not only $\overline{C}^*$ but also the relative intensity to the rms concentration fluctuation. Therefore the fluctuation intensity of the reactive system must be related to the fluctuation intensity of the non-reactive system. In the final step this can be done in terms of the stoichiometric factor provided the reactant means are equated to the mean of the reactants.

The difficulty in applying even this approach is that of determining $\bar{C}*$ and the probability distribution function which is a function of $\bar{C}*$ and C'. Clearly, it is desirable to devise experiments by means of which the probability distribution function, undoubtedly non-Gaussian in character, can be approximated by a suitable modelling of $\bar{C}*$. In general, the pdf should be obtained in the hot flame itself. However, it seems cold flow experiments could be conducted to determine the extent of the correspondence between the reacting and non-reacting systems.

5.3. Extension to Shear Flows

There has been an attempt recently (Lin and O'Brien, 1974) to extend the Toor analysis to shear flows. Other than the assumptions of incompressibility, dynamically passive reactants and a one-step, second-order, irreversible exothermic reaction, it is assumed that the heat generated due to chemical action is a constant and that it is the dominant term in the energy balance equation. The temperature fluctuations are taken to be small.

Considering a very fast reaction (species segregation and thin interface surfaces) and the non-negativeness of concentration, one can then write

$$L\,(\chi) = 0 \tag{5.3.1}$$

$$L\left[\Gamma_B + \Gamma_P\right] = 0 \tag{5.3.2}$$

$$L\left[\Gamma_B + T\right] = 0 \quad \text{and} \tag{5.3.3}$$

$$L\left[\Gamma_B\right] = W_A$$

where P is the product of the reaction between A and B, W_A is the rate of production of specie A in moles,

$$\chi(\underline{n},\, t) = \Gamma_B\,(\underline{n},\, t) - \Gamma_A\,(\underline{n},\, t) \tag{5.3.4}$$

and the operator L is given by

$$L = \partial/\partial t + \underline{V} \cdot \nabla - D\nabla^2.$$

It may be noted here that the turbulent mixing times are included in the operator L and there is no chemical kinetic scale in this formulation.

One can then set up an equation for the pdf in the following manner.

$$\lim_{R\to\infty} P_{\Gamma_B}\left[\Gamma_B(\underline{n},\ t)\right] = P_{\kappa}\left[\Gamma_B(\underline{n},\ t)\right]\ H\left(\Gamma_B\right) +$$

$$\delta\left(\Gamma_B\right)\int_{-\infty}^{0} P_{\kappa}[\kappa(\underline{n},\ t)]\ d\kappa \tag{5.3.6}$$

where R is the ratio of the reaction rate to the rate of turbulent diffusion and H is the Heaviside's unit function. Equation 5.3.6 thus replaces Eqn. 3.4.1 and the pdf for B is given now in terms of the pdf for κ.

The question next is how to solve the set of Eqns. 5.3.1-- 5.3.4 and 5.3.6 without the almost impossible recourse to the Navier-Stokes equations. The suggestion has been made that one can set up an analogy between the joint pdf of the coupled reactive system and the joint pdf of a double scalar mixing field. It is assumed in the analogy that the velocity fields are identical.

In order to understand the nature of the analogy, one can consider the two-dimensional mixing layer in the wake of a plate. If the fluids on either side of the plate are at the same tempera- ture but with concentrations Γ_A and Γ_B, the analogical flow becomes one with no concentration and a certain temperature on the upper- side and a different temperature and the experimental concentration on the lower side. Supposing then the directly measured pdf's of the temperature and the temperature-nonreacting concentration fields is available; it is shown that the pdf's of Γ_A, Γ_B and Γ_P can be obtained. Numerical integration of the experimental data will be required in all cases except for the mean concentration of the product.

In the foregoing example, if the original system under consider- ation has equal temperature fluids on either side, $P_{\Gamma_B T}[\Gamma_B,\ T]$ can be obtained from $P_{\Gamma_B \Gamma_P}[\Gamma_B,\ \Gamma_P]$ directly.

5.4.   Intermittency in Mixing Layers

As a final example of superposing the chemical field on the turbulence field, we shall discuss a theory (Spalding, 1970; Bilger and Kent, 1972; Libby, 1973) based on the role of intermittency in mixing layers.

Intermittency is a fundamental characteristic of mixing layers, especially when the central mixing layer is bound by interfaces separating the outer non-turbulent flows. It has been pointed out (Wygnanski and Fiedler, 1970) that intermittency is distributed throughout the mixing region. Turbulent fluctuations are produced by intermittency effects in addition to those produced by fully developed turbulence. A striking feature of turbulent shear flows is the existence of sharply defined interfaces (which are neither stream nor path surfaces) separating a region of (homogeneous) vorticity fluctuations from the rest of the fluid (Corrsin, 1943; Townsend, 1949; Corrsin and Kistler, 1954). The intermittency demonstrates a large scale superstructure upon a fine scale turbulent flow. It affects the measurement and interpretation of the pdf of the fluid mechanical quantities common to the regions on either side of the interfaces. A stretched lateral coordinate scale, for instance, requires an assumption regarding the distribution of the instantaneous turbulent front coordinate. Some interesting ideas can be developed concerning the average duration and length of a turbulent burst from a knowledge of the velocity at which the interface is convected downstream. The manner in which that velocity is related to the convection velocity of large scale eddies is unclear. However, one can observe the connection between the roughness or flatness behavior of the mixing layer and similar phenomena in turbulent boundary layers.

Another important finding in regard to the convoluted turbulent interface is that while intermittency has no significant effect on the velocity of the fluid in the nonturbulent region between the successive lateral excursions of the turbulent front, the scalar properties within the same excursions may remain uniform and display a significant change across the turbulent front. Thus, in the case of a reactive fluid, there will arise random fluctuations of stoichiometry which are related to the intermittency. If it is assumed that the instantaneous composition is completely determined by the instantaneous stoichiometry, it becomes necessary to have some knowledge of the pdf of a parameter indicating the departure from mean stoichiometry. The difference between the distributions that one can assume for the velocity fluctuations and the scalar property fluctuations in the outer parts of the mixing layer becomes one of the central issues in including intermittency considerations.

Consider a shear flow generated by the mixing of two incompressible, parallel streams with a one-step, very rapid, isothermal chemical reaction involving two reactants A and B under dilute conditions. Denoting the mass fractions of species by Y_i, the rapid reaction is defined by the relation

$$\overline{Y_A Y_B} = 0$$

and the mean rate of production of each species $\bar{w}_i$ being the value appropriate to the processes governing the system. Assuming highly diluted reactants, the velocity field is first represented by the mean velocity field and an assumed distribution of intermittency. The simplest way of doing this is to invoke similarity and employ the distribution of intermittency obtained in non-reactive shear layers. Thus, introducing the stream function $f(\eta)$ and the intermittency I, one can write

$$f'(\eta) = f'(0) + 1/2 \ G \ [erf \ (\sigma(\eta + \eta_0)) - erf(\sigma\eta_0)] \qquad (5.4.1)$$

$$f = \int_0^\eta f' \ d\eta \qquad (5.4.2)$$

$$\bar{I} = 1 - erf \ (\alpha\sigma^2\eta^2) \qquad (5.4.3)$$

where η is the stretched variable, x_2/x_1, G is the ratio of the mean velocities of the two streams, σ is a spreading parameter of the mixing layer and α is a constant of $O(1)$.

Considering next the concentration field, the similarity form for the distribution of the mass fractions is written as follows:

$$f \ Y_i' = \left[\overline{Y_i' \ u_2}/U\right]' - \left[\bar{w}_i \ x_1/U\right] \qquad (5.4.4)$$

where U is the velocity of the reference stream. Now it is argued that if we know $\bar{Y}_i$ by whatever means, then the mean specie generation $(\bar{w}_i x_1/U)$ can be found as a function of η from Eqn. 5.4.4.

Defining the mass fraction of element i by Z_i, the conservation equation for Z_i can be written from Eqn. 5.4.4 as follows:

$$f \ \bar{Z}_i = \left[\overline{u \ L_i'}/U\right]' \qquad (5.4.5)$$

Introducing again a conserved scalar quantity ζ defined by

$$\zeta = Z_B - \left[W_B/W_A\right] Z_A = Y_B - \left[W_B/W_A\right] Y_A \qquad (5.4.6)$$

where W is the molecular weight, it follows that at a particular space-time point,

$$\zeta > 0 \longrightarrow \zeta = Y_B$$

$$\zeta < 0 \longrightarrow \zeta = Y_A.$$

In other words, one can introduce conserved concentration variables such that at a given space-time point, $\zeta > 0$ indicates passage of eddies with reactant B and product and $\zeta < 0$ indicates passage of eddies with reactant A and product.

The mean concentrations can therefore be related to the <u>conditional</u> probability

$$\overline{Y}_B = E(\zeta|\zeta > 0)$$

$$\overline{Y}_A = - W_A/W_B \; E(\zeta|\zeta < 0)$$

$$(5.4.7)$$

where E is the expected value of ζ. The conditional events are given to the right of the vertical bar. The entire concentration field therefore can be found from a knowledge of Z_i and Eqns. 5.4.5 and 5.4.7.

The question then is how to establish by theory or experiment adequate information regarding the pdf of ζ, the mean values of Z_i and ζ and the first few moments of the fluctuations. It is precisely here that several models for the pdf of ζ have been examined incorporating the intermittency of the flow. However, there is not yet a convincing means of accounting for the intensity of the scalar variable fluctuations. The phenomenology of scalar quantities in intermittent flows is inadequate at present. Even the mean values of Z_i and ζ can be established only on heuristic grounds.

5.4.1 In this connection, on the basis of a model for the ensemble average concentration, it has been shown (Alber, 1973) that the mean and fluctuating species profiles can be developed for a two-specie, incompressible, infinitely fast chemical reaction, say

$$A + B \rightarrow C$$

where C is the product. Assuming equal diffusivities and defining

$$Z_1 = A - B, \quad \text{and}$$

$$Z_2 = A + C,$$

one can set up the diffusion equations governing Z_1 and Z_2 in a two-dimensional mixing layer formed between two streams with either A or B. In the fast reaction limit, as stated earlier, one can say

$$\text{if} \quad Z_1 > 0, \quad A = Z_1, \quad B = 0, \quad C = Z_2 - Z_1, \quad \text{and}$$

$$\text{if} \quad Z_1 < 0, \quad A = 0, \quad B = Z_1, \quad \text{and} \quad C = Z_2.$$

Thus, if the complete turbulent fields related to Z_1 and Z_2 are known in terms of their statistical properties, the distribution of the product in the shear layer can be established.

In order to establish the ensemble average concentration of the product, suppose we assume Z_1 and Z_2 are perfectly correlated. One can represent the ensemble average of the product in terms of a single variate probability distribution function, say

$$<C> = \int_{-\infty}^{\infty} C(Z_2) \, f(Z_2) \, dZ_2$$

where it is necessary to model the distribution function $f(Z_2)$. On the basis of other arguments one can treat Z_2 as a chemically passive scalar, say temperature.

The product of the reaction is related to Z_2 in the following manner. Suppose we assume that the time-averaged solutions for the diffusion equations are given by

$$\frac{<Z_1> + B_0}{A_0 + B_0} = \frac{<Z_2>}{A_0} = F(\eta)$$

where $(\ )_0$ refers to the two streams and $F(\eta)$ is linear in η, the stretched nondimensional coordinate normal to the shear layer. Suppose also we assume that the solution holds instantaneously. Then

$$\frac{Z_1}{A_0} = \left(1 + \frac{B_0}{A_0}\right) \frac{Z_2}{A_0} - \frac{B_0}{A_0} \ .$$

It follows then that

$$\frac{C}{A_0} = \frac{B_0}{A_0} \left(1 - \frac{Z_2}{A_0}\right), \quad Z_1 < 0, \quad \text{and}$$

$$\frac{C}{A_0} = \frac{Z_2}{A_0}, \quad Z_2 > 0 \ .$$

5.5. Bray's Analysis

An important question raised by Bray (1973, 1974) in turbulent flame modelling (see also Section 6.3) is related to the implications of neglecting reaction-rate dependent terms in modelling turbulence quantities and of starting with equations applicable to cold, incompressible and non-reacting flows. Bradshaw and Ferris (1971) already showed how the incompressible flow equation was not adequate for compressible turbulent boundary layers at low Mach numbers. Bray shows how an order of magnitude analysis of compressible, reactive flow reveals several terms which have to be taken into account, for example, the terms indicating the coupling between the velocity divergence and the reaction and heat release.

The order of magnitude analysis is based on the following simplified model: (a) A two-dimensional shear layer (low Mach number and high Reynolds number) is considered with a distributed time-average chemical reaction zone and under boundary layer approximations. (b) The fluctuating magnitudes of all quantities are estimated on the basis of the approximation

$$f_0'/f_0 = E^{\frac{1}{2}} \tag{5.5.1}$$

where $E \sim V/U$ and f_0 is any flow property in a reference state. (c) The pressure fluctuations are assumed to be proportional to $\bar{\rho}_0 U_0^2$. Further, it is assumed that the largest contributions to the pressure fluctuation terms arise on account of the appearance of discrete flamelets within the combustion zone and that the con- tributions can be taken to be well-correlated in view of the possible regular structure of the flamelets. The reasons for neglecting the pressure fluctuations in the final energy balance equation are thus heuristic. (d) The reaction rate fluctuations are considered large. Thus, it is assumed that

$$w_0'/\bar{w}_0 \sim 1$$

where $\bar{w}_0$ is the time-average rate of production of species given by

$$\bar{w}_0 = \bar{\rho}_0 U_0/\ell_0$$

(e) Two turbulence length scales are introduced, one characteristic of the energy-bearing eddy scales and the other being a Taylor microscale. They are related to the dissipation term in the energy balance equation. Two reference length scales are also introduced, one the width of the mixing region (a dissipation length scale) and another for the length of interest in the flow direction. (f) In order to obtain closure of the equations, the Reynolds stress is related to the turbulence energy (Rhodes and Harsha, 1972). (g) Finally, the specific heats are assumed to be constant and the Prandtl-Schmidt number assumed to be of order unity.

Thus the basic equations in two-dimensions become the following.

(1) Time-average divergence of velocity

$$\frac{\partial U}{\partial x} + \frac{\partial V}{\partial y} = -\frac{\gamma-1}{\gamma} \frac{1}{p} \sum_{i=1}^{n} H_i \bar{w}_i \tag{5.5.2}$$

where H_i is the heat of formation.

(2) The pressure-fluctuation terms

$$\overline{p'\,\frac{\partial u}{\partial x}} + \overline{p'\,\frac{\partial v}{\partial y}} + \overline{p'\,\frac{\partial w}{\partial z}} = \frac{\gamma-1}{\gamma}\frac{1}{\overline{p}}\left[\overline{k\,p'\,\frac{\partial^2 T'}{\partial x_\beta^2}}\right.$$

$$+ \sum_{i=1}^{n}\varepsilon_{pi}\,\overline{T}\,\overline{D}_i\,\overline{\rho}\,\overline{p'\,\frac{\partial^2 C_i}{\partial x_\beta^2}} - \sum_{i=1}^{n}H_i\,\overline{p'\,w_i'}\left.\right] \qquad (5.5.3)$$

where the three terms on the right hand side can be shown to be of
the order of

$$E^{\frac{1}{2}}/15, \quad E^{\frac{1}{2}}/15 \quad \text{and} \quad E$$

respectively.

(3) The turbulence kinetic energy equation

$$\overline{\rho}\,U\,\frac{\partial \overline{q^2}}{\partial x} + \overline{\rho}\,V\,\frac{\partial \overline{q^2}}{\partial y} = \frac{a_1\,\overline{\rho}\,\overline{q^2}}{\left|\frac{\partial U}{\partial y}\right|_m}\left[\frac{\partial U}{\partial y}\right]^2$$

$$+ a_4\,\overline{\rho}\,\overline{q^2}\,\frac{\gamma-1}{\gamma}\frac{1}{\overline{p}}\sum_{i=1}^{n}H_i\,\overline{w}_i + \overline{p}\,\frac{\partial}{\partial y}\left[\frac{a_1\overline{q^2}}{N_{q^2}\left|\frac{\partial U}{\partial y}\right|_m}\frac{\partial \overline{q^2}}{\partial y}\right]$$

$$+ \left[\frac{1}{N_{q^2}} + \frac{1}{N_p}\right]\frac{a_1\overline{q^2}}{\left|\frac{\partial U}{\partial y}\right|_m}\frac{a_1\overline{q^2}}{\left|\frac{\partial U}{\partial y}\right|_m}\frac{\partial \overline{q^2}}{\partial y}\frac{\partial \overline{p}}{\partial y}$$

$$- a_2\,\frac{\overline{p}\,\overline{q^2}}{\ell_{q^2}}\left[1 + \frac{a_3}{R_T}\right] \qquad (5.5.4)$$

where a_1, a_2, a_3, a_4 are constants of proportionality, $\left|\frac{\partial U}{\partial y}\right|_m$ is the
maximum value when $\frac{\partial U}{\partial y} \approx 0$,

$$R_T = \frac{\overline{\rho}\,\overline{q^2}^{\frac{1}{2}}\,\ell_q}{\mu} ,$$

$\ell_q \equiv$ the dissipation length

$$= a_2\,\overline{\rho}\,\overline{q^2}^{\frac{3}{2}}\left/\left\{\frac{1}{2}\,\overline{\mu}\left[\frac{\partial v_\alpha}{\partial x_\beta} + \frac{\partial v_\beta}{\partial x_\alpha}\right]\right\}\right.$$

and N is the Prandtl-Schmidt number.

The foregoing turbulence kinetic energy equation may be compared with a turbulence kinetic energy equation for incompressible flow (Harsha), namely

$$\overline{\rho} \, U \frac{\partial \overline{q^2}}{\partial x} + \overline{\rho} \, V \frac{\partial \overline{q^2}}{\partial y} = \frac{a_1 \, \overline{\rho} \, \overline{q^2}}{\left| \frac{\partial U}{\partial y} \right|_m} \left[\frac{\partial U}{\partial y} \right]^2$$

$$+ \, \overline{\rho} \, \frac{\partial}{\partial y} \left[\frac{a_1 \, \overline{q^2}}{N_{q^2} \left| \frac{\partial U}{\partial y} \right|_m} \frac{\partial \overline{q^2}}{\partial y} \right]$$

$$+ \frac{a_1 \overline{q^2}}{N_{q^2} \left| \frac{\partial U}{\partial y} \right|_m} \frac{\partial \overline{q^2}}{\partial y} \frac{\partial \overline{\rho}}{\partial y} - a_2 \frac{\overline{\rho} \, \overline{q^2}^{\frac{3}{2}}}{\ell_q}. \tag{5.5.5}$$

Bray has pointed out that, while the foregoing Harsha equation may be successful some distance downstream of the origin of the mixing layer, one should note (a) the removal of turbulence energy due to the effect of heat release on the velocity divergence, (b) the generation of turbulence by the shear produced within the flame and (c) the effect of density gradient on the diffusion of turbulence energy within the reaction zone. There is also the effect of an order of magnitude reduction in turbulence Reynolds number due to entry of new fluid into a flame zone. The manner in which one can incorporate the heat release and the density gradient in reactive flows into the incompressible, cold flow should therefore take into consideration the foregoing. This is especially important in regard to experimental studies.

6.   TURBULENT COMBUSTION

It is convenient to divide combustion systems into premixed systems and nonpremixed systems even when finite rate chemistry is considered. It is necessary to distinguish between combustion in laminar flows, turbulent flows and transitional flows from laminar, turbulent or transitional combustion in given flows. We have very little understanding of the influence of combustion on turbulence, transition or laminarization. That there should be mutual interaction between the chemical and the flow fields is obvious. However, if the reactants and products are assumed to be dynamically passive, we can restrict attention to the influence of the flow field on the reaction.

6.1. Some Features of Laminar Combustion

It is considered useful in several respects to recall certain features of laminar combustion as a basis for examining combustion in turbulent flows.

The general methods of attack for combustion in laminar flows seem to be well established (Williams, 1971). It is possible to take into account the nonisothermal reaction kinetics, variable properties and arbitrary orders of reaction within the limitations of regular and singular perturbation techniques and similarity assumptions.

Consider a reaction of order m with respect to a reactant A and n with respect to a second reactant B. One can write the species conservation equation for the coupling function y (of the mass fractions Y of the species) as follows.

$$L(y) = N_D \ [y + f \ (x)]^n \ [y - f(x)]^m \ F(Y) \qquad\qquad (6.1.1)$$

where N_D is the relevant Damkohler group, $k\ell_D^2/D$, ℓ_D being the thickness of the diffusion layer, $F(Y)$ is the Arrhenius rate function and x is a non-dimensional distance. The operator L depends upon whether we are considering a premixed or an unmixed situation, the premixed situation often requiring a convective term. It also depends upon the flow conditions, for example on whether shear (or flame stretch) is included or not. It may also be necessary to introduce a time-dependent term if unsteadiness is suspected and may lead to energy accumulation. It has been pointed out (Williams, 1974) that in the case of unmixed or diffusion flames, no steady state solution is possible in the absence of shear. Laminar flames can be shown (Bush and Fendell, 1970) to consist of two zones, an upstream convective-diffusive zone (with negligible reaction) and a reactive-diffusive zone (where convection may not be significant). In the convective-diffusive zone, the shear determines the evolution of the flame. For positive flame stretch, a convective-diffusive balance can be established on both sides of the reaction zone in a diffusion flame. A point of some importance here is that large positive strain rates will tend to cause extinguishment. In a diffusion flame, a stretched condition causes greater reactant flux into the reactant zone; this decreases the residence time therein compared to the chemical reaction time and hence there is insufficient time for heat release. If γ is the flame stretch indicating the rate of increase of the flame surface area and τ_{ch} is the chemical reaction time, it is significant to consider when $\gamma\tau_{ch} <$ the Damkohler number for extinction.

We shall mention two other aspects of laminar diffusion flames. First, there are the classical results (Burke and Schumann, 1928; Shvab and Zeldovich, 1948): consider a one-step irreversible very fast reaction. Combustion is confined mathematically to a thin surface. The finite rate heat generation included in Eqn. 6.1.1 requires a flame surface across which there is interpenetration of reactants. If very fast chemistry is assumed, even if the Lewis number is not unity, one can examine a variety of combustion problems in terms of conservation equations governing the convection-diffusion of chemically inert species, even though the solutions may not be uniformly valid.

We shall now consider the case of two streams of reactants which are coming into contact in a two-dimensional mixing layer (Linan, 1963; Clarke, 1967). An experimental study has been reported (Phillips, 1965) of such a configuration in a combustion tunnel with a stratified methane stream interacting with air. If some approximation is made concerning the initial mixing region and the point of ignition, one can calculate (Williams, 1971) the formation of the laminar premixed flame zone, its extension to the diffusion layer, its extinguishment and the downstream progress of the diffusion flame.

A related problem (Libby and Economos, 1963) is the interaction between a flat plate boundary layer flow of an oxidizer over a flat fuel surface. In both problems, the crucial question arises in the initial region of fuel-oxidizer contact. A tentative analysis of the premixing, homogeneous ignition and the gradual change from the premixed flames to diffusion flames has been made (Murthy, 1967) using the PLK method (which has been described by Crocco, 1973). The activation energy and the difference in temperature between the oxidizer and the fuel surface are two of the parameters which determine the nature of the solution.

The second problem in laminar flow combustion which becomes of some significance in turbulent flow combustion is the possibility of multiple diffusion flames arising for two-step chain reactions (Bush and Fendell, 1973). One case that is studied relates to the counter stagnation flow in which forced convection introduces a constant rate of strain along the flame. Both equilibrium burning with two spatially separated flames of zero thickness and near-equilibrium burning with two flames of small but finite thickness are studied. The influence of transient tangential straining on the rates of reactant consumption shows an upper limit to the strain before reactant consumption begins to fall.

6.2. Structure of Turbulent Combustion Zones

The structure of combustion zones in turbulent flow can be understood in many respects by examining the turbulence length scales and the strain rate. For example, the applicability of the wrinkled laminar flame model may be related to the flame thickness being less than the Kolmogoroff length scale. Flame stretch is a useful concept both in determining the smoothening and thickening of the flame profiles (negative strain rate) and the extinguishment of hot spots (large positive strain rates).

In the experiments on the mixing layer downstream of a splitter plate with two-dimensional, fully-developed, subsonic, chemically inert streams (Brown and Roshko, 1971), it has been found that (a) a large scale structure appears indicating that the interface between the two streams is a thin, smeared out, highly-convoluted vortex sheet and this remains in tact for a considerable distance from the point of initial contact and (b) the density traverses indicate either the one or the other density of the two streams and very little of a mixture density. The time-averaged transverse profiles for velocity and density remain the conventional shear layer profiles.

In the case of a reactive system of a similar configuration, one can consider (Bush and Fendell, 1973; Libby, 1972) that, for Prandtl and Lewis numbers equal to unity, the reactants coexist only in the vicinity of the vortex sheet (on an instantaneous basis) and under a fast reaction approximation one can identify the behavior of the diffusion flame in analogy with the behavior of the interface. Thus, the diffusion flame structure may be governed largely by the local strain rate and this stretching of the flame leads to a lengthening of the flame, an increase in reactant consumption and when the strain rate is too large, an annihilation of the hot spot.

The behavior of a diffusion flame under a finite tangential strain in an unbounded counter flow has been examined (Carrier et al., 1973) using a singular perturbation analysis; such an analysis permits a reaction zone of small but finite thickness.

6.3. Modelling of Turbulent Flames

We have referred earlier to various modelling schemes in turbulent flows including cases where scalars and chemical reaction are present.

6.3.1 One scheme that has been evolved for turbulent diffusion flames (Spalding, 1970) consists in predicting the complete flame

properties, under the assumption of very fast chemistry, by solving
six simultaneous equations for the six variables as follows: three
for the hydrodynamic field (time mean value of velocity, turbulence
energy, and length parameter) and three for reaction (stagnation
enthalpy, mixture fraction and concentration fluctuation squared).
The mixture fraction measures the fraction of all of the material
in a sample which has emanated from one of the supply sources; it
is not the same as the mass fraction of either stream. One of the
features of the model is that it yields a diffusion flame of finite
thickness, a consequence of the large fluctuations of concentration.
At any section along the flow, there is a radial distribution of
temperatures, the maximum, the mean and the minimum; that is, at
any section along the flow, in any radial position the mixture
ratio is stoichiometric for a finite fraction of time.

6.3.2 Another scheme for turbulent flows is based upon the
premise that, under appropriate conditions, laminar flow results
hold instantaneously in turbulent flows (Toor, 1962). We have
already discussed how, provided the flow field, the molecular
diffusion coefficient and the boundary-initial conditions are
identical, the Shvab-Zeldovich function for a turbulent diffusion
flame, the instantaneous, time-averaged and fluctuating values,
satisfy the same boundary value problem as that obeyed by chemically
inert scalar quantities. The two boundary streams should have no
fluctuations in velocity.

Suppose we consider a variable-property flow (low kinetic
energy, high Reynolds number) in a thin shear layer with Prandtl
and Lewis numbers equal to unity. It can be shown (Bush and
Fendell, 1973) that the Shvab-Zeldovich function is described by
a boundary value problem, for instance, equivalent to that which
describes a chemically inert specie provided certain conditions
are satisfied. The equivalence is particularly attractive in the
very fast reaction limit. For dilute systems involving only
slightly exothermic reactions, an incompressible constant-property
approximation may suffice. However, if the system is assumed to
be isothermal under those conditions, one can introduce the
Stewartson transformation.

One can then proceed by applying one of the standard closure
approximations.

6.3.3. It has been suggested (Spalding, 1971) that the major
effect of turbulence on a chemical reaction can be understood on
the basis of an "eddy break up" model. It is presumed that (i) the
rate of consumption of fuel is proportional to the rate at which
the fuel concentration fluctuations disappear and (ii) an analogy
can be set up between the rate at which unburned gas lumps are
broken down into smaller lumps and the energy transfer rate from

large to small eddies. The fuel concentration fluctuations are
assumed to be calculable from the mass fraction of fuel. The rate
of energy decay, when the local generation rate of turbulence
equals the local decay rate, is given approximately by
$0.35 \rho \, |\partial U/\partial y|$ where $\partial U/\partial y$ is the local velocity gradient.
Defining a reactedness R* by

$$R* = \left(m_f - m_{f,u}\right)\bigg/\left(m_{f,b} - m_{f,u}\right) \qquad\qquad (6.3.1)$$

where m_f is the mass fraction of fuel and the subscripts b and u
represent the burned and the unburned parts and if the mass rate
of consumption of fuel by chemical action is $\dot{m}_f$ and its maximum
value with respect to R* is $\dot{m}_{f,max}$, then the time-mean volumetric
rate of fuel consumption can be written as equal to

$$\left\{\left(R*\dot{m}_{f,max}\right)^{-1} + \left[C(1 - R*) \, \rho \, |\partial U/\partial y|\right]^{-1}\right\}^{-1} \qquad (6.3.2)$$

where C is a constant of the order of 0.35. The implication here
is that the volumetric rate is the sum of that due to chemical
kinetic processes and eddy-break-up processes. It will be observed
that $\dot{m}_f$ is zero both for the gas with R* equal to zero and one.
Also, at some interface between lumps of gas the reactedness may
be, it is presumed, such that $\dot{m}_f$ is the maximum value. The
reactedness R* is described as a measure of the fraction of time
at which high reactedness gas is present.

      Some suggestions for improving this model have also been made.
One is to set up some connection between the turbulence and the
reaction rate, perhaps through a local Reynolds number. Another
is to associate reactedness with the gas locally and set up a
transport equation for the root mean square fluctuation of the
reactedness.

      This model was developed originally for confined premixed
turbulent flames. The method of using multiple equations for
turbulent diffusion flames has already been discussed in Section
6.3.1.

      6.3.4 Lastly, we may mention a closure model (Rhodes and
Harsha, 1972, 1973) which is based on a boundary layer formulation
for a coaxial fuel-oxidizer configuration. On the basis of
heuristic arguments, a probabilistic model is developed for the
interaction of turbulence and chemical reactions. Several interesting features are also incorporated into the closure of the momentum
equation (see also Morel and Torda, 1973), but we shall concentrate
here on the chemistry model.

In the usual formulation of a chemically reactive problem, one sets up the specie continuity equation with an appropriate production term. One can replace that by means of an element concentration equation which also represents the mean total enthalpy if the Prandtl and Lewis numbers are taken to be unity. However, the average density at a point cannot be related directly to the average total enthalpy and the average element concentration. The central question then is how to specify the average density with turbulent equilibrium or frozen chemistry and with turbulent finite rate chemistry. In essence, the average density is determined not from the average of species and enthalpy at the point under consideration but from an average taken over the species and enthalpy values which contribute to the averages at the point. As a consequence of this procedure, one can model the diffusion and production of species without the necessity of including the species continuity equation; it is not necessary to define profiles for the species.

The mixing layer is divided simultaneously into a discrete number of zones and classes. A zone is an elemental physical region characterized by a time-averaged elemental composition. The distribution of the zones is determined by the spatial distribution of stoichiometry. A class is fluid located anywhere, which can be characterized by an instantaneous elemental composition.

When there is mass addition to a class, the added mass is of the same elemental composition as the class, by definition. The material of the class may be distributed throughout the mixing layer but is most abundant in the zone where the average stoichiometry is the same as the class stoichiometry. The majority of a class may or may not be formed in that zone.

It is assumed in the model that a single density may be assigned to a fluid in a class. The velocity assigned to a class is that of the zone with the same elemental composition as the class even though the correlation between species and velocity fluctuations is not unity. The class total enthalpy is assumed to be linearly related to the class elemental composition and this assumption is consistent with Lewis number being unity and the total enthalpy being large compared to the energy dissipation.

There is a finite probability, P_{ij}, in turbulent flow of finding class j in any zone i. If the pdf of concentration is P_{Ci}, then

$$P_{ij} = \int_{C_j - \frac{\Delta C}{2}}^{C_j + \frac{\Delta C}{2}} P_{Ci} \, dC \qquad (6.3.3)$$

where C_j is the instantaneous elemental composition. The time-averaged values may be defined by

$$\overline{C}_i = \sum_j P_{ij} \, C_j \qquad\qquad\qquad (6.3.4)$$

$$\overline{\rho}_i = \sum_j P_{ij} \, \rho_j \qquad\qquad\qquad (6.3.5)$$

In order to use Eqn. 6.3.4, it is necessary to make some approximation for P_{Ci}. In view of the lack of necessary data, this is done rather arbitrarily at the moment, for example, by relating P_{Ci} to the standard deviation and relating the standard deviation for concentration to that for the fluctuating velocity. One can then determine the time-averaged density in an equilibrium or frozen flow.

In order to extend this approximation to a situation with finite rate chemistry, each class is considered as a transient perfectly stirred reactor. However, this development is still in its initial stages.

6.3.5 Another model for turbulent flow combustion is based (Libby, 1973, 1974) on the concept of the "oscillation" of a flame sheet within the mixing zone formed between two parallel streams each containing one of the reactants. There is emphasis here on the intermittent nature of the turbulent flow and hence the structure of the turbulent-non-turbulent interfaces and also of the flame sheet. We shall note at the outset that the two reactants are assumed to be involved in a one-step, single product, fast reaction. Some of the background for this model has been provided in Section 5.4.

It is significant to note here that the reaction surface is assumed to be a convoluted oscillating surface, the dynamical nature of which is determined by the presence of the turbulent strain and mechanisms associated with it. However, the convoluted oscillating reaction surface or the outer interfaces of the mixing zone are not viewed as involving the large scale structure that has been observed (Brown and Roshko, 1972; Winant and Browand, 1973) in the development of the mixing layer. A distinction is made between the possibility of existence of contiguous large eddies (one with one reactant and an adjacent eddy with another reactant, with diluent and product in each) and the possibility of an oscillating reaction surface. Whether there is a connection between the mechanics of the large scale structure and the mechanics of the strained convoluted surface is not yet clear. The model is based on interpreting the oscillation of the reaction surface in

terms of the pdf of the mass fractions of the two reactants, which
in turn is related to the intermittency of the flow. The pdf are
not assumed to be Gaussian or near-Gaussian. Suppose we have the
concentration time-history at a point in the mixing layer. One
can develop a zero-one discriminating function to identify the
species and then obtain zone averages, crossing frequencies and
point statistics by the usual methods of conditional sampling.
One can also deduce the pdf based on an adequate number of traces.

In developing an analysis based on this model, one proceeds,
as discussed earlier, by establishing under appropriate conditions
a connection between the behavior of chemically passive scalars
and chemical reactants. Next, one has the problem of estimating
the pdf of a concentration variable at each location and computing
the conditional expectation (see Section 5.4). Information
pertaining to the mean values and fluctuation intensities (taking
into account intermittency) of the concentration variables then
becomes central to the modelling of the pdf.

6.3.6 Finally, there are the turbulence velocity divergence
(5.5.2) and the turbulence kinetic energy (5.5.4) equations which
have been developed by Bray (1973, 1974) and which can be employed
for modelling and computing turbulent flame structure and propagation. Those equations have been discussed in Section 5.5. Bray
discusses several experimental findings on open and ducted flames
both in the premixed and the diffusion cases, some of them with
variations in turbulence scale and intensity. It is necessary to
identify clearly definable turbulent reactive flows and experiments
on flames in order to assess the significance of the additional
terms and the associated assumptions in comparison with the
incompressible, nonreactive flow equations. One can then proceed
to examine the interaction between molecular mixing and turbulence.

7. CONCLUSIONS

One of the objectives throughout the review has been to
elucidate some of the problems which remain in the experimental and
analytical studies on mixing. In experimentation, it is necessary
to examine clearly defined flow configurations and measurement
techniques. The effects of compressibility and of chemical reaction
have not been examined in detail to date. It appears that apart
from data acquisition, data processing will play an important role
in the understanding of the structure of turbulence.

In analysis, it is of the greatest importance in practice to
advance modeling procedures. At the same time it seems that
attempts at incorporating the structural information directly into
the phenomenology of turbulence and turbulent mixing may not be

too far-fetched. Recent evidence on the presence of large-scale coherent structure may lead to completely new possibilities in the analysis of turbulent flows (see Phillips, 1972, for example, in regard to modelling the turbulent entrainment interface). If some of the possibilities appear incompatible from a dynamical point of view, one may also raise questions regarding the method of incorporating stresses and strains in the moment equations.

It may be said concerning shear flows that the coupling between the mean velocity field and the turbulent velocity field is fundamental in those flows. The turbulent velocity field is sustained and structured by the interaction between the mean velocity non-uniformities on the one hand and the Reynolds stresses and the vorticity and other fluctuations on the other. The behavior of dynamically passive scalars (with possibilities of chemical action in the case of chemically active species) is therefore perhaps most effectively studied in shear flows. Whether the problem of shear flow turbulence will be any simpler than that of homogeneous turbulence is not yet clear. If the Reynolds stresses in shear flow arise from the direct interaction of turbulent fluctuations and the mean shearing motion, the processes governing the distribution of Reynolds stresses can often be described by linear equations. If, on the other hand, the spectral cascade of energy plays a central role and the Reynolds stresses are a result of the nonlinear turbulent mode-mode interactions in a shearing field, then all the problems of homogeneous turbulence will remain in shear flow problems also.

We may also remark here that nonhomogeneity is a basic characteristic of transport and reactive systems. The determination of the interaction of turbulence and reaction with respect to correlations, spectral analysis and probability concepts such as intermittency is the central problem in both theory and experiment.

The evolution of a general theory for mixing layers, jets and wakes including transport and reaction rests largely in the future even though some advances seem to be possible on a phenomenological basis.

ACKNOWLEDGMENT

The author is appreciative of the support from Project SQUID (Office of Naval Research) during the preparation of this position paper on turbulent mixing.

8. BIBLIOGRAPHY

1. Abramovich, G. N. (1963): The Theory of Turbulent Jets,
    M.I.T. Press.

2. Alber, I. E. (1973): TRW Systems Report 18117-6019-RU-00,
    Redondo Beach, Calif.

3. Batchelor, G. K. (1959): J. Fluid Mech. 15, p. 113.

4. Batchelor, G. K., Howell, I. D. and Townsend, A. A. (1959):
    J. Fluid Mech. 15, p. 133.

5. Batchelor, G. K. (1953): Homogeneous Turbulence, Cambridge
    University Press.

6. Batt, R. G., Kubota, T. and Laufer, J. (1970: AIAA Paper
    No. 70-721.

7. Bilger, R. W. and Kent, J. H. (1972): Charles Kolling
    Research Lab. Tech. Note F-46, The University of Sydney,
    Sydney, Australia.

8. Birch, S. F. and Eggers, J. M. (1972): NASA SP-321, p. 11.

9. Bradshaw, P. et al. (1967): J. Fluid Mech. 28, (3), p. 593.

10. Bradshaw, P. (1971): An Introduction to Turbulence and its
     Measurements, Pergamon Press, London.

11. Bradshaw, P. (1972): Aeronautical Journal, 78, p. 403.

12. Bradshaw, P. (1972): I. C. Aero Report 72-21, Imperial
     College of Science and Technology, London. See also J. Fluid
     Mech. 63, p. 449, April 1974.

13. Bradshaw, P. (1973): Imperial College Aero Rep. 73-05.

14. Bray, K. N. C.: Paper presented for AGARD Propulsion and
     Energetics Panel Specialists' Meeting, Leige, Belgium,
     1-2 April, 1974; also AA Su Ref. No. 332, University of
     Southampton, England.

15. Brodkey, R. S. (1973): Article in Fluid Mechanics of Mixing,
     A.S.M.E. New York.

16. Brown, G. and Roshko, A. (1971): AGARD Conference Proceedings
     No. 93, London, p. 23-1 -- 23-12.

17. Brown, G. L. and Rebello, M. R. (1972): AIAA J. $\underline{10}$, p. 649.

18. Brown, G. L. and Roshko, A. (1974): To be published in
    J. Fluid Mech.

19. Burke, S. P. and Schumann, T. E. W. (1928): Industrial and
    Engineering Chem., $\underline{20}$, p. 998-1004.

20. Bush, W. B. and Fendell, F. E. (1970): Combustion Science
    and Technology, $\underline{1}$, p. 407.

21. Bush, W. B. and Fendell, F. E. (1972): Project SQUID Tech.
    Rep. TRW-2-PU.

22. Bush, W. B. and Fendell, F. E. (1973): Phys. Fluids, $\underline{16}$.

23. Bywater, R. J. and Chung, P. M. (1973): AIAA Preprint
    Paper No. 73-646.

24. Carrier, G. F., Fendell, F. E. and Marble, F. E. (1973):
    Project SQUID Tech. Rep. TRW-5-PU, Project SQUID Headquarters,
    Purdue University, West Lafayette, Indiana.

25. Chandrasekhar, S. (1943): Review of Modern Physics, $\underline{15}$,
    p. 1-89.

26. Chou, P. Y. (1945): Quart. App. Math. $\underline{3}$, p. 38.

27. Chu, B. T. and Kovasznay, L. S. G. (1958): Journal of Fluid
    Mechanics, $\underline{5}$, (3), p. 494.

28. Chung, P. M. (1969): AIAA J., $\underline{7}$ (10), p. 1982-1991.

29. Chung, P. M. (1970): Physics of Fluids, $\underline{13}$, p. 1153.

30. Chung, P. M. (1972): Physics of Fluids, $\underline{15}$ (10), p. 1735-
    1746.

31. Chung, P. M. (1973): Physics of Fluids, $\underline{16}$, p. 1646.

32. Clarke, J. F. (1967): Proc. Roy. Soc. London, $\underline{A296}$, p. 519.

33. Corrsin, S. (1951): J. App. Phys., $\underline{22}$, p. 469.

34. Corrsin, S. and Kistler, A. (1954): NACA TR 1244.

35. Corrsin, S. (1958): Physics of Fluids, $\underline{1}$, p. 42.

36. Corrsin, S. (1964): Phys. Fluids $\underline{7}$, p. 1156.

37. Daly, B. J. and Harlow, F. H. (1970): Physics of Fluids, <u>13</u> (11).

38. Davidov, B. I. (1961): Soviet Physics - Doklady, <u>4</u> (5), p. 10-12.

39. Davies, P. O. A. L. et al. (1963): J. Fluid Mech. <u>15</u>, p. 337.

40. Donaldson, C. duP. (1971): <u>AGARD Conference Proceedings</u> No. 93, London, p. B-1 -- B-24.

41. Donaldson, C. duP. and Hilst, G. R. (1972): <u>Proc. of the 1972 Heat Transfer and Fluid Mechanics Institute</u>, Stanford University Press, Stanford, Calif., p 256-261.

42. Dopazo, C. and O'Brien, E. E. (1973): Paper presented at the 4th International Colloquium on Gas Dynamics of Explosions and Reactive Systems, La Jolla, Calif.

43. Edelman, R. and Fortune, O. (1968): AIAA Paper No. 68-114.

44. Edwards, S. (1964): J. Fluid Mech. <u>18</u>, p. 239.

45. Edwards, S. and McComb, W. D. (1971): Proc. Roy. Soc. London, <u>A325</u>, p. 313.

46. Edwards, S. F. and McComb, W. D. (1972): Proc. Roy. Soc. London, <u>A330</u>, p. 495.

47. Eschenroeder, A. Q. (1965): AIAA J. <u>3</u> (10), p. 1839.

48. Favre, A. (1965): J. App. Mech. <u>326</u>, p. 241.

49. Favre, A. et al. (1967): J. Fluid Mech. <u>10</u>, p. 138.

50. Fendell, F. E. (1971): Project SQUID Tech. Rept. TRW-1-PU.

51. Ferri, A. (1973): <u>Annual Review of Fluid Mechanics, Vol. 5</u>, ed. M. Van Dyke et al., Annual Reviews, Palo Alto, Calif., p. 301.

52. Gibson, C. H. (1968): Physics of Fluids, <u>11</u> (11), p. 2305-2315.

53. Gibson, C. H. and Libby, P. A. (1973): Combustion Science and Technology, 6, p. 29-35.

54. Graham, S. C. et al. (1973): Submitted for publication in AIAA J.

55. Grant, A. J. et al. (1973): Combustion Institute European
     Symposium, Academic Press, London.

56. Harlow, F. H. and Nakayama, P. I. (1967): Phys. Fluids,
     10 (11), p. 2323.

57. Hawthorne, W. R. et al. (1949): Third Symposium on Com-
     bustion, Flame and Explosion Phenomena, Williams and Wilkins,
     Baltimore, p. 266.

58. Heisenberg, W. (1949): Z. Physik, 124, p. 628.

59. Henjalic, K. and Launder, B. E. (1970): Journal of Fluid
     Mechanics, 52, p. 689.

60. Heskestad, G. (1965): Journal of Applied Mechanics, 32,
     p. 721-734.

61. Hill, J. C. (1969): Goddard Space Flight Center, Preprint
     No. X-641-69-108, Greenbelt, Md. See also: Hill, J. C.
     (1970): Phys. Fluids 13, p. 1394-6.

62. Hilst, G. R. (1973): AIAA Preprint Paper 73-101.

63. Hopf, E. (1952): Journal of Rational Mechanical Analysis,
     1, p. 87-123.

64. Hottell, H. C. and Hawthorne, W. R. (1949): Third Symposium
     on Combustion, Flame and Explosion Phenomena, Williams and
     Wilkins, Baltimore, Md., p. 254.

65. Jones, B. G., et al. (1973): AIAA Paper No. 73-225.

66. Kaplan, R. E. and Laufer, J. (1968): Proc. Twelfth Inter-
     national Congress of Mechanics.

67. Kodomtscv, B. B. and Kostomarov, D. P. (1972): Phys. Fluids,
     15 (1), p. 1.

68. Kolmogorov, A. N. (1941): C. R. Academy of Sciences, USSR,
     30, p. 301.

69. Kovasznay, L. S. G. (1956): Jet Propulsion, 26, p. 485.

70. Kovasznay, L. S. G. (1962): see The Mechanics of Turbulence,
     Ed. A. Favre, Gordon and Breech.

71. Kovasznay, L. S. G. et al. (1970): J. Fluid Mech. 41,
     p. 283.

72. Kraichnan, R. H. (1959): J. Fluid Mech. 5, p. 497.

73. Kraichnan, R. H. (1962): Proc. Symp. Appl. Math. 13,
     p. 199.

74. Landahl, M. T. (1967): J. Fluid Mech., 29, p. 441.

75. Laufer, J. (1969): AIAA J. 7, p. 706.

76. Launder, B. E. and Spalding, D. B. (1972): Mathematical
     Models of Turbulence, Academic Press, London.

77. Launder, B. E. et al. (1973): Rept. No. HTS/73/31.
     Imperial College of Science and Technology, London.

78. Libby, P. A. (1973): Combustion Science and Technology,
     6, p. 23-28.

79. Libby, P. A. (1973): To appear in Combustion Science and
     Technology.

80. Libby, P. A. (1974): Paper presented at the AGARD
     Conference, Apr. 1-2, Liege.

81. Libby, P. A. and Economos, C. (1963): Int. J. Heat and
     Mass Transfer 6, p. 113.

82. Liepmann, H. and Laufer, J. (1949): NACA TN 1257, NACA.

83. Lin, C.-H. and O'Brien, E. E. (1972): Astronautica Acta,
     17, p. 771-781.

84. Lin, C.-H. and O'Brien, E. E. (1974): Private Communication,
     to be published in the Journal of Fluid Mechanics, 1974.

85. Linan, A. (1963): Ph.D. Thesis, California Institute of
     Technology.

86. Liu, C. Y. and Lees, L. (1961): in Rarefied Gas Dynamics,
     Academic Press, New York, p. 391.

87. Lundgren, T. S. (1969): Physics of Fluids, 12, p. 485.

88. Markovin, M. V. (1964): in The Mechanics of Turbulence,
     Ed. A. Favre, Gordon and Breach.

89. Mellor, G. L. and Herring, H. J. (1973): AIAA Journal,
     11 (5), p. 590-599.

90. Mollo-Christensen, E. (1973): Annual Reviews of Fluid
     Mechanics 5, Annual Reviews, Palo Alto, Calif., p. 101.

91. Murthy, S. N. B. (1967): Unpublished notes.

92. Murthy, S. N. B. (1971): ARL Rep. No. 71-0244.

93. Nee, V. W. and Kovasznay, L. S. G. (1969): Physics of
     Fluids 12, p. 473.

94. O'Brien, E. E. (1966): Phys. Fluids 9, p. 1561.

95. O'Brien, E. E. (1968): Physics of Fluids, 11, p. 1883.

96. O'Brien, E. E. (1969): Physics of Fluids, 12, p. 1999.

97. O'Brien, E. E. (1971): Physics of Fluids, 14, p. 1326.

98. O'Brien, E. E. (1973): Paper presented at the Second
     IUTAM-IUGG Symposium on Turbulent Diffusion in Environmental
     Pollution, Charlottesville, Va.

99. Obukhoff, A. M. (1941): Compt. Rend. Acad. Sci. URSS, 32,
     p. 19.

100. Obukhov, A. M. (1949): Compt. Rend. Acad. Sci. URSS, 66,
     p. 17.

101. Pao, Y. H. (1965): Physics of Fluids 8, p. 1063.

102. Patel, R. P. (1973): AIAA J. 11, p. 67.

103. Pearson, J. R. A. (1963): App. Sci. Res., A11, p. 321-340.

104. Phillips, H. (1965): Tenth Symp. Intern. Combustion, The
     Combustion Institute, Pittsburgh.

105. Phillips, O. M. (1972): J. Fluid Mech. 51, 1, p. 97-118.

106. Reynolds, W. C. (1970): Rept. No. MD-27, Thermosciences
     Div., Stanford University, Stanford, Calif.

107. Rhodes, R. P. and Harsha, P. T. (1972): AIAA Paper No. 72-68.

108. Rhodes, R. P. et al. (1973): Paper presented at the 4th
     International Colloquium on Gas Dynamics of Explosions and
     Reactive Systems, San Diego, California.

109. Rodi, W. and Spalding, D. B. (1970): Warme-und-Stoffubertragung,
     3, p. 85.

110. Rotta, J. (1951): Z. Physik, 129, p. 547.

111. Rotta, J. C. (1962): Progress in Aeronautical Sciences,
      Vol. 2, Pergamon Press, London.

112. Rotta, J. C. (1971): AGARD Conference Proceedings, No. 93,
      London.

113. Saffman, P. G. (1963): J. Fluid Mech. 16, p. 545.

114. Saffman, P. G. (1970): Proc. Roy. Soc. London A317, p. 417.

115. Shackelford, W. L., et al. (1973): AIAA Preprint Paper
      73-640.

116. Shvab, V. and Zeldovich, L. (1948): See Combustion Theory,
      F. A. Williams, Addison-Wesley, Reading, Mass.

117. Spalding, D. B. (1970): VDI-Berichte Nr. 146, VDI Verlag,
      Dusseldorf, p. 25.

118. Spalding, D. B. (1971): 13th Symposium (International) on
      Combustion, The Combustion Institute, Pittsburgh, p. 649.

119. Spencer, B. W. and Jones, B. G. (1971): AIAA Preprint Paper
      No. 71-613.

120. Stanford, R. A. and Libby, P. A. (1973): To be published in
      Phys. Fluids.

121. Stuart, J. T. (1971): in Annual Review of Fluid Mechanics
      3, Annual Reviews, Palo Alto, Calif., p. 347.

122. Ting, L. (1959): J. Math. and Phys. 38, p. 153.

123. Toor, H. L. (1962): A.I.Ch.E. J., 8, p. 70-78.

124. Townsend, A. (1951): Proc. Roy. Soc. London, 209A, p. 418.

125. Townsend, A. A. (1956): The Structure of Turbulent Shear
      Flow, Cambridge University Press.

126. van Driest, E. R. (1956): J. Aero. Sci. 23, p. 1007.

127. Way, J. and Libby, P. (1971): AIAA J. 9, p. 1567.

128. Williams, F. A. (1971): Annual Review of Fluid Mechanics,
      Vol. 3, Annual Reviews, Palo Alto, Calif., p. 171-188.

129. Williams, F. A. (1974): Paper presented at the AGARD Conference, Apr. 1-2, 1974, Liege.

130. Winant, C. D. and Browand, F. K. (1973): To appear in J. Fluid Mech. in 1974.

131. Wygnanski, I. and Fiedler, H. E. (1970): Journal of Fluid Mechanics, 41, (2), p. 327-361.

132. Yule, A. J. (1972): AIAA J. 10 (5), p. 686.

TURBULENCE MODELING: SOLVED AND UNSOLVED PROBLEMS

D. B. Spalding

Imperial College of Science and Technology

London, England

ABSTRACT

The paper is a status report on turbulence modelling, with special emphasis on practical achievements. The need for turbulence modelling is reviewed. It is asserted that the computational problem has been satisfactorily solved, and that available models permit satisfactory predictions to be made of many hydrodynamic, mixing and chemical-reaction processes in turbulent flow. The need for further knowledge of low-Reynolds-number turbulence, and of turbulence ~ reaction interactions, is emphasized. A proposal is made for a new way of handling turbulent diffusion, based on ideas from radiative transfer theory. Finally, suggestions are made for experiments which would be especially informative.

1.   INTRODUCTION

The purpose of this paper is to review, in so far as it is possible to do so within a short presentation, the current status of turbulence modelling as it appears to me. In choosing topics to include and to omit, I have been guided by a desire to provide something of interest for each member of a non-homogeneous audience; and I have sought to leave this balanced impression: much of practical value has already been achieved; but much remains to be done.

The subject matter may be classified under the following headings.

- Practical problems (Sections 2 and 3). Here I affirm that my interest in turbulence modelling is motivated by a desire to make useful predictions of real phenomena.

- The turbulence-model approach (Sections 4 through 6). Here is given a brief outline of the nature and style of fluid-flow prediction by way of turbulence models.

- Solved problems (Sections 7 through 14). My aim in this part of the lecture is both to inform and to encourage.

- Unsolved problems (Sections 15 through 17). Here I draw attention to two of the most urgent of these.

- Transport of a scalar (Sections 18 through 21). In this part of the lecture, I return to a topic that has long concerned me: how to take account of the fact that, in turbulent flow, the distances over which mixing occurs are not small compared with the distances over which time-average gradients vary.

- Some informative experiments (Sections 22 through 25). Many people say that we need more experimental information before we can advance further our knowledge of turbulence models. This is my view also; but I think that some experiments are likely to be much more informative than others. At the end of the lecture, I propose some of these.

2. TURBULENT FLOWS IN THE NATURAL ENVIRONMENT

The following table lists some of the practical problems involving turbulence modelling which concern fluid flow, heat transfer and mass transfer in the environment.

Table 1.   Turbulent Flows in the Natural Environment

Process	Mathematical Type
Thermal plume in river	(3D, Steady, Parabolic)
Smoke plume in atmosphere	(3D, Steady, Parabolic)
Pollutant dispersal in ocean	(3D, Unsteady, Elliptic)
Fire spread by wind	(2 or 3D, Unsteady, Elliptic)
City meteorology	(3D, Various)
Weather prediction	(3D, Unsteady)

The mere mention of the names of the processes, without detailed description, will be enough to remind this audience of their physical nature and their importance. I shall therefore here concentrate attention on the right-hand column, which provides a mathematical classification.

Nearly all the processes are three-dimensional in character; but whether the associated <u>computer storage</u> is 2D or 3D depends on whether or not the flow is steady or unsteady; and, if the former, on whether the phenomenon is to be classified as "parabolic" or "elliptic," to use the words which are commonly employed. The 3D-parabolic processes require only 2D computer storage; the 3D-elliptic ones require 3D computer storage, and so are more expensive to solve.

Both the thermal plume and the smoke plume are 3D parabolic. This means that prediction can proceed by way of marching integration from upstream to downstream; at each section across the plume, there are two-dimensional fields of velocity, temperature and concentration to predict; then attention passes to the next-downstream section; and so on. Methods of solution have been provided and reported by: Gosman and Spalding (1971); Caretto, Curr and Spalding (1972); Patankar and Spalding (1972); Patankar, Pratap and Spalding (1974a, 1974b).

The other problems mentioned in the table are of such a character that three-dimensional storage is required in the computer; for the unsteadiness of the process, or the fact that all three velocity components may take values of either sign, ensures that influences from any point in the domain can extend, at least in principle to any other point. Procedures are available for calculating these flows also, for example those of: Harlow and Amsden (1968); Amsden and Harlow (1970); Chorin (1968); Caretto, Gosman, Patankar and Spalding (1973); and Patankar and Spalding (1972). A valuable bibliography has been published by Harlow (1969).

Since methods are available for solving equations of both these types, it behoves us to make sure that we can write the equations correctly. It is this matter that turbulence-modellers are concerned with.

3.   TURBULENT FLOWS IN MAN-MADE DEVICES

Practical problems of engineering, in which the ability to predict turbulent flows quantitatively is of high importance, are exemplified in the following table. Here the list of equipment items is supplemented by reminders of the special processes which require consideration in addition to those of hydrodynamics.

Table 2.   Turbulent Flows in Man-Made Devices

Equipment	Special Features
Combustion in diesel engine	(Particles, Reaction)
Combustion in furnace	(Particles, Reaction)
Heat transfer to turbine blade	(Low-Re)
Aluminum smelter	(Magnetic)
Gas-mixing device	(Concentration fluctuations)
Liquid-cooled nuclear reactor	(Buoyancy)

The first two examples involve the contribution of fuel droplets, and so cause the turbulence-modeller to ask how exo-thermic chemical reaction influences the rate of turbulence-energy generation, how the reaction rate depends on the turbulence macro- and micro-scales, and how particles of dense fluid are transported by a lighter one and interact with its fluctuating velocities. Quantitative knowledge is required about these effects.

The turbine-blade example is one in which the turbulence in the main stream diffuses into a boundary layer of such low Reynolds number that, left to itself, it would be laminar. Experience shows that this diffusing mainstream turbulence sufficiently disturbs the boundary layer to cause a practically-important enhancement of heat transfer; it is desirable to be able to compute the magnitude of this enhancement.

The aluminum smelter usually involves a large bath of turbu-lent molten aluminum, stirred by electromagnetic forces. The presence of eddy currents, etc., presumably influences the rate of decay of turbulence energy; and perhaps it influences the length-scale also. Such effects require to be incorporated into turbu-lence models quantitatively.

The table, which could have been greatly extended, gives further examples in which buoyancy and concentration fluctuations are of importance. It is clear that the man who seeks to predict turbulent flows in all the circumstances of practical engineering must make allowance for a much larger range of physical processes than are taken into account by most academic researchers on turbulence.

4.   THE NATURE OF THE TURBULENCE-MODEL APPROACH

As has already been stated, methods exist for solving the equations of unsteady three-dimensional motion; such methods are in principle capable of predicting the details of any turbulent flow, by starting from the highly reliable Navier-Stokes equations. However, the practical possibility of doing this is very small because of the great disparity of scale between the size of most practical flow domains and the sizes of the eddies which exist, and exert their influences, within them.   Some authors, notably Schumann (1973), have sought to compute simple turbulence phenomena in this way; but the cost is great, and the success so far small. Phenomena of engineering or environmental practice could not possibly be computed in this way with current, or currently planned, computers.

It is for this reason that, in recent years, all persons seriously intending to compute turbulent flows have employed "turbulence models." These are sets of equations, purporting to describe the convective transport, the diffusion, the generation and the decay of certain statistical properties of a turbulent fluid, the so-called "correlations." Examples of these are: $\overline{u_i u_j}$, where u_i and u_j are instantaneous fluctuating components of velocity in a Cartesian coordinate system; $\overline{u_i c'}$, where c' is the instantaneous fluctuating component of concentration; or $\overline{\Sigma_{i,j}\mu(\partial u_i/\partial x_j)^2}$, where μ is the fluid viscosity and x_j stands for one of the three Cartesian-coordinate distances.

The equations having these correlations as dependent variables contain "constants" or "functions" which are hopefully supposed by the turbulence-modeller to be universally valid.   Usually they are obtained from experimental evidence; but their origin is immaterial. The most important thing is that they should be known.

Once supplied with the equations, the constants (or functions) and the solution procedures, the analyst can make the desired prediction. He is not faced with a scale-disparity problem, and so is not required to employ an impractically fine grid, because the scale of variation of the correlations is the same order as that of such properties as time-average velocity or temperature.

Of course, predictions based on such solutions must be employed warily; for the presumed constants do not truly have universal values. However, the predictions are usually accurate enough to justify the expense of making them.

General works on turbulence models include: Harlow (1973); Launder and Spalding (1972a, 1972b, 1974); Launder, Spalding and Whitelaw (1973); Mellor and Herring (1973).

5. THE TURBULENCE-MODEL APPROACH; THE TYPICAL EQUATION

Consider the equation

$$\frac{\partial}{\partial t}(\rho\phi) + \text{div}(G\phi - \Gamma \text{ grad } \phi) = S_+ - S_- \qquad (1)$$

Here ϕ stands for an extensive fluid property, t for time, ρ for density, G for the mass-velocity vector, Γ for an exchange coefficient, and S_+ and S_- for positive and negative source terms.

This equation is typical of those which are solved by turbulence-modellers, whether the variable ϕ is one of their correlations or one of the more-usually-solved-for quantities like time-average velocity (i.e. momentum per unit mass in a particular direction) or specific enthalpy.

In the former case, ϕ might stand for the mean square concentration fluctuations, $\overline{c'^2}$; then Γ would be a corresponding turbulent-exchange coefficient, taken as proportional to the square root of the local turbulence energy, to the time-average density ρ, and to the turbulence macro-scale ℓ. Thus:

$$\Gamma = \text{const } (\overline{u_i u_i})^{\frac{1}{2}} \rho\ell \qquad (2)$$

The source of concentration fluctuations would be taken (Spalding, 1971) as being given by:

$$S_+ = \text{const} \cdot \Gamma \ (\text{grad } \overline{c} \cdot \text{grad } \overline{c}); \qquad (3)$$

and the sink is then taken (loc. cit) as being given by:

$$S_- = \text{const } \frac{\overline{c'^{\frac{1}{2}}} \cdot (\overline{u_i u_i})^{\frac{1}{2}}}{\ell} \qquad (4)$$

It can be seen that, provided other equations are available for the simultaneous computation of ρ, G, $\overline{u_i u_i}$, $\overline{c}$, and ℓ, and provided suitable boundary conditions have been prescribed, the field of $\overline{c'^2}$ can be computed. Of course the values of the constants must be known also.

6. SOME EXAMPLES OF TURBULENCE MODELS

The following table lists the principal authors of some of the turbulence models from which the analyst can make his choice, together with their dates of publication, and an indication of the dependent variables of the differential equations that are incorporated.

Table 3.   The Turbulence-Model Approach; Some Examples

Originator	Date	Dependent Variables
Kolmogorov	1942	$\overline{u_i u_i}$, $\overline{u_i u_i}^{\frac{1}{2}}/\ell$
Prandtl	1945	$\overline{u_i u_i}$
Rotta	1951	$\overline{u_i u_i}$, $\overline{u_i u_j}$, $\overline{u_i u_i}\,\ell$
Bradshaw	1967	$\overline{u_i u_j}$
Harlow	1967	$\overline{u_i u_i}$, $\overline{u_i u_i}^{\frac{3}{2}}/\ell$
Kovasznay	1969	$\overline{u_i u_i}^{\frac{1}{2}}\,\ell$
Spalding	1969	$\overline{u_i u_i}$, $\overline{u_i u_i}/\ell^2$

Kolmogorov (1942) may be regarded as the father of turbulence modelling; for, in a little-known publication, he proposed that turbulent-flow phenomena should be computed by way of the solution of two equations, like equation (1), the dependent variables of which would be, first, the energy of the turbulent motion, and secondly, its "frequency." In the table, the first of these is represented by $\overline{u_i u_i}$ ($\equiv$ twice the energy); and the second by $(\overline{u_i u_i})^{\frac{1}{2}}/\ell$, where ℓ is the length scale, which obviously has the right dimensions. The hydrodynamic equations were of course to be solved in addition; there the crucial assumption, since made by the majority of workers in the field, was that the shear stresses could be computed by way of the "effective-viscosity" concept of Boussinesq (1877), the value of which, like that of Γ, was equal to a constant times $\rho\,\overline{u_i u_i}^{\frac{1}{2}}\,\ell$.

Prandtl's (1945) proposal was similar, in respect of the energy equation; but the distribution of the length scale ℓ was to be prescribed beforehand, not computed from a second differential equation. The length scale was needed by Prandtl, because he used the same effective-viscosity concept as Kolmogorov (although he must have been unaware of the latter's work), and because S_-, in equation (1) for energy, was to be computed from a constant times $\rho\,\overline{u_i u_i}^{\frac{3}{2}}/\ell$. Kolmogorov, of course, obtained ℓ by dividing $\overline{u_i u_i}^{\frac{1}{2}}$ by $\overline{u_i u_i}^{\frac{1}{2}}/\ell$.

Rotta (1951) took the matter further, pointing out that one could derive differential equations for the shear-stress components $\overline{u_i u_j}$ themselves, thus obviating the use of the effective-viscosity concept, and also other "moments" of the fluctuation spectrum, resulting in other combinations of $\overline{u_i u_j}$ and ℓ.

Bradshaw, Ferris and Atwell (1967) made an independent proposal for a one-equation model; this was similar to that of Prandtl (1945) in requiring specification of the length-scale distribution, yet different in supposing the transfer rates of both momentum and energy to be proportional to the energy itself. Another one-equation model was that of Nee and Kovasznay (1969) who employed the effective-viscosity itself as the dependent variable.

Harlow and Nakayama (1968) and Spalding (1969) made independent proposals for two-equation models. The energy equation was used in both cases, but the former authors proposed that the quantity $\overline{u_i u_j}^{3/2}/\ell$ should be the dependent variable of the second equation, while the latter proposed the use of $\overline{u_i u_j}/\ell^2$; this is of course the square of Kolmogorov's variable, although it was thought of as being vorticity fluctuations. Saffman (1970) made a similar proposal quite independently. (One remarkable feature of the literature of turbulence modelling is preponderance of independent invention over transmission. Knowledge of turbulence models at first, like the most primitive of the models themselves, has been dominated by the processes of local generation and dissipation; diffusion and convection of this knowledge are only now beginning to play important roles.)

7. SOLVED PROBLEMS

1. How to Solve the Equations

I now begin the review of the successes of turbulence model-ling, and start by clearing away the computational question: there are nowadays no difficulties about solving the differential equa-tions once they have been set up; adequate algorithms, computer codes and computers, already exist. Of course, all three can be and should be improved; but this is not a matter that the turbu-lence modeller need worry about.

Already in Section 2 some of the numerical methods were men-tioned. Since this lecture has a different focus of attention, I will here simply state that my personal preferences, based upon some years of experience, are for methods characterized by the following.

(i)   the "staggered-grid" arrangement of Harlow and
      Amsden (1968);

(ii) the "hybrid" or "mixed central and upwind" difference
      scheme of Spalding (1972a);

(iii) an implicit, conservative formulation of the difference
      equations;

(iv) the SIMPLE (semi-implicit method for pressure-linked
      equations) algorithm of Patankar and Spalding (1972a)
      for handling the continuity and momentum equations;

(v)   line-by-line use of TDMA (tri-diagonal matrix
      algorithm) for iterative solution of all finite-
      difference equations;

(vi) where possible the use of non-dimensional streamline
      coordinates, as in the 2D boundary-layer method of
      Patankar and Spalding (1967, '70).

However, other authors have other preferences; and time will
tell which combinations of particular features are best adapted to
the majority of practical problems.

8.   EXAMPLES OF SUCCESSFUL COMPUTATIONS

(a)   3D Steady Parabolic

Figure 1 illustrates what it is now possible to do by combining
modern computing methods with up-to-date turbulence models. My
student J. McGuirk (1974) has employed the method of Patankar and
Spalding (1972), and the k ~ ε model of turbulence as described by

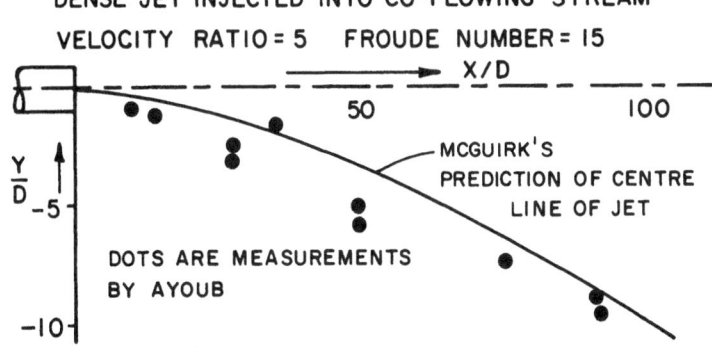

Figure 1. Computational Examples; 3D, Steady, Parabolic.

Launder and Spalding (1974), to compute the sinking of a jet of
dense polluted water in a co-flowing stream of clear water in a
rectangular channel. The computer program used is called STABLE
(Sharma, 1974), standing for steady three-dimensional analysis of
boundary-layer equations. Further information is contained in a
paper by Spalding (1974a).

The abscissa of the graph is downstream distance x divided by
jet diameter D; the ordinate is the distance from the jet center
line y, divided by D, at which the density of fluid has its maximum.
The points are measurements by Ayoub (1971); the curve is McGuirk's
prediction. The jet velocity at entry is five times the stream
velocity; the Froude number (≡ jet velocity squared divided by D
times gravitational acceleration) is 15.

Evidently, the agreement is rather satisfactory suggesting
that the turbulence model, which contains no term for buoyancy-
turbulence interactions, is quite satisfactory in this case.
Other equally good results are found for other velocity ratios
and Froude numbers.

9. EXAMPLES OF SUCCESSFUL COMPUTATIONS

(b) 3D, Steady, Elliptic

Figure 2 provides some information about computations carried
out by Patankar and Spalding (1973) to demonstrate how turbulence
models (among other things) might be applied to practical combustion-
chamber flows.

The computer program TROG (the paper was read at Trogir,
Yugoslavia) solved twelve differential equations. Their variables

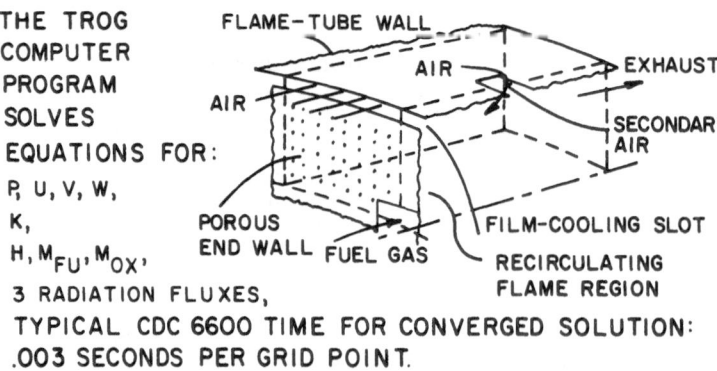

THE TROG COMPUTER PROGRAM SOLVES EQUATIONS FOR: P, U, V, W, K, H, M_{FU}, M_{OX}, 3 RADIATION FLUXES,

FLAME-TUBE WALL — AIR — EXHAUST — AIR — SECONDARY AIR — POROUS END WALL — FUEL GAS — FILM-COOLING SLOT — RECIRCULATING FLAME REGION

TYPICAL CDC 6600 TIME FOR CONVERGED SOLUTION: .003 SECONDS PER GRID POINT.

Figure 2. Computational Examples; 3D, Steady, Elliptic

were: the three velocity components u, v and w; the pressure, p;
the fluctuation energy and dissipation rate of turbulence k and ε;
the specific enthalpy, h; the mass fractions of fuel and oxygen,
m_{fu} and m_{ox}; and the radiation-flux sums in the three coordinate
directions. The flow was of course three-dimensional, and
exhibited a complex recirculation pattern; therefore the computer
storage was also three-dimensional.

Calculations were carried out, for lack of funds, with only a
coarse $(7 \times 7 \times 7)$ grid; so no quantitative accuracy could be claimed.
Nor were measurements available for comparison. However, the computer time was modest (about one minute on a CDC 6600 computer);
and the predictions entirely plausible.

The conclusion that may reasonably be drawn from this example
is that only two things stand in the way of large-scale use of
programs like TROG for engineering design: the costs of fine-grid
calculations; and the prevailing uncertainties about the validity
of the incorporated models of turbulence, etc. Both these
obstacles can be removed by detailed development.

10. EXAMPLES OF SUCCESSFUL COMPUTATIONS

(c) 3D Unsteady with Free Surfaces

Figure 3 provides the third and last example. This computer-
generated output records the results of applying the computer
program TRIC (three-dimensional recirculating flow in Cartesian
coordinates), from which TROG was derived, to a two-fluid phenomenon.

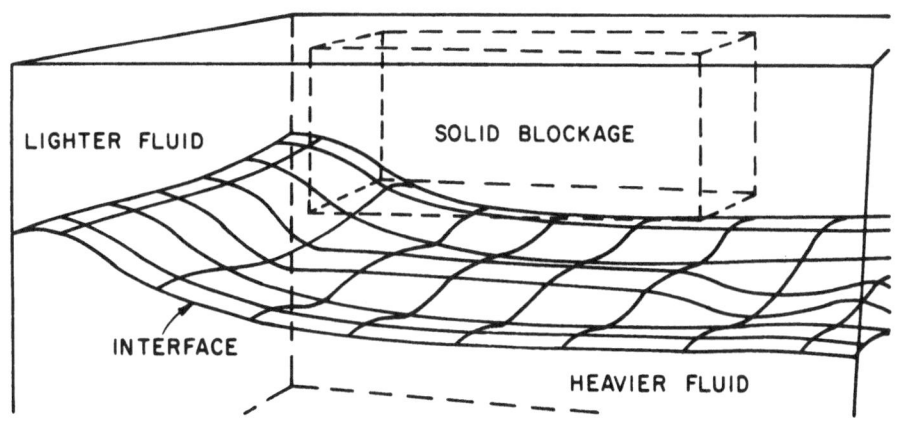

Figure 3. Computational Examples; 3D-Unsteady.

The two fluids are immiscible and of slightly different density; both are turbulent; the upper fluid is obstructed by a rectangular obstacle, shown dotted; and gravity and electromagnetic body forces affect the flow. The work was contributed to by several of my colleagues at Imperial College (A. D. Gosman, M. Koosinlin, S. V. Patankar, D. Rafinejad), and is concerned with an industrial process.

It is not necessary to say more about the computations than that they have been made, that the results they produce are in qualitative agreement with the (alas, non-quantitative) experimental observations, and that they provide so much insight into the relevant process that the mathematical model is now being extensively exercised in design studies. Although one would of course be grateful for confirmation or correction of the turbulence model, even its present uncertainties are preferable to the previously prevailing ignorance.

11.   SOLVED PROBLEMS

2.   The k~ε Turbulence Model for Large Reynolds Number

Let us now review in more detail the extent to which the two-equation turbulence model first proposed by Harlow and Nakayama (1968) has subsequently been tested and developed. Since a recent paper by Launder and Spalding (1974) has concerned itself with this matter, the present account can be brief.

We refer to the model here as the $k \sim \varepsilon$ model. k stands for $\frac{1}{2} \overline{u_i u_i}$, i.e. the kinetic energy of the fluctuating motion per unit mass. ε stands for $\sum_{i,j} \overline{\mu(\partial u_i / \partial x_j)^2} / \rho$ where ρ is the density, i.e. for the rate of dissipation of this energy per unit mass. ε is also taken as constant times $k^{\frac{3}{2}}/\ell$, where ℓ is the length scale of the turbulence.

Rodi (1972) and Launder, Morse, Rodi and Spalding (1972) have presented the results of many comparisons of numerical predictions based on the $k \sim \varepsilon$ model with experimental data culled from the world's literature. Agreement with experiment varies from good to very good, although in some cases this achievement is at the cost of inelegant amendments to the model which are of doubtful generality. I shall refer to one of these later.

The data just referred to related to steady two-dimensional "parabolic" flows, such as: plane jets, wakes and mixing layers; wall jets on planes and cones; boundary layers of various kinds on solid walls; "fan" jets; and flows in pipes and annuli. Fewer comparisons have been made with two-dimensional "elliptic" flows; however agreement is satisfactory here also. The work of Elghobashi (1974) serves as example.

Most of the comparisons have been made in situations of uniform density, for lack of other data; and it must be admitted that these other data, when they do exist, exhibit rather less good agreement with predictions. Nevertheless, the k ~ ε model still permits fairly accurate predictions to be made even when density variations are very large, as in combustion phenomena. From the practical point of view, its utility is very great indeed.

12.   SOLVED PROBLEMS

A 2D Elliptic Application of the k ~ ε Model

Figure 4 provides confirmation of the assertion just made that 2D recirculating turbulent flows can be adequately predicted using the k ~ ε model. The geometrical situation is that of two coaxial jets, mixing in a cylindrical duct. The curves represent Elghobashi's predictions, based on the k ~ ε model; and the points represent the experimental data of Barchilon and Curtet (1964).

13.   SOLVED PROBLEMS

3.   Prediction of Concentration Fluctuations

Another minor success is illustrated by Figure 5, which relates to the distribution of concentration fluctuations in an axisymmetrical jet. The points represent the experimental data of Becker, Hottel and Williams (1967); the curves result from

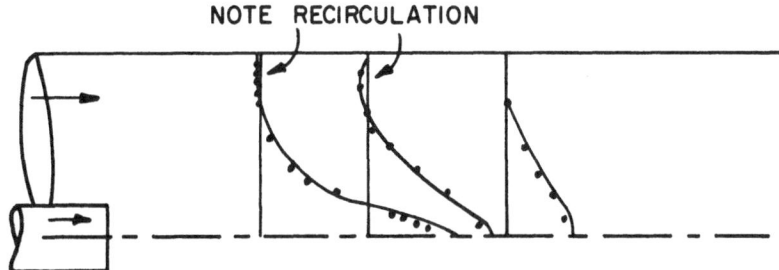

VELOCITY PROFILES IN CONFINED JET, DOTS ARE EXPERIMENTAL MEASUREMENTS, CURVES ARE PREDICTIONS OF ELGHOBASHI,USING K~ TURBULENCE MODEL.

Figure 4. Solved Problems 2; 2D, Steady, Elliptic.

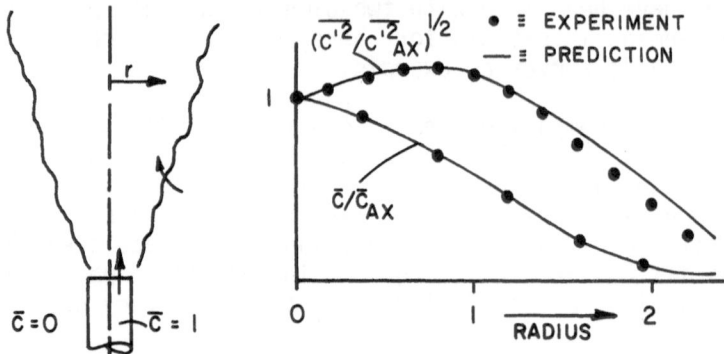

Figure 5.   Solved Problems 3; Concentration Fluctuations.

predictions of a three-equation model of turbulence (Spalding, 1971), the three dependent variables of differential-equations being the fluctuation energy, the mean-square "vorticity fluctuations," and the mean-square concentration fluctuations. Evidently, the agreement is quite satisfactory in this case.

14.   SOLVED PROBLEMS

General Remarks

In summary, three points need to be grasped about the successful side of turbulence modelling. They are:

    (i) It is now easy to solve the turbulence-model equations for nearly all practically interesting problems.

    (ii) As far as the hydrodynamics of the processes are concerned, it suffices to employ turbulence models possessing differential equations for just two quantities, for example k and ε.

  (iii) The successful prediction of concentration fluctuations, by the use of a third differential equation, opens the door to the calculation of interactions between turbulence and chemical reactions, a subject of especial concern in this conference.

Concerning the second item, it needs to be emphasized that:

(a) At least two differential equations must be employed, because otherwise it is necessary to provide as an algebraic input the length-scale distribution; and no knowledge of this exists for complex flows.

(b) Which two-equation model is used is still somewhat a matter of preference. For flows remote from walls, they all give approximately the same result; it is just for flows near walls that the k ~ ε model proves superior. The reason is explained by Launder and Spalding (1974).

Concerning the third item, the door is only just open, as will be indicated in section 16 of this lecture. For what we often need to know is not just the mean-square value of the concentration fluctuations, but also their shape. However, even limited progress is better than none.

15.   UNSOLVED PROBLEMS

1.   Low-Reynolds-Number Turbulence

"Constants" have appeared in various places in the discussion so far; and they are highly desirable features of turbulence models. For example, most modellers employ some form of the effective-viscosity assumption of Kolmogorov (1942), which, in terms of k and ε, we write as:

$$\mu_{eff} = C_\mu \rho \frac{k^2}{\varepsilon} .$$ (5)

Can C_μ retain a constant value even when the fluctuations are small, i.e. when $(\mu_{eff} - \mu)/\mu$, which we can regard as the local Reynolds number of turbulence, is less than unity? It seems unlikely; for we must then expect that $\mu_{eff} \approx \mu$ and that ε is proportional to k rather than to k^2. We must conclude that C_μ is a function of the local Reynolds number $k^{\frac{1}{2}}\ell/\mu$ at least.

What kind of function? Preliminary suggestions were made by Glushko (1965) and Wolfshtein (1967), within the framework of one-equation turbulence models. Jones and Launder (1972) have made further suggestions. My own view is that the question is still unresolved, and that what is needed is, ideally:

(i) guidance from fundamental turbulence theory about the relation between μ_{eff}, μ, ρ, k and ε, as to the form of a C_μ ~ Re function.

(ii)   comprehensive examination of existing experimental data
        with a view to establish whatever constants the new
        functional relation is thought to contain.

(iii)   performance of new experiments, designed to test the
        resulting function in new circumstances and so to
        establish its validity or to point to further improve-
        ments.

The trouble about item (i) is that the persons who are
interested in fundamental turbulence theory dislike sullying their
minds with so non-fundamental a concept as μ_{eff}. Item (ii) involves
more work than most turbulence modellers are prepared to undertake.
And item (iii) necessitates an adequate grasp, on the part of the
deviser of the experiments, of what questions truly need to be
answered. It is therefore easily understood that progress in this
direction is slow.

16.   UNSOLVED PROBLEMS

2.   Interactions between Turbulence and Chemical Reaction

We know that the instantaneous reaction rate per unit volume
of gas depends on the instantaneous pressure, temperature and
composition of the gas; and, in some cases, we can even ascribe
correct quantitative expressions for the dependences. In turbulent
flows however we need to know the time-average value of the reaction
rate; and this is certainly not related to the time-average values
of pressure, temperature and concentrations by the same expressions.
The time-average reaction rate must depend upon the fluctuating
concentrations as well, and indeed upon their correlations. No
significant knowledge exists of these at present.

Despite this lack of knowledge, some formulae do exist which
enable the time-average reaction rates to be computed; but they
connect these rates with hydrodynamic quantities at least as much
as with concentrations.

For turbulent diffusion flames, from which chemical-kinetic
limitations are absent, and in which it is the inter-diffusion of
eddies containing separated fuel and oxygen that controls the rate
of burning, the appropriate theory was developed five years ago
(Spalding, 1970); but it has been little developed since that date.
The essential idea is that the decay of concentration fluctuations,
calculable as described in the paper mentioned in Section 13 above
(Spalding, 1971), coupled with a presumption as to the concentration
distribution function, provide all the information that is required
if kinetic influences are insignificant. The quantitative predictions are in fair agreement with the available experimental data.

For <u>confined pre-mixed flames</u>, although the detailed roles of
chemical reaction and inter-diffusion must be entirely different,
the practical result is still remarkably similar: the time-average
reaction rate depends primarily on the rate of break-up of large
eddies.   The theory was put forward in the Thirteenth Combustion
Symposium (Spalding, 1971b), and has been followed up by Mason and
Spalding (1973).

Despite initial successes, defects in the theories remain,
both at the level of concept and at that of agreement with experiment.   The conceptual difficulty appears when it is sought to
combine the two models, i.e. to answer the question: what governs
the reaction rate when chemical kinetics influences a flame to
which fuel and air are supplied separately?   This, after all, is
the most important practical situation.

The defects in prediction accuracy appear as an inability
correctly to foretell the effect of a change of fuel type on the
length of a diffusion flame, and in an over-estimate of the influence of upstream turbulence characteristics on the rate of spread
of a confined pre-mixed flame.   These defects, and their possible
causes and remedies, are the subject of a separate lecture
(Spalding, 1974).

17.   UNSOLVED PROBLEMS

3.   A Further List

One of the most tiresome features of even the best turbulence
models is that none of them so far can predict both the plane and
the axi-symmetrical turbulent jet, in stagnant surroundings, with
the use of a single set of turbulence constants.   The work of Rodi
(1972) makes this clear.   My own view is that the reason is that,
near the axis of the "round" jet, the "gradient assumption" breaks
down; for the length scale of turbulence is no longer small compared with the radius of curvature.   However, I have not had time
to make a comprehensive study of the matter; and Rodi's enquiries
led him in a different direction.   I shall return to this topic
below.

Then there is the question: when the "gradient assumption"
is plausible for diffusion of turbulence quantities, which of
these three quantities satisfies it most closely: $k^{3/2}/\ell$ (i.e. ε),
$k\ell$ (suggested by Rotta (1951)), k/ℓ^2 (suggested by Spalding,
1969a)), or simply ℓ (suggested by Spalding (1969b)).   One thing
is certain: they cannot <u>all</u> do so; rigorous mathematical investigation of a particular type <u>could</u> decide the matter, as described
in Section 23 below.

Many turbulent flows of practical importance contain particles: droplets of water and particles of sand are conveyed by the natural wind; droplets of liquid fuel are sprayed into the combustion chambers of diesel engines, furnaces and gas turbines; and the two-phase mixtures in steam generators and some nuclear reactors take the form, in some regions, of suspensions of droplets in a gaseous phase. What effects will the presence of the dense-phase material have on such processes as dissipation of turbulence energy and transfer of momentum between adjacent streams? Despite the importance of the question, there is no firm knowledge, so far as I am aware, and there are few guesses. If I am wrong about this, I shall be glad to be corrected.

Finally, body forces may have a great influence on the generation, damping and transport of turbulence quantities. The most familiar demonstration of this phenomenon is provided by the interaction of a vertical temperature gradient and the earth's gravitational field: when the warm air is down below one may have a bumpy airplane ride; when it cools in the evening, smoke plumes spread as laminar layers. Something is known of this matter; for example, the Richardson number is certainly an important dimensionless parameter. But who can tell us how all the various "constants" in the turbulence-model equations depend upon this parameter?

18.   THE TRANSPORT OF A SCALAR; ANALOGY WITH RADIATIVE TRANSFER

I turn now from surveys and generalities to a conceptual novelty which, I believe, deserves someone else's attention; for several years have passed since I first advanced the idea (Spalding, 1972b) without my having done anything with it.

The problem is this: how can we relate the turbulent diffusion flux $\overline{G'C'}$ to local properties of the fluid? Of course, we have the gradient-diffusion approximation:

$$\overline{G'c'} = - \Gamma \ grad \ \overline{c} \ ; \tag{6}$$

where G' is the fluctuating density-velocity product, and c' and $\overline{c}$ are respectively the fluctuating and time-average components of concentration, and Γ is an "effective exchange coefficient" of the turbulent fluid; yet experience shows us that $\overline{G'c'}$ and - grad $\overline{c}$ do always even have the same sign.

What can be the reason for this? One possibility is that the length scale of turbulence, to which Γ is proportional, is not always small enough in comparison to the distance over which grad $\overline{c}$ varies appreciably.

The situation is similar in this regard to that encountered
in radiative transfer; for often the "mean free path of radiation"
is of the same order of magnitude as the dimensions of the apparatus.
Let us see therefore whether we can apply to turbulent transfer
phenomena ideas such as those which Schuster (1905) and Hamaker
(1947) have developed for radiation.

19. TRANSPORT OF A SCALAR; THE TWO-FLUX MODEL

Consider for simplicity the axi-symmetrical situation of
Figure 6. Turbulent diffusion is supposed to take place in the
radial direction as a result of random fluid motion. Specifically,
if the net rate of fluid movement in the <u>positive</u>-r direction per
unit area is g (and the net rate in <u>reverse</u> direction is also g,
for continuity of mass), the turbulent diffusion is supposed to be
the result of this fact: that the average c - value of outgoing
material is c_+ while the average value of ingoing material is c_-.

It follows that the diffusion flux is given by:

$$\overline{G_r'c'} = g(c_+ - c_-) \tag{7}$$

where

$$\overline{G_r'^2} \equiv g^2 \tag{8}$$

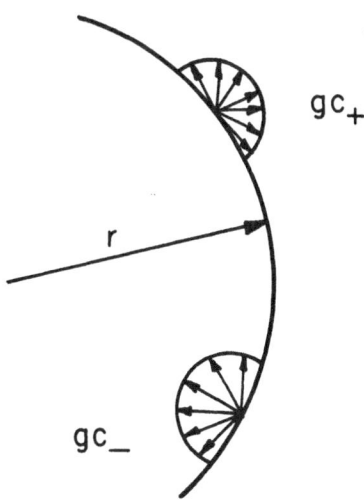

Figure 6.   Illustration of the Two-Flux Model.

Let us now define an "eddy-absorption coefficient" a, such that in the steady state and in the absence of sources and sinks of the material whose concentration is measured by c, the quantities c_+ and c_- are governed by the differential equations:

$$\frac{d}{dr} (rgc_+) = - rga(c_+ - \bar{c}) + gc_- \,, \tag{9}$$

$$\frac{d}{dr} (rgc_-) = rga(c_- - \bar{c}) + gc_- \tag{10}$$

The significance of the terms in these equations are as follows:

(i)   The left-hand sides express the changes with radius of the net outgoing and ingoing fluxes of material.

(ii) The first terms on the right-hand side express the fact that the concentration in an eddy tends to become the same as that of the time-average material at the layer in question, $\bar{c}$.

(iii) The third term on the right-hand side expresses the fact that, because of the curvature of the surface in question, a certain proportion of the material carried randomly in the inward direction, i.e. that following a nearly tangential direction, passes directly into the outgoing stream.

It is on these two first-order differential equations that the two-flux model is based.

20.   TRANSPORT OF A SCALAR; THE RESULTING SECOND-ORDER DIFFERENTIAL EQUATION

Let us now define c_* as the arithmetic mean of c_+ and c_-; and let J stand for the outward-moving diffusive flux. Then:

$$c_* \equiv \tfrac{1}{2} (c_+ + c_-) \,, \tag{11}$$

and

$$J \equiv \overline{G_r' c'} \,, \tag{12}$$

Then addition of equations (9) and (10), with introduction of (11) and (12) leads to:

$$2r \frac{d}{dr} (gc_*) = - J(1 + ar) \; ; \tag{13}$$

and subtraction leads similarly to

$$\frac{d}{dr} (Jr) = - 2 \; rag \; (c_* - \bar{c}) \tag{14}$$

These two equations connect $\bar{c}$, c_* and J.   There is normally another one also, representing the conservation law for the chemical species in question.   In the general case of an axi-symmetrical flow, this might run as follows:

$$\frac{1}{r} \frac{\partial}{\partial r} (Jr) = - S \quad , \tag{15}$$

where:

$$S \equiv \frac{\partial}{\partial t} (\rho \bar{c}) + \frac{\partial}{\partial x} (\rho u \bar{c}) \tag{16}$$

Here t stands for time, x for distance in the axial direction, and u for the fluid-flow velocity in that direction.

21.   TRANSPORT OF A SCALAR; THE DIFFUSION APPROXIMATIVE

Let us now suppose that the "source term" S is a constant, as might well be true over a restricted region.   Then equation (15) yields:

$$J = - \frac{Sr}{2} \; ; \tag{17}$$

and substitution into (14) thereafter entails:

$$S = 2 \; ag(c_* - \bar{c}) \tag{18}$$

Let us further suppose that a and g can be regarded as independent of radius; then we deduce:

$$\frac{dc_*}{dr} = \frac{d\bar{c}}{dr} \tag{19}$$

Finally, substitution in equation (13) yields:

$$\boxed{J = - \frac{2g}{(a+\ell/r)} \frac{d\bar{c}}{dr}} \tag{20}$$

This is of course the gradient-diffusion approximation for J, with the interesting features:

> (i) that the exchange coefficient Γ tends to 2g/a at large ar; and

> (ii) that it tends to 2gr, i.e. to zero, at low ar.

Here then is a possible explanation of the anomaly noted in Section 17, concerning the apparent change in turbulence constants near an axis of symmetry. If we adopt the symbol Γ_∞ for 2g/a, we learn that

$$\Gamma = \Gamma_\infty/\{\ell + \ell/(ar)\}$$

$$= \Gamma_\infty/\{\ell + \ell_\infty/(2gr)\} \qquad (21)$$

So $\Gamma_\infty/(2gr)$ can be regarded as the dimensionless argument of which the "constants" are likely to be functions. Since $\Gamma_\infty/(2g)$ can be regarded as a measure of the length scale of turbulence ℓ, we conclude that this dimensionless quantity is nothing but ℓ/r.

The matter will be dropped at this point; but it seems to me that further exploration and development of the model is worthwhile.

22.   SOME INFORMATIVE EXPERIMENTS

1.   Low-Reynolds-Number Turbulence

I turn now to my last topic, the illustration of what experiments would be particularly illuminating from the point of view of the turbulence modeller; and the first example concerns the important subject of low-Reynolds-number turbulence.

Figure 7 illustrates what, in my opinion, would be both easily practicable and extremely informative. In a steady low-Reynolds-number channel flow, turbulence of controlled energy and scale is produced by a grid. The two walls of the duct are at different temperatures; and the fluid upstream of the grid may be at a third. Measurements are made of the time-average temperature and velocity profiles downstream of the grid; from these are deduced the fields of effective viscosity and effective thermal conductivity.

Even from these time-average measurements, it would be possible to discriminate between alternative formulations of the low-Re turbulence-model equations. If measurements of turbulence energy, Reynolds stresses and turbulent heat fluxes were made simultaneously, still finer discrimination would be possible. Until this is done,

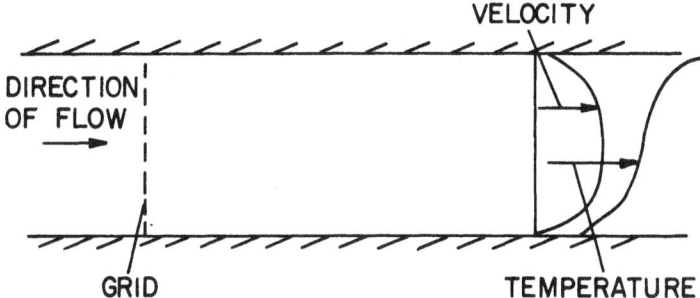

Figure 7. Apparatus for the Investigation of Low-Reynolds-Number
           Turbulence.

however, it is possible that highly non-universal models will
remain in the literature without their inadequacies being detected.

23. SOME INFORMATIVE EXPERIMENTS

2. What is it that Diffuses?

Let us now consider the question: what is the power of n
which causes the variable $(k\ell^n)$ to obey most closely a gradient-
diffusion equation, in the first instance at high Reynolds number?

The following suggestion is perhaps somewhat naive; but I
should like to know why it would not work.

Consider the grid of graded blockage and spacing shown in
Figure 8. It seems to me possible to arrange that the fluid
emerging from it has a uniform time-average velocity, and linear
variations of temperature, turbulence energy and turbulence scale.
That must surely be just a matter of patience and ingenuity.

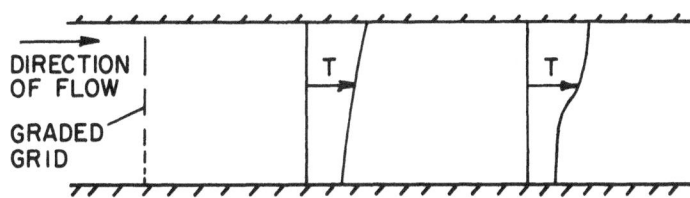

Figure 8. Apparatus for the Determination of n.

Then measurements of the temperature profiles, and of those of turbulence quantities if available, at downstream sections, can be compared with the predictions of turbulence models having k and $k\ell^n$ as dependent variables of transport equations, with various values ascribed to n. Presumably <u>one</u> value of n will fit the data better than another. That is the one n to use until further notice.

Of course, it is not impossible that the appropriate n value will depend upon Reynolds number and upon other quantities; but at least such dependences will be revealed, and can be investigated, by the apparatus that I have in mind. I should be very pleased to learn if anyone has already made such an investigation.

24. SOME INFORMATIVE EXPERIMENTS

3. Mixing Layer with Chemical Reaction

Excellent experimental studies have been made of the plane turbulent mixing layer in an inert fluid. When will someone carry out such studies of mixing layers in which chemical reaction occurs? When they do, we shall truly begin to learn something about turbulent reacting flows.

What would be the arrangement? Figure 9 illustrates what is wanted: two streams at least one of which is sufficiently hot to cause chemical reaction, with a variety of combinations of fuel, oxygen and inert concentrations. What would be measured would be the self-similar distributions, at downstream localities, of time-average concentrations and temperature, as well as the distribution of mean-square fluctuating quantities when appropriate.

What is especially interesting about such flows is the fact that, after an initial development region, all quantities should be constant along lines of constant y/x. Or rather <u>many</u> quantities would be; but the ratio of the eddy size to the micro-scale of turbulence would steadily enlarge. What, if anything, would this do to the concentration profiles? Different models of turbulence

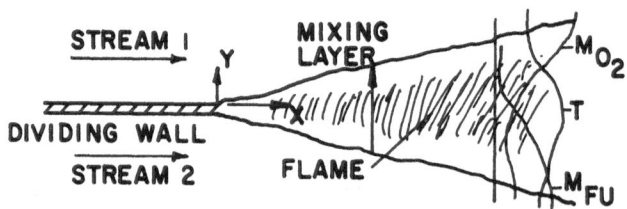

Figure 9. An Apparatus for Turbulent-Combustion Studies.

will predict different answers; therefore experiments on the
mixing layer with reaction can provide useful discrimination.

25.   SOME INFORMATIVE EXPERIMENTS

4.   The Influence of Particles

This item I will illustrate more briefly. Suppose that we
have an inert plane mixing layer, but that at least one of the
streams contains a suspension of solid particles. Certainly the
rate of spread of the mixing layer will be influenced, especially
in the upstream region in which the ratio of turbulence scale to
particle size is relatively small. Will the particles have an
influence also in the downstream region? On some views of what
happens in turbulence, the answer is negative; but it would be
useful to find out. In any case, comparisons could be made of
the distributions of velocity, concentration, etc. with the predictions of various theoretical models. The results could not
fail, in my opinion, to lead to improvements to these models.

In concluding these suggestions about experiments, I want to
remark that they are just a selection of a long series that would
greatly assist the modeller. How interesting it is that so few
really valuable experiments are performed nowadays, despite the
great improvements that have been made to methods of instrumentation! The reason appears to be that our education divides us into
experimentalists and theoreticians; and the former, though they
like measuring things, do not care enough for the needs of theory
to perform the right experiments; while the latter are oppressed
by the thought that experiments are difficult, and so make no
suggestions.

26.   CONCLUDING REMARKS

Having now arrived at the end of the lecture, I must underline
the main points which I would like the audience to be persuaded of.

The first is that the turbulence-model approach, coupled with
modern computing techniques, is already successful in predicting,
with speed and economy, turbulent-flow phenomena in many fields of
engineering and of the natural environment.

Secondly, this is the only hopeful approach in prospect. It
is therefore fruitless for purists to press on us their certainty
that our turbulence models are too crude to comprehend all that is
known. We are aware of this; yet, until computers are larger and
experimental researches more systematic, we shall have simply to

ignore the more depressing knowledge, on the grounds that it is unusable.

Thirdly, there is much more for further research, both original and routine. Original experiments are needed, as well as original theoretical notions. Perhaps the most difficult feat of all, in this area of study, is to maintain the proper mean between naive optimism and naive despair. What is needed is simultaneously a willingness to use boldly the achievements that have already been made, and a readiness to criticize the results realistically and to search untiringly for improvements of concept and method.

Lastly, it must be remembered that serious turbulence-model research has been in progress for less than ten years. We are still in the stage of rapid discovery, and the practical impact on engineering has scarcely started yet.

REFERENCES

1. Amsden, A. A. and Harlow, F. H. "A simplified MAC technique for incompressible fluid flow calculations." Los Alamos Report Number LA-DC-11272, 1970.

2. Ayoub, G. M. "Dispersion of Buoyant Jets in a Flowing Ambient Fluid." Imperial College, London, Civil Engineering Department, Ph.D. Thesis, March 1971.

3. Barchilon, M. and Curtet, R. "Some details of the structure of an axi-symmetric confined jet with back flow." Journal of Basic Engineering, Vol. 86, p. 777, 1964.

4. Becker, H. A., Hottel, H. C. and Williams, G. C. "The nozzle-fluid concentration field of the round turbulent free jet." Journal of Fluid Mechanics, Vol. 30, p. 285, 1967.

5. Boussinesq, J. "Theorie de l'ecoulement tourbillant." Mem. Acad. Sci., Vol. 23, no. 46, 1877.

6. Bradshaw, P., Ferriss, D. H. and Atwell, N. P. "Calculation of Boundary-Layer Development Using the Turbulent Energy Equation." Journal of Fluid Mechanics, Vol. 28, part 3, pp. 593-616, 1967.

7. Caretto, L. S., Curr, R. M. and Spalding, D. B. "Two Numerical Methods for Three-Dimensional Boundary-Layers." Computer Methods in Applied Mechanics and Engineering. Vol. 1, pp. 39-57, 1972.

8.   Caretto, L. S., Gosman, A. D., Patankar, S. V. and Spalding,
     D. B. "Two Calculation Procedures for Steady, Three-Dimensional
     Flows with Recirculation." Proceedings of the Third Inter-
     national Conference on Numerical Methods in Fluid Mechanics.
     Vol. II, pp. 60-68. Published by Springer-Verlag Heidelberg.
     Edited by J. Ehlers, K. Hepp, H. A. Weidenmüller, 1973.

9.   Chorin, A. J. "Numerical Solution of the Navier-Stokes
     Equation." Mathematics of Computation, Vol. 22, pp. 745-762,
     number 104, 1968.

10. Elghobashi, S. E. "Characteristics of Gaseous Turbulent
     Diffusion Flames in Cylindrical Chambers - A Theoretical and
     Experimental Investigation." Imperial College, London,
     Mechanical Engineering Department, Ph.D. Thesis, 1974.

11. Glushko, G. S. "Turbulent boundary layer on a plane plate
     in an incompressible fluid." Izv. Akad. Nauk SSSR Ser. Mech.
     Number 4, pp. 13-23, 1965.

12. Gosman, A. D. and Spalding, D. B. "The prediction of confined
     three-dimensional boundary layers." Symposium on Internal
     Flows, University of Salford. Published by Imperial College,
     London, Mechanical Engineering Department, 1971.

13. Hamaker, H. C. "Radiation and heat conduction in light-
     scattering material." Philips Res. Rep. 2, pp. 55-67, p. 103,
     112, 420, 1947.

14. Harlow, F. H. "Numerical methods for fluid dynamics; an
     annotated bibliography." Los Alamos Laboratory Report Number
     LA - 4281, 1969.

15. Harlow, F. H. "Turbulence Transport Modelling." AIAA
     Selected Reprint Series, Vol. XIV, AIAA, New York, 1973.

16. Harlow, F. H. and Amsden, A. A. "Numerical calculation of
     almost incompressible flow." Journal of Computational
     Physics 3, 1, 1968.

17. Harlow, F. H. and Nakayama, P. I. "Transport of turbulence
     energy decay rate." LA Report Number 3854, Los Alamos Sci.
     Lab., University of California, February 1968.

18. Jones, W. P. and Launder, B. E. "The prediction of laminarisa-
     tion with a two-equation model of turbulence." International
     Journal of Heat and Mass Transfer, Vol. 15, p. 301, 1972.

19.   Kolmogorov, A. N.   "Equations of turbulent motion of an
      incompressible fluid."   Izv. Akad. Nauk SSSR Ser. Phys.
      Vol. 6, No. 1/2, pp. 56-58, 1942. (Translated into English
      at Imperial College, London, Mechanical Engineering Department,
      Report Number ON/6, 1968)

20.   Launder, B. E., Morse, A. P., Rodi, W. and Spalding, D. B.
      "The prediction of free shear flows - A comparison of six
      turbulence models."   NASA Free Shear Flows Conference,
      Virginia.   NASA Report Number SP-311, July 1972.

21.   Launder, B. E. and Spalding, D. B.   "Turbulence models and
      their application to the prediction of internal flows."
      Heat and Fluid Flow, Vol. 2, No. 1, pp. 43-54, 1972.

22.   Launder, B. E. and Spalding, D. B.   "Mathematical models of
      turbulence."   Published by Academic Press, London and New
      York, 1972.

23.   Launder, B. E. and Spalding, D. B.   "The numerical computa-
      tion of turbulent flows," Computer Methods in Applied
      Mechanics and Engineering, Vol. 3, pp. 269-289, 1974.

24.   Launder, B. E. Spalding, D. B. and Whitelaw, J. H.   "Turbulence
      models and their experimental verification."   A course of
      lecutures at Imperial College London, Mechanical Engineering
      Department.   Recorded in Heat Transfer Section Reports, Numbers
      HTS/73/16, 17, 18, 19, 20, 21, 22, 23, 24, 25, 26, 27, 28,
      1973.

25.   Mason, H. B. and Spalding, D. B.   "Prediction of reaction
      rates in turbulent pre-mixed boundary-layer flows."   Combustion
      Institute European Symposium, pp. 601-606, 1973.

26.   McGuirk, J. J.   Private communication, 1974.

27.   Mellor, G. and Herring, H.   "A survey of the mean turbulent
      field closure models."   In - AIAA Journal, Vol. 11, p. 590,
      1973.

28.   Nee, V. W. and Kovasznay, L. S. G.   "Simple phenomenological
      theory of turbulent shear flows." Physics of Fluids, Vol. 12,
      no. 3, pp. 473-484, March 1969.

29.   Patankar, S. V., Pratap, V. S. and Spalding, D. B.   "Prediction
      of laminar flow and heat transfer in helically coiled pipes."
      Journal of Fluid Mechanics, Vol. 62, part 3, pp. 539-551, 1974.

30. Patankar, S. V., Pratap, V. S. and Spalding, D. B. "Prediction of turbulent flow in curved pipes." Imperial College, London, Mechanical Engineering Department, Report Number HTS/74/1, 1974.

31. Patankar, S. V. and Spalding, D. B. "A finite-difference procedure for solving the equations of the two-dimensional boundary layer." International Journal of Heat and Mass Transfer, Vol. 10, pp. 1389-1411, 1967.

32. Patankar, S. V. and Spalding, D. B. "Heat and mass transfer in boundary layers." Published by Intertext Books, London, 2nd Edition, 1970.

33. Patankar, S. V. and Spalding, D. B. "A calculation procedure for heat, mass and momentum transfer in three-dimensional parabolic flows." International Journal of Heat and Mass Transfer, Vol. 15, pp. 1787-1806, Published by Pergamon Press, 1972.

34. Patankar, S. V. and Spalding, D. B. "Numerical prediction of some three-dimensional fluid flows." Imperial College, London, Heat Transfer Section Report Number HTS/72/4, 1972.

35. Patankar, S. V. and Spalding, D. B. "Simultaneous predictions of flow pattern and radiation for three-dimensional flames." "Heat Transfer from Flames" - Lecture delivered at the International Centre for Heat and Mass Transfer Seminar, Trogir, Yugoslavia. Imperial College, London. Mechanical Engineering Department Report Number HTS/73/39, 1973.

36. Prandtl, L. "Uber ein neues Formelsystem für die ausgebildete Turbulenz." Nachr Akad. der Wissenschaft in Göttingen. Göttingen: van den Loeck und Ruprecht, pp. 6-19, 1945.

37. Rodi, W. "The prediction of free turbulent boundary layers by use of a two-equation model of turbulence." Ph.D. Thesis, Imperial College, London, Mechanical Engineering Department Heat Transfer Section Report Number HTS/72/24, 1972.

38. Rotta, J. "Statistische Theorie Nichthomogener Turbulenz." Z Physik, Vol. 129, pp. 547-572 (1951) and Vol. 131, pp. 51-77 (1953). Translated into English by W. Rodi as - Imperial College, London, Mechanical Engineering Department Technical Notes, Numbers TWF/TN/38 and TWF/TN/39.

39. Saffman, P. G. "A Model for Inhomogeneous Turbulent Flow." Proceedings of the Royal Society, London, Vol. A317, pp. 417-433, 1970.

40. Schumann, V. "Results of a numerical simulation of turbulent channel flows." Reactor Heat Transfer, Gesellschaft für Kernforschung, Karlsruhe, pp. 230-251, 1973.

41. Schuster, A. Astrophysics Journal, Vol. 21, pp. 1-22, 1905.

42. Sharma, D. The STABLE Program, CHAM Technical Report, 1974.

43. Spalding, D. B. "The prediction of two-dimensional steady, turbulent, elliptic flows." International Seminar on Heat and Mass Transfer in Flows with Separated Regions, Herceg-Novi, Yugoslavia, September, 1969. Imperial College, London, Mechanical Engineering Department Report Number EF/TN/A/16, 1969.

44. Spalding, D. B. "The calculation of the length scale of turbulence in some shear flows remote from walls." Progress in Heat and Mass Transfer, Vol. 2, pp. 255-266, edited by T. F. Irvine et al., Oxford, Pergamon Press, 1969.

45. Spalding, D. B. "Mathematische Modelle Turbulenter Flammen." Vorträge der VDI-Tagung, Karlsruhe, 1969. "Verbrennung und Feuerungen." VDI-Berichte, nr 146, Düsseldorf: VDI-Verlag, pp. 25-30, 1970.

46. Spalding, D. B. "Concentration fluctuations in a round turbulent free jet." Chemical Engineering Science, Vol. 26, pp. 95-107, 1971.

47. Spalding, D. B. "Mixing and chemical reaction in steady confined turbulent flames." Proceedings of the Thirteenth Symposium on Combustion, The Combustion Institute, Pittsburgh, p. 649, 1971.

48. Spalding, D. B. "A novel finite-difference formulation for differential expressions involving both first and second derivatives." International Journal for Numerical Methods in Engineering, Vol. 4, pp. 551-559, 1972.

49. Spalding, D. B. "Mathematical models of free turbulent flows." Instituto Nazionale di Alta Matematica Symposia Mathematica, Vol. IX, pp. 391-416, 1972.

50. Spalding, D. B. "The mathematical modelling of rivers." Imperial College, London, Mechanical Engineering Department Report Number HTS/74/4, January 1974.

51. Spalding, D. B. "Mathematical models of chemical reaction
     in turbulent flow." Imperial College, London, Mechanical
     Engineering Department, Heat Transfer Section Report Number
     HTS/74/30, 1974.

52. Wolfshtein, M. "Convection processes in turbulent impinging
     jet." Ph.D. Thesis, University of London, Imperial College,
     London, Mechanical Engineering Department Report Number
     SF/R/I/2, 1967.

DISCUSSION

BRADSHAW: (Imperial College, London, England)

I think we would all agree with Professor Spalding's closing
remarks, anyway. At an early stage, Professor Spalding dismissed -
and rightly, I think we also agree - a solution of the time
dependent Navier-Stokes equations as a means of obtaining engineering solutions.

What are the speaker's views about the possibilities of using
a few Navier-Stokes solutions for determining the constants
required in turbulence models? In principle, these solutions can
be used to derive any measurable quantity.

SPALDING:

I think that is a very helpful way. I do know that Schumann
at Karlsruhe has made a very impressive step forward along that
line and I am hoping that we are going to get answers, which would,
for example, tell us, on theoretical grounds, that the von Karman
constant should be 0.41 or thereabout.

KOVASZNAY: (Johns Hopkins University, Baltimore, Maryland)

I would like to make some comments and also ask some questions.
The question is first. You have not mentioned much about three-
dimensional flows. What are your views of the approach concerning
such flows? By three-dimensional flows, of course, I mean, not
the "degenerated" case namely axisymmetric flow, but flows with
different behavior in the different directions. My comment is
about the low Reynolds number flows. The low Reynolds numbers
have two difficulties that are far more fundamental than just
practical. One is, that you may have upstream turbulence, where
the scale or frequency varies, and one becomes extremely dependent
on the rather non-universal manner. The other difficulty is the
classical assumption, the use of gradient diffusion, whenever
transport is called for, and this is getting worse all the time

with decreasing Reynolds numbers. The largest and the smallest of
the eddies are becoming about the same size.

SPALDING:

As to the first point, the testing of turbulent models has
been mainly confined to two-dimensional plus low Reynolds numbers.
It is proper; there were many experimental data available and the
computations were easier. The present position is that, for our
part, we are just beginning to make comparisons with truly three-
dimensional flows. There was one, incidentally, in the lecture
and that was the flow in a river; it appears as though the same
turbulence model as for 2D flows does predict the phenomena. So,
my approach to the three-dimensional turbulent flows is one of
wary hope. We know now we can solve the equations. We make the
predictions first of all without modification to the turbulence
model and then compare with experiment. At present, we have no
evidence that any drastic change will be required, but as evidence
comes along, we may have to introduce those changes.

As to the low-Reynolds-number matter, I agree with what you
say. It may well be that we have to recognize that in all the
lower-Reynolds-number flows, it is simply not adequate to characterize our turbulence by just two quantities. But the question
will always be: with how few can we get away? I have no solutions
here; I just see this as an important practical problem requiring
experimental and theoretical study.

LIBBY: (University of California, La Jolla, California)

In the spirit of the workshop, I would like to ask two questions. First, you made the comment with respect to passive
scalars, that the present modeling is quite satisfactory and
that, of course, is a very important conclusion with respect to
the reacting case. But it was my understanding from private
communications with both Launder and Bradshaw that they are
uncomfortable with the present modeling of passive scalars. I
would like to have some discussion, perhaps from you and from
Bradshaw, about that situation. Certainly your comparison with
respect to the circular jets into a coaxial domain was quite
satisfactory.

The second question concerns the matter of the variable
density problem. I notice you had a small comment on the bottom
of one of your slides about the problem, but you did not say very
much at all about it. The particular thing that I would like to
hear discussed at this workshop is the question of whether or not
we are supposed to be using Favre averaging or do the same averaging
for the density as we do for the velocity. That seems to be an

important question. At the Liege meeting,* Ken Bray--he is also
here--showed us how complicated the resulting equations become
with decomposition of the density into mean and fluctuating parts,
but I presume he has some arguments in favor of doing that, and
I personally have some arguments for not doing it that way. So
I'd like to have your comments and perhaps that of others, on
that point as well.

SPALDING:

     Well, I can also be uncomfortable about the calculation of
passive scalars! Perhaps I can give two answers. If one will
accept the idea of an empirical Schmidt number then one can be
comfortable. What I would say is this, that if the calculation
of the point average scalar quantity is adequately characterized,
then the concentration fluctuations can also be characterized.
The point which I am making is really related to the concentration
fluctuations. But if we back away from inserting an empirical
Schmidt number, then I agree that the situation is uncertain. As
to the other matter, I believe that I favor Fav're averaging,
except that I've almost forgotten which side of the fence he is
on!

KOVASZNAY:

     When you speak of effective Schmidt number you mean effective
turbulence, don't you?

SPALDING:

     I think so.

DONALDSON: (ARAP, Princeton, New Jersey)

     There is one problem with the dispersal of passive scalar
which does not show up in the particular problem that you discussed. It is when one has a passive scalar that is a thin thread
that introduces the turbulence. This might be the release of a
plume in an atmosphere which is turbulent and has a scale much
different from the plume. One has a problem then deciding what
scale to put into the model computations. Now, we have tried a
few things. I have no slides, but in the spirit of the workshop,
I will tell you what the results are. I have no idea whether they
are correct, but there should be very interesting experiments that
might be done to try to resolve whether the results we have found
are correct.

---

*   AGARD Specialists Meeting on Analytical and Numerical Methods
    for Investigation of Flow Fields with Chemical Reactions, especi-
    ally related to combustion, April 1-2, 1974, Liege, Belgium.

If you choose the scale that you put into the diffusion calculation to be the scale of the turbulence itself, then you get a hopelessly wrong answer. If you put in the scale of the mean concentration profile, you get a hopelessly wrong answer. On the other hand, if the scale used to model diffusion in the transport correlation is the scale of the mean profile but you correct the turbulent energy in the term so that it is the energy contained in the total turbulence spectrum for scales at and below the scale of the concentration profile, the solutions come out in a very interesting fashion. Whether they are right or not, I do not know. When you do the above for an unstably stratified flow, you find that you do not get a nice Gaussian profile. The profiles you get are bimodal in the vertical direction. Our interpretation of this is that the big scale turbulence is responsible in the solution for a flapping of the plume while the small scale is responsible for the normal diffusion seen in a typical plume.

KOVASZNAY:

What is the variable?

DONALDSON:

The mean concentration. You find that the mean concentration $\bar{c}$ has a maximum, a minimum, and a maximum in the vertical direction (see Figure α). The maximum fluctuation c'^z tends to be at the center (see Figure β). And what that is, is a picture of a flapping jet. It spends somewhat more time at the points A and A' in the sketch than it does at the center. It is interesting that these equations do have that kind of solution inherent in them. Whether or not it is a proper description of the kind of phenomenon that has been seen when a small plume flaps in a larger scale flow is open to question. I think a very interesting experiment might be carried out to resolve this and give us a great deal of information on how to treat this problem.

LIBBY:

Is not the ideal experiment which represents what you are talking about the small heated wire behind a grid?

DONALDSON:

Sure.

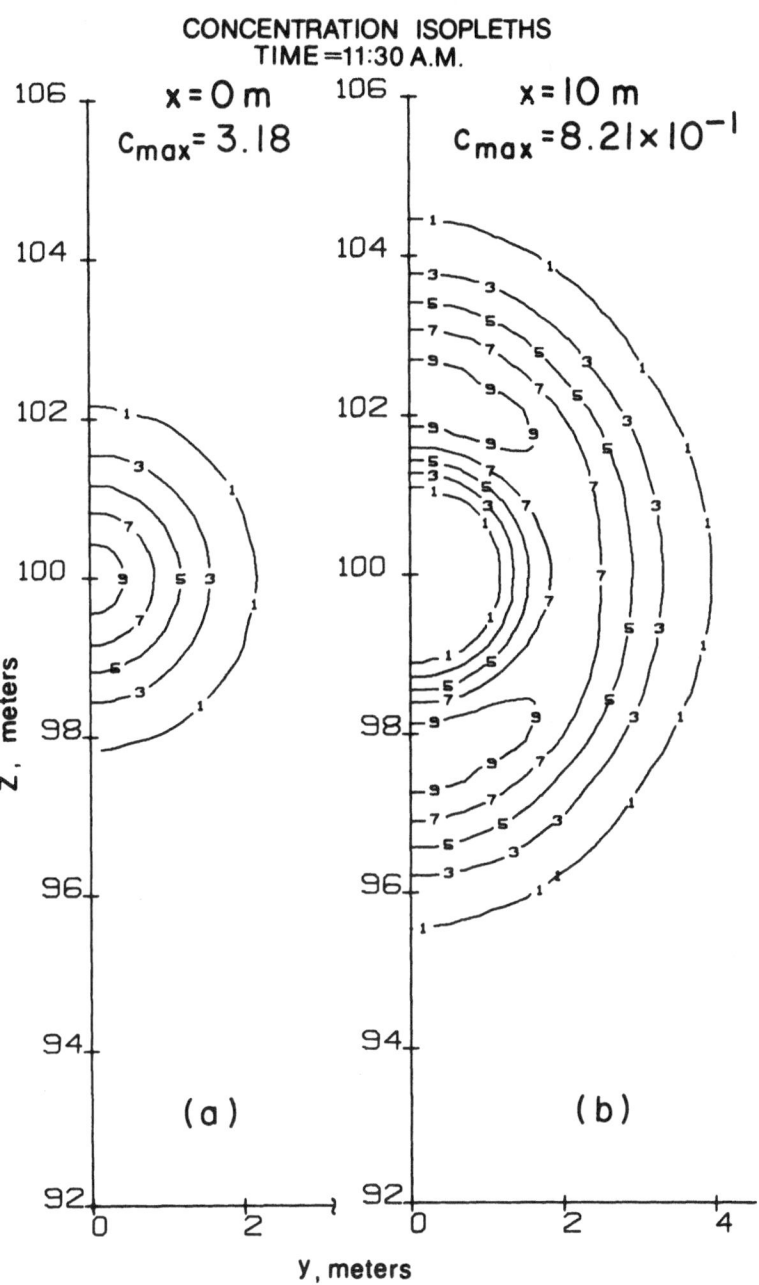

CONCENTRATION ISOPLETHS
TIME = 11:30 A.M.

x = 0 m c_{max} = 3.18

x = 10 m c_{max} = 8.21 × 10⁻¹

(a)

(b)

Z, meters

y, meters

Figure α

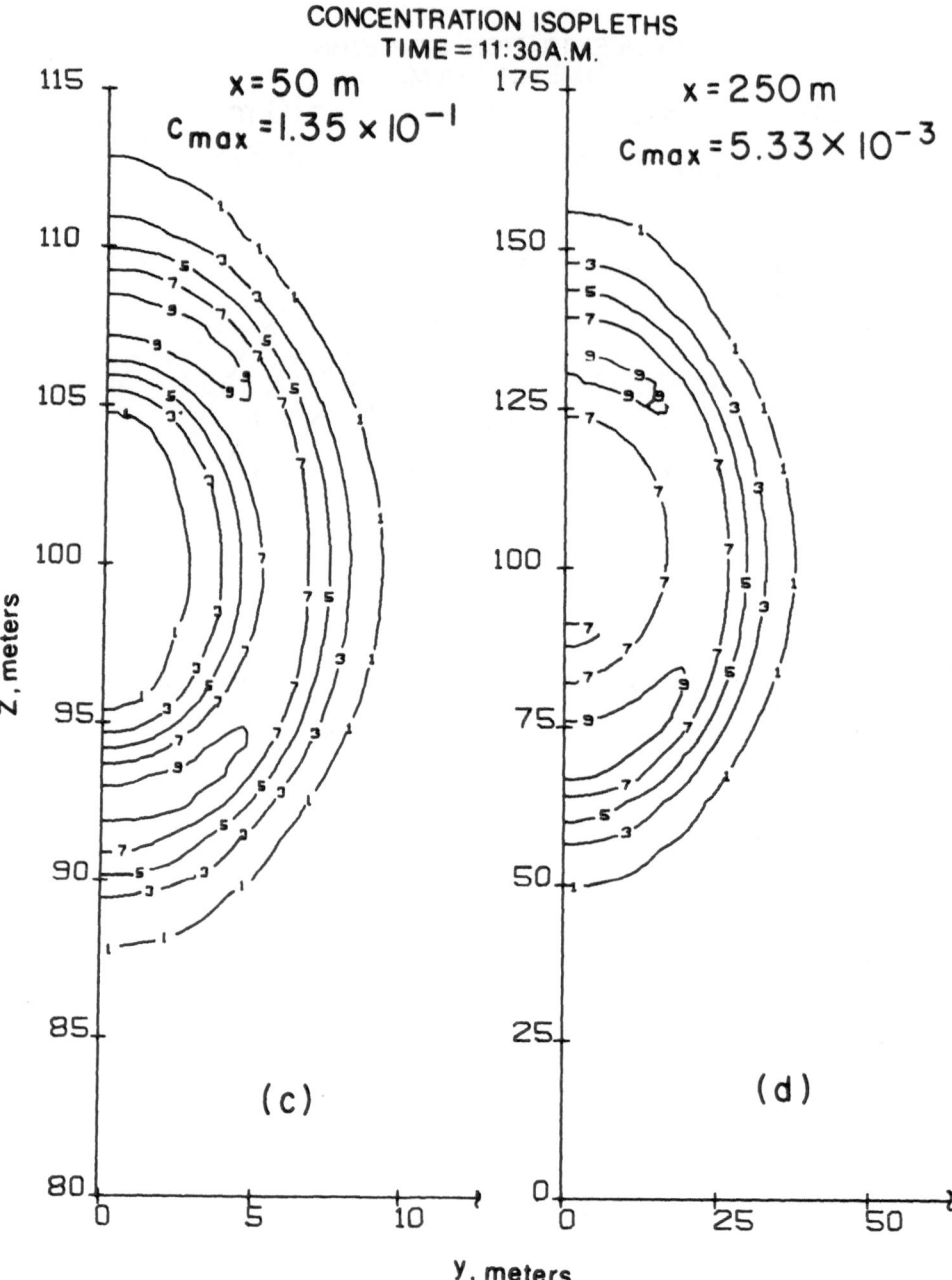

Figure α.   (Continued)

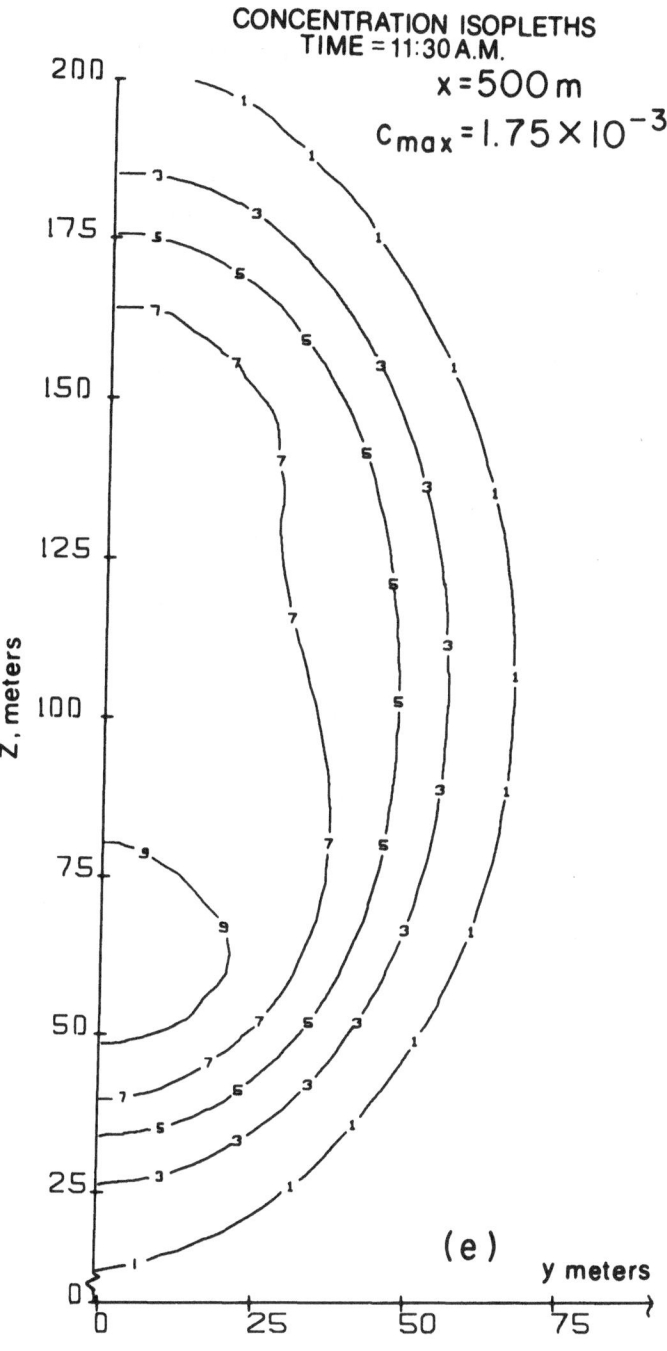

Figure α.   (Concluded)

ISOPLETHS OF MEAN CONCENTRATION FLUCTUATIONS
TIME = 11:30 A.M.

Figure β

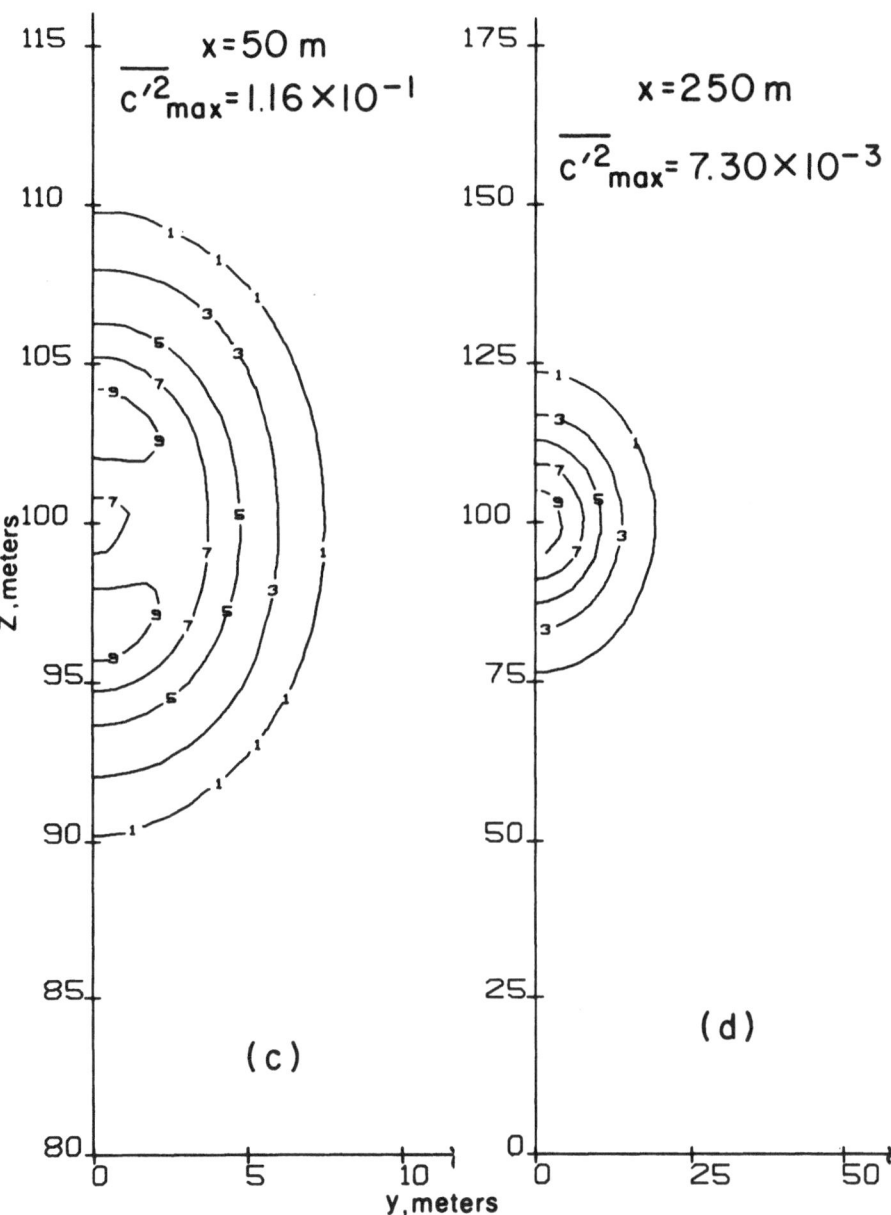

Figure β.   (Continued)

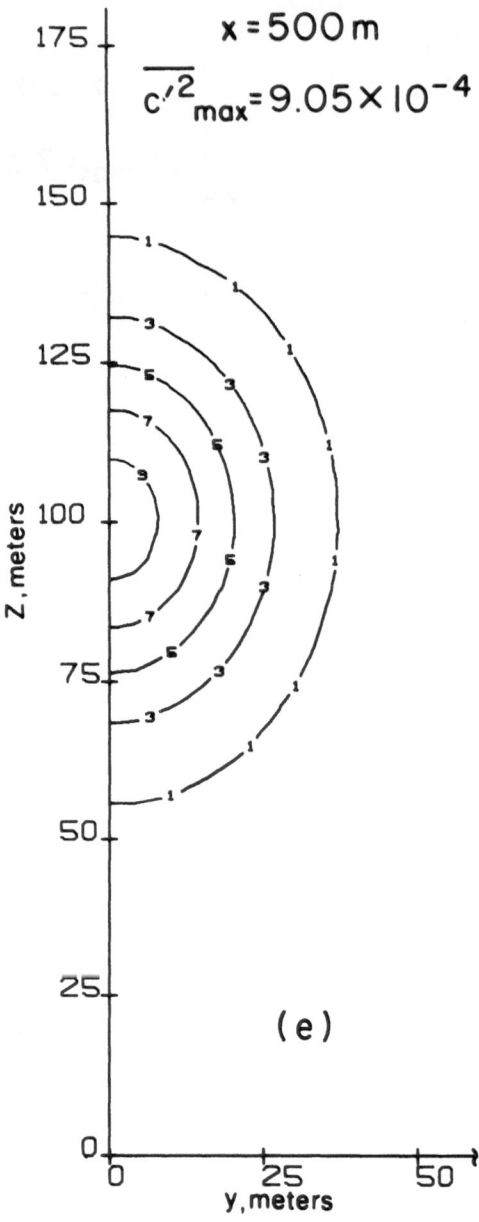

ISOPLETHS OF MEAN CONCENTRATION FLUCTUATIONS
TIME = 11:30 A.M.

$x = 500\,m$

$\overline{c'^2}_{max} = 9.05 \times 10^{-4}$

(e)

Figure β. (Concluded)

LIBBY:

I had a note* in the AIAA Journal on that last year. We
took the data of Corrsin and Uberoi, and also the old data of
Townsend and were able to put it on one nice graph (see Figure γ)
by using a length scale which is associated with the temperature,
not with the grid turbulence. I think this supports what you are
talking about.

DONALDSON:

That is the model experiment along the line that I am talking
about. It is two dimensional and it leaves out some of the
features which this stratified experiment does because, in the
stratified experiment, the scales are really different in the
lateral and the longitudinal directions. This would be a very
hard test of any model.

NAGIB: (Illinois Institute of Technology, Chicago, Illinois)

We just finished some experiments very similar to those of
the two-dimensional heated source or heated wire which Professor
Libby referred to, except that we used a short section of the
heated wire and we followed its temperature and velocity wakes.
Taking that very short heated segment of a wire which is like a
heated spot and putting it in different turbulence conditions,
we actually followed the spreading of the heat in the different
turbulence conditions. The results are different depending on
the turbulence condition.**

ALBER: (TRW Systems, Redondo Beach, California)

The problem of wind-wave generation, the flow of air over a
wavy boundary, represents another practical problem of interest
in turbulence that has different types of scales and has really
not been treated adequately by current turbulence theories. The
initial work on this problem by Miles[†] lets you predict how much
power goes into a wave from the perturbed pressure component 90
degrees out of phase with the wave height. One finds that in

---

*   Libby, P. A. and Scragg, C., "On the Diffusion of Heat from a
     Line Source Downstream of a Turbulence Grid," AIAA J., 11, 4,
     562-563, (April 1973).

**   Wigeland, R. 1974, Diffusion from a Periodically Heated Line-
     Source Segment and Its Application to Measurements in Turbulent
     Flows. M. S. Thesis, Illinois Institute of Technology.

†   Miles, J. W. (1957) On the Generation of Surface Waves by
     Shear Flows, JFM 3 (10)

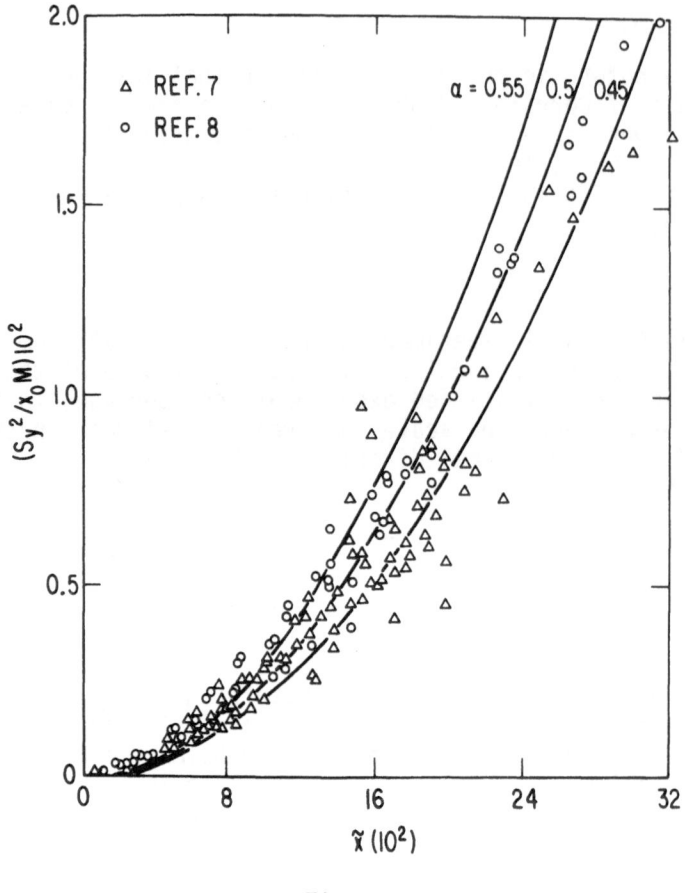

Figure γ

practical applications, for example, in looking at the wave
generation in the ocean, the Miles' theory grossly underestimates
the wave growth. It has been indicated by some recent laboratory
experiments that perhaps the difference results from the change in
the Reynolds stress field caused by the wave-induced velocity and
pressure perturbation. There is an effort now being made to try
to understand the connection between the wave induced field and
the Reynolds stress field associated with the incoming wind shear
flow. That coupling is quite an unsolved problem that is going to
require further study. I do not think it has been adequately
treated by any of the current turbulent models.

KOVASZNAY:

     There are the experiments of J. M. Kendall (JFM 41, 259 (1970))
on the boundary layer over a moving wave.

MORKOVIN: (Illinois Institute of Technology, Chicago, Illinois)

One comment with respect to the low Reynolds number experiment. I have a suspicion that the particular experiment you outlined will suffer from intermittency which will be somewhat different from what you are expecting. In other words, these flow fields, at least from the experience of other people who studied low Reynolds (slightly over 2000) number flows in pipes that have been disturbed on purpose so as to have turbulence generated--they are highly intermittent things. There are at least two types of turbulence structures involved. One that is coupled to the wall conditions and presumably the bursting phenomenon. The other one is much more in the center of the pipe, and it is dispersed with regions of laminar flow in which the turbulent stuff is completely restrained. In other words, I think that one would have to look for generation and relaminarization during periods of time, and it would have to be basically a conditional-type effort. In other words, in low Reynolds number phenomenon the intermittency situation can make life even more complicated than I think you have assumed.

What I am saying is that the grid generated case should not be too different in its development from their different ways of disturbing it. In the experiments which I referred to, they have generated turbulent flows in the center of the pipe by smaller orifices (different orifice plates), by using low pressure pulse upstream in the pipe,* by single cylinders across the inlet, by centered circular disk wakes, etc. They have had small and large turbulence from upstream as well. Anyway, that part of the experiment would at least be a guide. From the evidence of any one of the pipe people (Don Coles has had some experience in the pipe as well; he can tell us a little bit more about the evidence of the intermittency) I think, every mode of the flow, no matter how you created the turbulence, has displayed some intermittency at these Reynolds numbers.

KLINE: (Stanford University, Palo Alto, California)

I do not know the full implications of the exact experiments you are asking for but there are some other ones which have similar problems and they tend to bear out what Morkovin is saying. W. M. Kays' student, Cannon,** some years ago introduced a turbulent flow into a pipe, ordinary turbulent pipe flow, which then

---

* This part of the experiment was presented at the March 1974 Symposium on Large Turbulent Structures at Southampton and is contained in Tel-Aviv University Report TAU/SOE-94/74 by Wygnanski, Sokolov and Friedman.

** Cannon, J. "Heat Transfer from a Fluid Flowing inside a Rotating Cylinder," Ph.D. Dissertation, Stanford (1965).

goes into a section of pipe which rotates. The rotating pipe
generates a stable vorticity layer at the wall. This layer diffuses
inward but leaves a non-rotating flow in the middle, the turbulence
decays very quickly at the wall. Farther downstream one then
observes the second thing that Morkovin is talking about. Retransi-
tion occurs along the interface of the diffusion layer, between the
vorticity layer and the middle 'core' of the flow. That definitely
can occur, but whether it always occurs, of course, is not clear.
Also, in the experiments of Halleen and Lezuis, two of Jim Johnston's
students, the quite complicated effects of Coriolis forces on a
channel now have been studied. There are certain zones in these
flows in which you get disruption of the stabilized wall, creating
turbulent spots, but the spots are damped and they subsequently
decay rather than grow. Again you see the intermittency factor.
So I think, although we do not have exactly the experiment which
you are asking for, one needs to be able to include the kinds of
things that Morkovin is asking about if the modelling is to be
adequate for general flows.

COLES: (California Institute of Technology, Pasadena, California)

    Well, I think the experiment Professor Spalding suggested is
probably new. I do not know what he meant by low Reynolds number,
whether he meant below 1900.

SPALDING:

    The Reynolds number, such that turbulence will undoubtedly
decay as long as heat is present.

COLES:

    Right, I think that would be a clean experiment. However, it
would tell you more about the turbulence generated at the grid than
anything else. But it would be a useful experiment to study damping
of turbulence due to a lack of production. I would not expect
intermittency at that Reynolds number.

MORKOVIN:

    If you are talking about very low Reynolds numbers for pure
decay, there are two more experiments that could be mentioned for
guidance in this direction. These experiments go from a small pipe
to a larger pipe. That was done by Laufer* and by Sibulkin.**
I think you get the decaying mode if you wish.

---

*   J. Laufer: "Decay of a nonisotropic turbulent field," Miszel-
    laneen der Angewandte Mechanik, W. Tollmien Festschrift (60th
    birthday), M. Schäfer, Editor, Akademie-Verlag, Berlin, 1962.

** M. Sibulkin: "Transition from turbulent to laminar pipe flow,"
    Physics of Fluids, 5, 280-284, 1962.

SPALDING:

You say small pipe to larger pipe, then there is sudden enlargement?

MORKOVIN:

That's right, sudden enlargement by means of which the Reynolds number actually goes down. This is in the literature, both Sibulkin and Laufer have it, and I think that would be close to your grid initial conditions. The case of relaminarization by heating in pipes has been reported by C. A. Bankston: "The transition from turbulent to laminar gas flow in a heated pipe," Journal Heat Transfer, ASME TRANS. Series C, Vol. 92, 569-579, 1970.

BRADSHAW:

With reference to your exposition on numerical methods, one of the things that worries one about upstream-difference methods is the appearance of the pseudo-viscosity which, as I understand it, gives you a Reynolds number of order unity based on the grid size, through-flow velocity and the pseudo-viscosity, if the streamlines are not aligned with the grid lines. I noticed in one of your slides that you suggested using orthogonal, rectangular coordinates, and merely said "streamline coordinates if necessary." Would you regard it as necessary to use streamline coordinates to cut the pseudo-viscosity down?

SPALDING:

I think it is always   desirable to align one's grid with streamlines as much as possible. However, it is certainly right to state all numerical procedures are liable to introduce false viscosity with an order of magnitude such that the false Reynolds number based upon the velocity on the diagonals is unity.

BUSHNELL: (NASA-LaRC, Langley AFB, Virginia)

In regard to the windward differencing, have you ever thought of using second order windward differences, as opposed to first order with high artificial viscosity--just second order spatial differences?

SPALDING:

I do not think that I know what you mean and it would probably take us too long to discuss this. I have thought of doing things of this kind and found no satisfactory way of doing so.

BRODKEY: (Ohio State University, Columbus, Ohio)

In chemical engineering we must often deal with chemical kinetic problems that involve liquid flows. There is a vast difference in the Schmidt number between the gas and liquid flows, and I wonder if anybody has had experience in using computational methods for the decay of the scalar field under such conditions. Because of the vast difference in Schmidt number of three to four orders of magnitude, the concentration fluctuation wave numbers extend far beyond those of the velocity.

We are interested in the decay of concentration fluctuations, because that is what controls higher order chemical reactions.

SPALDING:

The results I spoke of were for gases, with Schmidt numbers of the order of unity. I do believe that the Schmidt number is important. High Schmidt numbers cause concentration fluctuations to persist.

ON THE MODELING OF THE SCALAR CORRELATIONS NECESSARY TO CONSTRUCT

A SECOND-ORDER CLOSURE DESCRIPTION OF TURBULENT REACTING FLOWS

Coleman duP. Donaldson

Aeronautical Research Associates of Princeton, Inc.

50 Washington Road, Princeton, New Jersey 08540

This workshop is concerned with the nature of turbulent mixing and the description of the rate at which chemical reactions proceed when turbulent mixing takes place. I am a member of that school of engineers that attempts to describe turbulent flows by what have come to be called second-order closure schemes. I believe that such schemes are useful, provided they are invariant closures and provided that, until one knows from experience that simplified schemes of second-order closure are adequate, one keeps equations for all the independent second-order correlations in the calculations. Of course, in keeping equations for all the independent second-order correlations, one adds a great deal of computational effort to the description of any given flow. However, until one understands the complete closure computations well enough to know that simplification will not destroy that portion of the physics of the problem that a complete second-order closure can give, this writer feels that one should make the effort necessary to cope with the equations for the complete set of independent second-order correlations.

It is true that a second-order closure scheme for computing turbulent flow can never be a complete description of the physics of the problem, and I often think the motto of the community that practices this art should be, "Lord, forgive us for we know not what we do." Be that as it may, I believe the method to be a powerful engineering tool, at this point in time, for the analysis of many important problems involving turbulent momentum, transport, and mixing. With this thought in mind, I would like to present to you today some thoughts on what I and my colleagues are doing to try to construct a second-order closure model of turbulent reacting flow. The scheme I will discuss is one that is, at present,

tentative. I put it forward, in the spirit of this workshop, as
an attempt to see what second-order closure schemes might provide
in the way of illumination of some of the complexities of turbulent
diffusion flames.

First let me try to describe the basic problem. Consider
Figure 1. The figure shows, in a rather conventional way, the
turbulent mixing between two streams, one of species α and one of
species β. The two species can react to form a third species γ
according to the simple reaction

$$\alpha + \beta \rightarrow \gamma \tag{1}$$

If one focuses attention on a particular spot in the turbulent
flow between the streams of α and β, say, on the point labelled P,
one might, in some particular interval of time, observe that the
mass fractions of the various species, namely, c_α, c_β, and c_γ
would have the values shown in Figure 1. Here we have shown a
hypothetical case where the observer would note first the passage
of a bit of species α, then a bit of species α mixed with species
β, then a bit of α, β, and γ mixed together, then a bit of mixed α
and β again, then a bit of β by itself, then a bit of β mixed with
γ, etc. In this thought experiment, the next interval of time
might bring a whole new set of bits or threads of various concentrations by the point in question. If one were to observe this
steady turbulent flow process long enough, one could, from such
observations, construct the probability functions and joint probability functions for finding c_α, say at some value; i.e.,

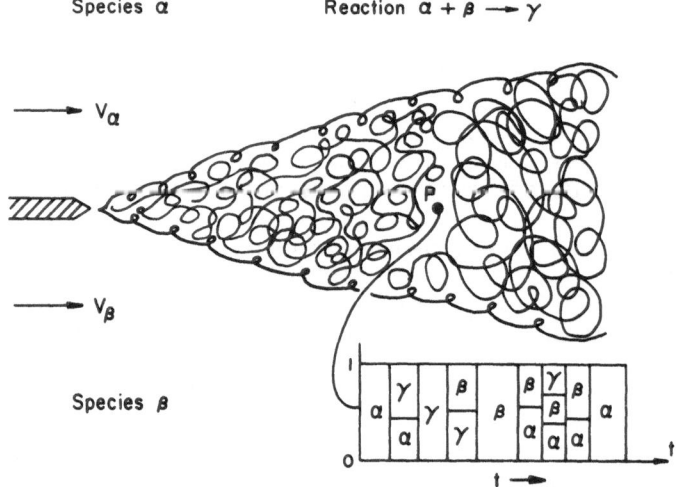

Figure 1.   Hypothetical Example of the Time Dependence of Species
            Mass Fractions c Passing a Point in a Turbulent Flame.

$p_\alpha = p(c_\alpha)$, or for finding c_α at one value and simultaneously
finding c_α at another value; i.e., $p_{\alpha\beta} = p(c_\alpha, c_\beta)$, etc. If the
streams of α and β were at different temperatures or if there was
heat released in the chemical reaction, then the various threads
passing the point in question might be alternately hot and cold,
and one could construct probability distribution functions such
as $p_{\alpha T} = p(c_\alpha, T)$, $p_{\beta T} = p(c_\beta, T)$, etc.

Probability distribution functions such as those just described
can be used to describe the rate at which chemical reactions can be
used to describe the rate at which chemical reactions proceed in
the turbulent flow in question. The question we ask here is, "Can
a complete second-order closure be used to approximate these distribution functions in some way?" Let us examine this question.

Consider for a moment that the series of threads that pass the
point P in Figure 1 is always the same and is always just the
series depicted in Figure 2. We might think of this as a "most
typical" occurrence of threads of various concentrations of α, β,
and γ and consider that this "most typical" series of threads
passes the point in question again and again and again. What,
then, would the distribution functions $p(c_\alpha)$ and $p(c_\beta)$ look like?
They would, of course, consist of five delta functions in each
case. Since there are threads or cells that contain no α, then
there would be a Dirac spike in the distribution function $p(c_\alpha)$
at $c_\alpha = 0$. Since there are cells which contain all α, there would
be an appropriate spike in $p(c_\alpha)$ at $c_\alpha = 1$. Since there are cells
which contain α in other amounts, but only cells which contain α
in three other very specific amounts, there will be three more
Dirac spikes in $p(c_\alpha)$ at the appropriate values of c_α. A similar
behavior can be deduced for $p(c_\beta)$. Indeed, all the probability
density functions can be derived from this very specialized model.
The spikey distribution functions for $p(c_\alpha)$ and $p(c_\beta)$ obtained
from the model just discussed are shown in Figure 3. Of course,
in real flows, no such thing happens. The distribution functions

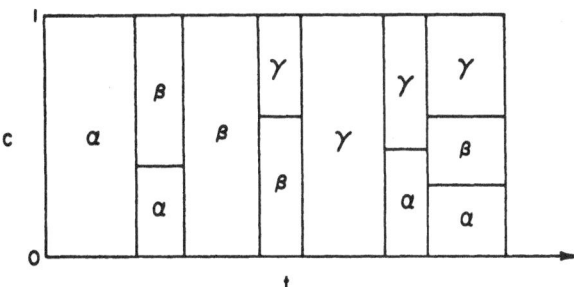

Figure 2. Hypothetical "Most Typical" Occurrence of Species Mass
           Fractions. All seven possibilities of mixtures of α,
           β, and γ are considered.

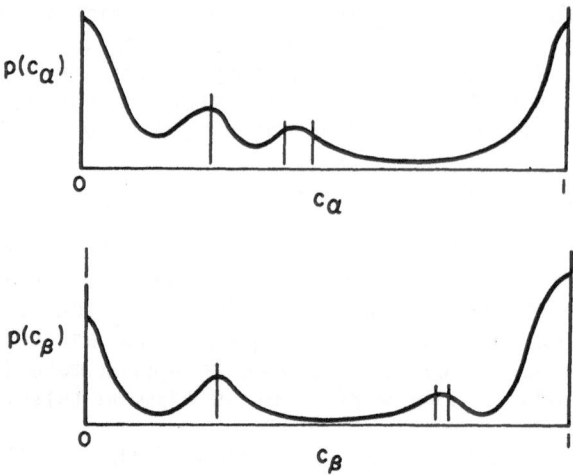

Figure 3.   Hypothetical Distribution Functions $p(c_\alpha)$ and $p(c_\beta)$
            Derived from Model of Figure 2 and the Possible Look
            of such Distribution Functions in a Real Turbulent Flame.

will look more like the smooth distribution functions plotted in
the figure.

     Before going on to discuss how one might construct a "typical
thread pattern" or "typical eddy" model of the probability distribution functions for a turbulent flow, let us look for the sake of
instruction, at how chemical reactions might proceed in a turbulent
flow when a typical eddy was followed in a hypothetical example.
Consider the "typical eddy" shown in Figure 4. It consists of
three cells when it is first observed and these three cells keep
going by the point in question again and again and again. Somewhere
downstream, this typical eddy goes by an observation point at a later
time, but it is not the same because chemical reactions have taken
place in each cell of the "typical eddy" as it passed downstream.
Since, in Figure 4, we have shown not only the concentrations in
the "typical eddy" but also have given the temperatures and, hence,
the reaction rate constants k_1 for the various cells of the eddy,
we may compute in detail the chemistry within each cell and, hence,
describe the mean characteristics of the flow consisting of these
eddies at any point downstream of the initial observation station.
These means may be computed by time-averaging the values of the
variables in each cell of the "typical eddy" to determine mean
values of c_α, c_β, $c_\alpha' c_\beta'$, etc. The mean values of c_α, c_β, and c_γ
found in this way are written $\bar{c}_\alpha$, $\bar{c}_\beta$, and $\bar{c}_\gamma$. Likewise, mean
values may be found for T and k, namely, $\bar{T}$ and $\bar{k}$. If the

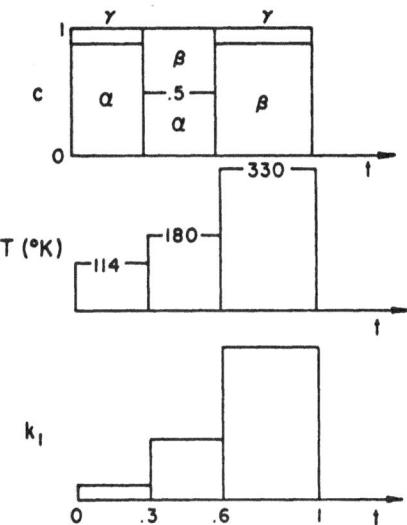

Figure 4.   An Example of a Rather Bizarre "Typical Eddy" which has
            been Chosen to Show the Effects of Nonmixedness in an
            Easily Seen Manner.

fluctuations from these mean values of any variable are denoted by
a prime, then one can compute the means of correlations of fluctuat-
ing quantities such as $\overline{c'_\alpha c'_\beta}$, $\overline{k' c'_\alpha}$, etc. It is found when this is
done that, for example, the rate of change of the mean value of c_α
may be written

$$\frac{D\overline{c_\alpha}}{Dt} = -\overline{k_1}\left(\overline{c_\alpha}\overline{c_\beta} + \overline{c'_\alpha c'_\beta}\right)$$

$$- \left(\overline{k'_1 c'_\alpha}\overline{c_\beta} + \overline{k'_1 c'_\beta}\overline{c_\alpha} + \overline{k'_1 c'_\alpha c'_\beta}\right) \tag{2}$$

This equation follows from the equation for the instantaneous
change in c_α, namely,

$$\frac{Dc_\alpha}{Dt} = -k_1 c_\alpha c_\beta \tag{3}$$

For the particular example we have chosen, what are the effects of
the various terms in Eq. (2)? The answers are given in Figure 5.

Consider first the term $\overline{k_1}\overline{c_\alpha}\overline{c_\beta}$. Since there is a considerable
amount of α and β around but it is only mixed on the molecular level
in the central cell, the expression $\overline{c_\alpha}\overline{c_\beta}$ does not in any way

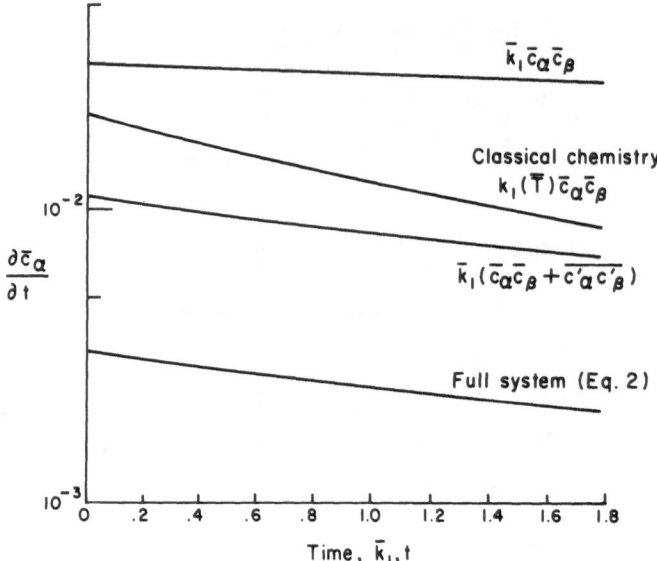

Figure 5.   Effects of Various Terms in the Rate Equation (2) on the
            Rate at which α is Disappearing in the Reaction $\alpha + \beta \to \gamma$
            for the "Typical Eddy" Shown in Figure 4.

properly measure the amounts of α and β that are capable of reacting.
Worse than that, the third or last cell, where α and β are not mixed
at all, has the highest temperature and by far the fastest reaction
rate. Thus the value of $\bar{k}$ is far too high since it is dominated by
the reaction rate available in the third cell. It is clear then
that the expression $\bar{k}_1 \bar{c}_\alpha \bar{c}_\beta$ can in no way be an adequate expression
for reaction rate in an arbitrary turbulent flow. For the case
under consideration here, the error may be evaluated by comparing
the two curves for $\partial \bar{c}_\alpha / \partial t$ versus $\bar{k}_1 t$ marked $\bar{k}_1 \bar{c}_\alpha \bar{c}_\beta$ and "full
system," the latter being the value of $\partial c_\alpha / \partial t$ obtained using all
the terms on the right-hand side of Eq. (2).

     Now consider the term $\bar{k}_1 (\bar{c}_\alpha \bar{c}_\beta + \overline{c'_\alpha c'_\beta})$. This gives the reaction
rate that would be achieved using the average value of k_1, namely,
$\bar{k}_1$, but correcting for the degree of molecular mixedness by adding
the term $\overline{c'_\alpha c'_\beta}$ to $\bar{c}_\alpha \bar{c}_\beta$. The result of making the molecular mixedness
correction may also be seen in Figure 5. The remaining error is, of
course, due to the fact that one has not used the proper reaction
rate which, in this very special case, happens to be that appropriate
to the second or center cell of our hypothetical flow. To correct
for the fact that one must use the appropriate reaction rate in the
appropriate cell, one needs to add the terms

$$\overline{k'_1 c'_\alpha}\, \bar{c}_\beta + \overline{k'_1 c'_\beta}\, \bar{c}_\alpha + \overline{k'_1 c'_\alpha c'_\beta}$$

as is shown by the right-hand side of Eq. (2).

It may be of some interest to compare, for the hypothetical case we have chosen here, the reaction rate that might be obtained by using the conventional or classical expression that has been used widely in the calculation of turbulent reactions in the past, namely,

$$\frac{\partial \overline{c}_\alpha}{\partial t} = - k_1(\overline{T})\overline{c}_\alpha\overline{c}_\beta \tag{4}$$

This result is also shown in Figure 5.

We note in the example we have chosen that no expression other than the complete equation (Eq. (2)) does an adequate job of predicting turbulent reaction rates properly. This is not to say that the expression $k(\overline{T})\overline{c}_\alpha\overline{c}_\beta$ is not a useful approximation to turbulent chemical reaction rates in many instances. Indeed, for most devices that depend on efficient chemical reaction to be considered at all, the hardware must be designed to achieve a high degree of molecular mixedness (and, since it is governed by the same general physical phenomena, thermal mixedness), and so the turbulent chemistry of "good" devices can generally be approximated by the simple expression

$$\frac{\partial \overline{c}_\alpha}{\partial t} = - k_1(\overline{T})\overline{c}_\alpha\overline{c}_\beta \tag{4}$$

Unfortunately, there are many practical problems in today's world where Eq. (4) is not good enough and where Eq. (2) should be used. We have illustrated the nature of this problem with a hypothetical case which is not meant to be typical but was chosen, as was said previously, because it illustrates the problem in a particularly clear way.

The problem we now wish to address is whether a complete second-order closure scheme for computing turbulent flow might not allow one to deal with all the terms in Eq. (2) and also those that arise in the equations for the complete set of second-order correlations. I will not present here the detailed equations for all the mean variables and second-order correlations for a compressible turbulent reacting flow in all their detail. I will, however, list the variables for which one must solve simultaneously. We will assume a low Mach number diffusion flame. We will have equations for the following mean variables:

$\overline{u}$ from the momentum equation

$\overline{v}$ from the continuity equation

$\overline{\rho}$ from the equation of state

$\overline{T}$ from the energy equation

$\overline{p}$ in free mixing, a given function of the streamwise
       coordinate

$\left.\begin{array}{c}\overline{c_\alpha} \\[6pt] \overline{c_\beta}\end{array}\right\}$ from the mass fraction equations.

Note that because this is a low Mach number flame, we may, in
computing ρ, $\overline{\rho}$, or ρ' from the general equation of state

$$p = \rho T R \sum_\alpha \frac{c_\alpha}{M_\alpha} , \qquad\qquad (5)$$

set p' equal to zero. This is because the low Mach number assumption tells us that the fluctuations in density are to be attributed
to the changes in temperature and concentration that are present and
not to the local fluctuations in pressure. Note also that $\overline{c}_\gamma$ is not
an independent mean variable because, by definition, $\overline{c}_\alpha + \overline{c}_\beta + \overline{c}_\gamma \equiv 1$.
The second-order correlations that can be constructed are

$\overline{u_i' u_k'}$ from double use of momentum equation

$\overline{u_i' T'}$ from use of momentum and energy equations

$\overline{u_i' \rho'}$ from use of momentum and continuity equations

$\left.\begin{array}{c}\overline{u_i' c_\alpha'} \\[6pt] \overline{u_i' c_\beta'}\end{array}\right\}$ from momentum and mass fraction equations.

Notice that $\overline{u' c_\gamma'}$ is not an independent second-order correlation
because, by definition, $c_\alpha' + c_\beta' + c_\gamma' \equiv 0$.

$\left.\begin{array}{c}\overline{c_\alpha' c_\beta'} \\[6pt] \overline{c_\alpha' c_\gamma'} \\[6pt] \overline{c_\beta' c_\gamma'}\end{array}\right\}$ from double use of the mass fraction equations.

Note again that $\overline{c_\alpha'^2}$, $\overline{c_\beta'^2}$, and $\overline{c_\gamma'^2}$ are not independent second-order
correlations because of the restraint $c_\alpha' + c_\beta' + c_\gamma' \equiv 0$.

$\left.\begin{array}{c}\overline{\rho' c_\alpha'} \\[6pt] \overline{\rho' c_\beta'}\end{array}\right\}$ from mass fraction and continuity equations.

Again, $\overline{\rho'c_\gamma'}$ is not an independent variable.

$\overline{\rho'^2}$ from double use of the continuity equation

$\overline{T'^2}$ from double use of the energy equation

$\left.\begin{array}{c} \overline{c_\alpha'T'} \\[1em] \overline{c_\beta'T'} \end{array}\right\}$ from use of mass fraction and energy equations.

Again, $\overline{c_\gamma'T'}$ is not an independent variable.

At this point we stop, with our minds reeling at the thought of the number of simultaneous equations to be solved. We have left out a couple of second-order correlations that are surely important, for they appear directly in Eq. (2). These are the correlations $\overline{k_i'c_\alpha'}$ and $\overline{k_i'c_\beta'}$ and their ilk. In what follows, we hope that we can show that these correlations, within what can be done with a second-order closure, are not independent of the large set given above.

When one writes the equations for all the second-order correlations given above, the really new thing that is added that was not there in previous closures is a valid scheme for modeling the correlations of fluctuations of scalar quantities up through the fourth order. A few typical examples of such correlations will suffice. They are of the form $\overline{k'c_\alpha'c_\beta'}$ (taken from Eq. (2)), $\overline{c_\alpha'^2c_\beta'}$, $\overline{c_\alpha'c_\beta'^2}$, $\overline{T'^3}$, $\overline{\rho'^3}$, and $\overline{k'c_\alpha'^2c_\beta'}$. Not only must one model these terms, but the models must be consistent with many constraints. For example, the already noted constraints,

$$\overline{c}_\alpha + \overline{c}_\beta + \overline{c}_\gamma \equiv 1 \tag{6a}$$

$$c_\alpha' + c_\beta' + c_\gamma' \equiv 0 . \tag{6b}$$

In addition, suppose no chemical reaction were possible although both α and β existed; then

$$c_\alpha(t)c_\beta(t) \equiv 0 \tag{7}$$

or

$$\overline{c}_\alpha\overline{c}_\beta + \overline{c}_\alpha c_\beta' + c_\alpha'\overline{c}_\beta + c_\alpha'c_\beta' \equiv 0 . \tag{8}$$

Averaging this expression, one finds the well known expression for the impossibility of reaction, namely,

$$\overline{c}_\alpha\overline{c}_\beta + \overline{c_\alpha'c_\beta'} \equiv 0 . \tag{9}$$

Subtracting this expression from Eq. (8) yields

$$\bar{c}_\alpha c_\beta' + c_\alpha' \bar{c}_\beta + c_\alpha' c_\beta' - \overline{c_\alpha' c_\beta'} \equiv 0 \tag{10}$$

as valid when α and β are not at the same place at the same time. One may multiply this expression by k_i' and average and obtain, under these conditions,

$$\overline{k_i' c_\beta'}\,\bar{c}_\alpha + \overline{k_i' c_\alpha'}\,\bar{c}_\beta + \overline{k_i' c_\alpha' c_\beta'} \equiv 0 \tag{11}$$

and this is a constraint that _must_ be applied to a modeling of the third-order correlation $\overline{k_i' c_\alpha' c_\beta'}$ that will be met when $\overline{c_\alpha' c_\beta'} = -\bar{c}_\alpha \bar{c}_\beta$.

By multiplying Eq. (10) by any other fluctuation or product of fluctuations and then averaging, all sorts of constraints are developed that must exist between the various correlations when $\overline{c_\alpha' c_\beta'} = -\bar{c}_\alpha \bar{c}_\beta$.

The method we have adopted to solve this modeling problem is the following. We try to determine the structure of the "most probable" or "typical" structure that passes each point in the flow that we are calculating. We have no idea what this "typical" structure looks like but we have a lot of information that is continuously being given to us as we solve for the means and second-order correlations listed above. What then is the information available to us? I will list the information in two groups. The first group assumes that the temperature of the flow is constant. We then have available to us the following:

$$\bar{c}_\alpha = \qquad\qquad \bar{c}_\beta =$$

$$\overline{c_\alpha' c_\beta'} = \qquad , \ \overline{c_\alpha' c_\gamma'} = \qquad , \ \overline{c_\beta' c_\gamma'} =$$

$$\overline{\rho' c_\alpha'} = \qquad , \ \overline{\rho' c_\beta'} = \tag{12}$$

$$\overline{\rho'^2} =$$

We are given $p = \bar{p}$ which can be turned into an expression for $\bar{\rho}$ for use in solving $\bar{v}$ when all the appropriate correlations required to compute $\bar{\rho}$ are known. The expression for $\bar{\rho}$ is

$$\bar{\rho} = \frac{\dfrac{\bar{p}}{R} - \bar{T}\sum_\alpha \dfrac{\overline{\rho' c_\alpha'}}{M_\alpha} - \overline{\rho' T'}\sum_\alpha \dfrac{\bar{c}_\alpha}{M_\alpha} - \sum_\alpha \dfrac{\overline{\rho' T' c_\alpha'}}{M_\alpha}}{\bar{T}\sum_\alpha \dfrac{\bar{c}_\alpha}{M_\alpha} + \sum_\alpha \dfrac{\overline{T' c_\alpha'}}{M_\alpha}} \tag{13}$$

Next we list the information available at a point in our solution which contains information concerning temperature. They are

$$\bar{T} = \qquad , \quad \overline{T'^2} = \qquad , \quad \overline{\rho'T'} =$$

$$\overline{c_\alpha'T'} = \qquad , \quad \overline{c_\beta'T'} = \qquad \tag{14}$$

Let us turn now to the task of constructing a "typical" structure or eddy. Suppose for the moment that the temperature is constant and is known. How many kinds of cells should there be in our eddy and for what percentage of the time of the passage of a typical eddy should the flow have what characteristics? Here we must introduce a physical model. The model is shown in Figure 6. We take time as the abscissa. As shown, it has been made nondimensional by means of the time of passage of a "typical" structure or eddy. The ordinate is mass fraction. We permit the following possibilities:

1. For the fraction of the time of the passage of a typical eddy ε_1, the mass fraction of α is one.

2. For the fraction of the time of the passage of a typical eddy ε_2, the mass fraction of β is one.

3. Likewise for the fraction of typical passage time ε_3, the mass fraction of γ is one.

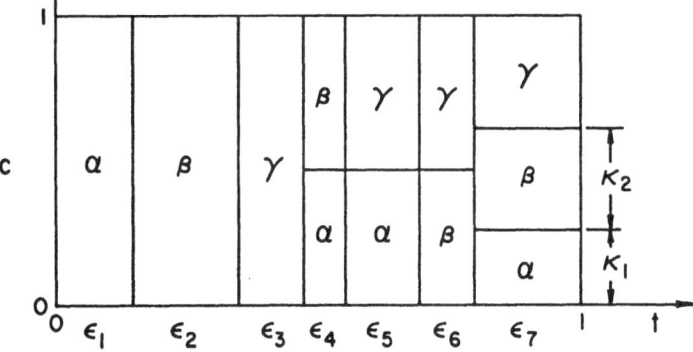

Figure 6.   Form of the "Typical Structure" or "Typical Eddy" Passing each Point in a Flow. The parameters ε_1, ε_2, ..., ε_7, and κ_1 and κ_2 must be evaluated from the values of the known quantities that have been solved for at the point in question.

4. We now consider that there are three cells that represent
    mixtures of the three substances. The mass fraction of
    each substance in each of these cells is 0.5. There is,
    thus, a cell containing α and β only for a fraction of
    the total time ε_4; there is a cell containing α and γ
    only for a fraction of the total time ε_5; and there is
    a cell containing only β and γ for a fraction of the total
    time ε_6.

5. If reaction is slow and diffusion rapid, then there needs
    to be a cell containing all three species. We set up a
    final or seventh cell. The fraction of the total time
    of the passage of a typical eddy that is taken up by such
    a cell is ε_7. We do not know how much α should be in
    this cell or how much β, so we let the mass fraction of
    α in this cell be κ_1, and the mass fraction of β be κ_2.
    The mass fraction of γ is, of course, $1 - \kappa_1 - \kappa_2$.

We have now proposed a model of a typical eddy. Do we have enough
information to determine the typical structure we have hypothesized?
The answer is yes. We have written down nine parameters (ε_1, ε_2,
ε_3, ε_4, ε_5, ε_6, ε_7, κ_1, κ_2). However, given the values of the
eight variables specified in Eqs. (12) and the fact that
$\varepsilon_1 + \varepsilon_2 + \varepsilon_3 + \varepsilon_4 + \varepsilon_5 + \varepsilon_6 + \varepsilon_7 = 1$, we have enough information to
construct the eddy or structure at each point in the flow as we
go along. If, indeed, we can do this for a model for the temperatures as well, it will be possible to construct all the higher
correlations of scalar quantities necessary to enable us to make
the calculations of the second-order quantities and means that we
have assumed to be at our disposal.

The model we assume for the temperature is shown in Figure 7.
We will assume that mass fractions in the various cells are given
as we have just described. Each cell has a basic temperature
associated with it, namely, T_1 for the α cell ε_1, T_2 for the β
cell ε_2, etc. In addition to this, since there must, in a single
species from a single source at a fixed temperature, be fluctuations
in temperature if this uniform fluid is brought in contact with a
colder or hotter surface, we add to each cell a fixed ΔT, i.e., the
same ΔT for all cells, so we assume that each cell spends one-half
its time at its basic temperature T_i and one-half its time at
temperature $T_i + \Delta T$.

To evaluate the various T_i and ΔT, we proceed in the following
fashion. We assume the α cell ε_1 to have an unknown temperature T_1
appropriate to a thread of α at that point in the fluid. We assume
the same for the β and γ cells, allowing them to have unknown
temperatures T_2 and T_3, respectively. For the cells ε_4, ε_5, and
ε_6 which are by mass half one substance and half another substance,

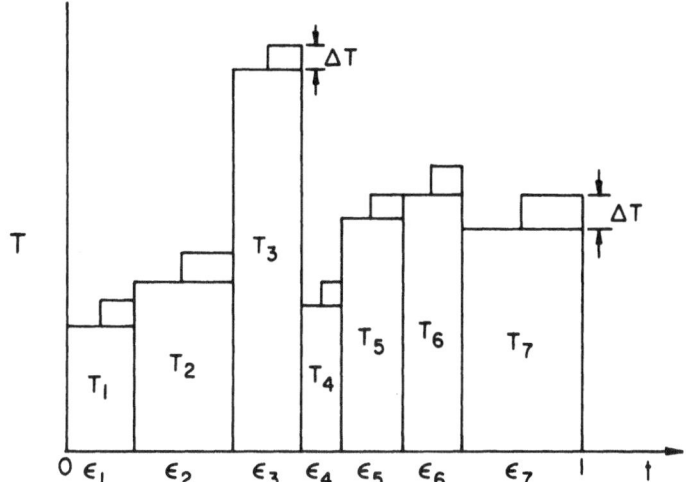

Figure 7.   Form of Temperature Distribution in a "Typical Structure"
             or "Typical Eddy." We assume we know the ε_i and we must
             now find expressions for the seven temperatures and ΔT
             used in the model.

we compute the appropriate average temperatures in each cell from
the basic temperatures T_1, T_2, and T_3 and the enthalpies of the
various substances at these temperatures. For example, the temperature in the cell containing α and β would be

$$T_4 = \frac{c_{p_\alpha} T_1 + c_{p_\beta} T_2}{c_{p_\alpha} + c_{p_\beta}} \tag{15}$$

The temperatures of cells 5 and 6 are found in the same way. The
temperature of the seventh cell we will not specify as it is going
to represent a final mixed condition and may have been affected by
all the other cells as well as its own constituents and reactions.
To construct our temperature model then, we will need to know the
five quantities $T_1 = T_\alpha$, $T_2 = T_\beta$, $T_3 = T_\gamma$, $T_7 = T_{mixed}$, and ΔT.
To solve for these five unknowns, we have the values of the five
quantities given in Eq. (14). The model is now complete.

        To make the model somewhat clearer, perhaps an example of how
one relates the ε_i 's, the κ 's to the mean values of $\bar{c}_\alpha$, $\bar{c}_\beta$, $\bar{T}$,
etc. and the second-order correlations $\overline{c'_\alpha c'_\beta}$, $\overline{c'_\alpha T'}$, etc. will be
helpful.

Consider Figure 6. Let us compute the mean value of c_α, namely, $\bar{c}_\alpha$. It is, quite clearly,

$$\bar{c}_\alpha = 1 \cdot \varepsilon_1 + 0 \cdot \varepsilon_2 + 0 \cdot \varepsilon_3 + \frac{1}{2} \cdot \varepsilon_4 + \frac{1}{2} \cdot \varepsilon_5 + 0 \cdot \varepsilon_6 + \kappa_1 \cdot \varepsilon_7 \qquad (16)$$

Similarly,

$$\bar{c}_\beta = 0 \cdot \varepsilon_1 + 1 \cdot \varepsilon_2 + 0 \cdot \varepsilon_3 + \frac{1}{2} \cdot \varepsilon_4 + 0 \cdot \varepsilon_5 + \frac{1}{2} \cdot \varepsilon_6 + \kappa_2 \cdot \varepsilon_7 \qquad (17)$$

Also from Figure 6, we have

$$
\begin{array}{llll}
c'_\alpha = 1 - \bar{c}_\alpha & \text{in } \varepsilon_1 & \quad c'_\alpha = 0.5 - \bar{c}_\alpha & \text{in } \varepsilon_5 \\[4pt]
c'_\beta = - \bar{c}_\beta & \text{in } \varepsilon_1 & \quad c'_\beta = - \bar{c}_\beta & \text{in } \varepsilon_5 \\[10pt]
c'_\alpha = - \bar{c}_\alpha & \text{in } \varepsilon_2 & \quad c'_\alpha = - \bar{c}_\alpha & \text{in } \varepsilon_6 \\[4pt]
c'_\beta = 1 - \bar{c}_\beta & \text{in } \varepsilon_2 & \quad c'_\beta = 0.5 - \bar{c}_\beta & \text{in } \varepsilon_6 \\[10pt]
c'_\alpha = - \bar{c}_\alpha & \text{in } \varepsilon_3 & \quad c'_\alpha = \kappa_1 - \bar{c}_\alpha & \text{in } \varepsilon_7 \\[4pt]
c'_\beta = - \bar{c}_\beta & \text{in } \varepsilon_3 & \quad c'_\beta = \kappa_2 - \bar{c}_\beta & \text{in } \varepsilon_7 \\[10pt]
c'_\alpha = 0.5 - \bar{c}_\alpha & \text{in } \varepsilon_4 \\[4pt]
c'_\beta = 0.5 - \bar{c}_\beta & \text{in } \varepsilon_4
\end{array}
\qquad (18)
$$

Using the information set forth in Eqs. (18), we find

$$\overline{c'_\alpha c'_\beta} = - \left(1 - \bar{c}_\alpha\right)\left(\bar{c}_\beta\right)\varepsilon_1 - \bar{c}_\alpha\left(1 - \bar{c}_\beta\right)\varepsilon_2 + \bar{c}_\alpha \bar{c}_\beta \varepsilon_3$$

$$+ \left(\frac{1}{2} - \bar{c}_\alpha\right)\left(\frac{1}{2} - \bar{c}_\beta\right)\varepsilon_4 - \left(\frac{1}{2} - \bar{c}_\alpha\right)\bar{c}_\beta \varepsilon_5$$

$$- \bar{c}_\alpha\left(\frac{1}{2} - \bar{c}_\beta\right)\varepsilon_6 + \left(\kappa_1 - \bar{c}_\alpha\right)\left(\kappa_2 - \bar{c}_\beta\right)\varepsilon_7 \qquad (19)$$

It is by the use of equations such as (16), (17), and (19) that the ε_i, the κ_i, and the T_i and, hence, the local eddy structure are found. The algebra is formidable, but the principle is straightforward.

By following the examples just presented, it is clear that $\bar{c}_\alpha$, $\bar{c}_\beta$, $\overline{c'_\alpha c'_\beta}$, $\overline{c'_\alpha c'_\gamma}$, and $\overline{c'_\beta c'_\gamma}$ can be computed. It should also be clear from the examples, since T in every cell is given, how $\overline{T}$, $\overline{T'^2}$, $\overline{c_\alpha' T'}$, and $\overline{c_\beta' T'}$ might be computed. Nine of the equations necessary for the computation of our thirteen-parameter "typical" eddy model are available. A little further explanation may be in order in regard to the expressions containing ρ, namely, $\overline{\rho'^2}$, $\overline{\rho' T'}$, $\overline{\rho' c_\alpha'}$, and $\overline{\rho' c_\beta'}$. First of all, we have assumed that the pressure in every cell is the same and is given by $\overline{p}$. Since the pressure, temperature, and mass fractions of each species are given in each cell, the density in each cell is known. One can then compute $\overline{\rho}$. If $\overline{\rho}$ is known, the ρ' in each cell is known so that $\overline{\rho'^2}$, $\overline{\rho' T'}$, $\overline{\rho' c_\alpha'}$, and $\overline{\rho' c_\beta'}$ can be calculated.

We may now observe, since all physical quantities are known in each cell, that it is possible to calculate any reaction rate constant in any cell, say, $(k_1)_i$. One can therefore calculate $\overline{k_1}$ and, hence, k_1' in each cell. With k_1' given in each cell, it is possible to calculate $\overline{k_1' c_\alpha'}$, $\overline{k_1' c_\beta'}$, $\overline{k_1' c_\alpha' c_\beta'}$, and $\overline{k_1' c_\alpha'^2 c_\beta'}$, etc.

In fact, since the fluctuation of any scalar qunatity is known in each cell, once the local "typical" structure or eddy has been determined, the correlation of scalar quantities to any order can be calculated. Of course, the higher the order of the correlation that one computes in this way, the further from reality he, in all probability, strays. Nevertheless, within the capabilities of a second-order closure scheme for computing chemically reacting turbulent flow, a model close to that proposed here is probably the most that can be expected and the best one can do. After all, for a fluid flow consisting of three species, it is a thirteen-parameter model. A bit of reflection on the part of the reader will show that the method may, in principle, be extended to any number of species.

It should be pointed out at this time, that because of the way the model has been constructed, it satisfies all the required constraints we have pointed out previously. Of particular interest, for example, might be the fact that when one does go to the trouble of calculating all the third-order correlations necessary to compute $\sum_\alpha \left(\overline{\rho' T' c_\alpha'} / M_\alpha \right)$ one finds that Eq. (13) is identically satisfied. One will, of course, also find that relations such as

$$\overline{k' c_\alpha' c_\beta'^2} + \overline{k' c_\beta'^3} + \overline{k' c_\gamma' c_\beta'^2} = 0 \qquad (20)$$

which is required by the fact that $c_\alpha' + c_\beta' + c_\gamma' \equiv 0$, are always satisfied.

How does one test whether the closure scheme proposed here which, as we have seen earlier, is just an attempt to obtain from

the information at hand from a turbulent calculation in progress,
a set of Dirac-function models of the probability distribution
functions and joint probability distribution functions (that
actually describe the problem) that are sufficiently accurate to
allow evaluation of the higher-order scalar correlations required
to continue the calculation? The way we have attempted to evaluate
our scheme at A.R.A.P. is to perform many, many numerical experiments of the kind described earlier. We assume a known turbulent
cell structure such as that shown in Figure 4 and say that this
flow goes by our observation point again and again and again.
Thus we <u>know</u> everything about the flow and can calculate what the
true chemistry for this thought experiment would be. To check our
model, we give our model only the information on the mean and
second-order correlations that exist at time t = 0. Given only
this information, we calculate the chemical behavior of the system
according to our second-order closure model. We judge the adequacy
of the closure model (for describing mixedness effects only in
these experiments) by how well the model chemistry follows the
known or true chemistry for each numerical experiment.

    To illustrate some of these results, we will make use of a
closure model somewhat simpler than that just described which has
been under investigation at A.R.A.P. for some time. Rather than
being a thirteen-parameter model, it is a nine-parameter model.
It makes use of the following nine mean and second-order quantities:

$$\bar{c}_\alpha = \qquad , \quad \bar{c}_\beta =$$

$$\overline{c'_\alpha c'_\beta} = \qquad , \quad \overline{c'_\alpha c'_\gamma} = \qquad , \quad \overline{c'_\beta c'_\gamma} = \tag{21}$$

and

$$\bar{T} = \qquad , \quad \overline{T'^2} =$$

$$\overline{c'_\alpha T'} = \qquad , \quad \overline{c'_\beta T'} = \tag{22}$$

This model eliminates the use of $\overline{\rho'^2}$, $\overline{\rho'T'}$, $\overline{\rho'c'_\alpha}$, and $\overline{\rho'c'_\beta}$. The
use of these four correlations enormously complicates the algebra
and, for our preliminary research, we were interested in how well
the general idea might work. We therefore investigated the
somewhat constrained model shown in Figure 8. Since we have had
to give up some information, we have to make more assumptions.
The model is the same as the general model shown in Figure 6 except
we assume we know the values of c_α, c_β, and c_γ in cell seven.
Indeed since this cell must be there to represent the fully diffused
solution in one limit of possible flow regimes, we set $c_\alpha = \bar{c}_\alpha$,
$c_\beta = \bar{c}_\beta$, and $c_\gamma = \bar{c}_\gamma$ in cell seven. We now have six unknowns
required to complete our cell model but only five equations
(eqs. 21). Remember there are seven ε_i but $\sum_i \varepsilon_i = 1$. We must

therefore choose the size of one cell. We elect to choose the
size of cell seven. Since we know that ε_7 must equal 1 if there
are no fluctuations and must equal 0 if $\overline{c'_\alpha c'_\beta} = -\overline{c}_\alpha \overline{c}_\beta$ or $\overline{c'_\alpha c'_\gamma} =$
$-\overline{c}_\alpha \overline{c}_\gamma$ or $\overline{c'_\beta c'_\gamma} = -\overline{c}_\beta \overline{c}_\gamma$, we might choose

$$\varepsilon_7 = \left(1 + \frac{\overline{c'_\alpha c'_\beta}}{\overline{c}_\alpha \overline{c}_\beta}\right)\left(1 + \frac{\overline{c'_\alpha c'_\gamma}}{\overline{c}_\alpha \overline{c}_\gamma}\right)\left(1 + \frac{\overline{c'_\beta c'_\gamma}}{\overline{c}_\beta \overline{c}_\gamma}\right) \tag{23}$$

This equation, however, will obviously not work well for those
cases (and there are such cases) when the second-order correlations
are positively correlated with the means. So we modify (23) to

$$\varepsilon_7 = \left(1 - \frac{|\overline{c'_\alpha c'_\beta}|}{|\overline{c}_\alpha \overline{c}_\beta|}\right)\left(1 - \frac{|\overline{c'_\alpha c'_\gamma}|}{|\overline{c}_\alpha \overline{c}_\gamma|}\right)\left(1 - \frac{|\overline{c'_\beta c'_\gamma}|}{|\overline{c}_\beta \overline{c}_\gamma|}\right) \tag{24}$$

when the numerical values of the correlation ratios are between 0
and 1; otherwise, we set $\varepsilon_7 = 0$.

There are now five unknowns and five equations, so that the
typical concentration structure of an eddy may be calculated.

To modify the temperature model of Figure 7 to adjust for our
restricted information, we no longer solve for the temperature of
the seventh cell. By analogy with what we have done for the mass
fractions in this cell, we put $T_7 = \overline{T}$. We then have four unknowns:
T_1, T_2, T_3, and ΔT, and we have four equations to determine these
unknowns, namely, Eqs. (22).

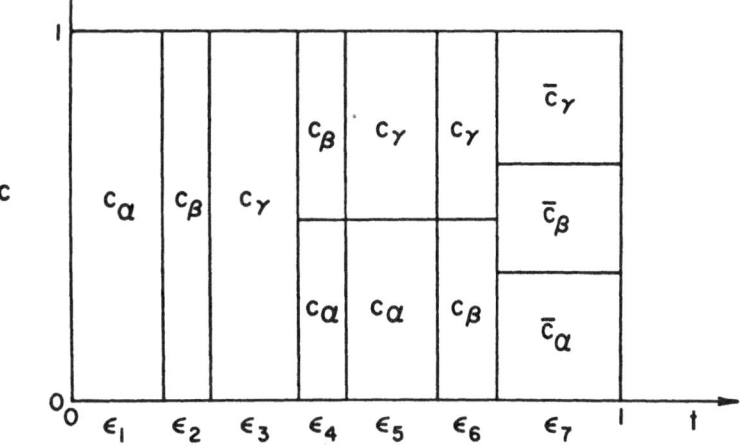

Figure 8.   Simplified "Typical Eddy" Model Used to Date in A.R.A.P.'s
            Studies of Second-Order Closure Methods for Turbulent
            Reacting Flows Containing Three Species.

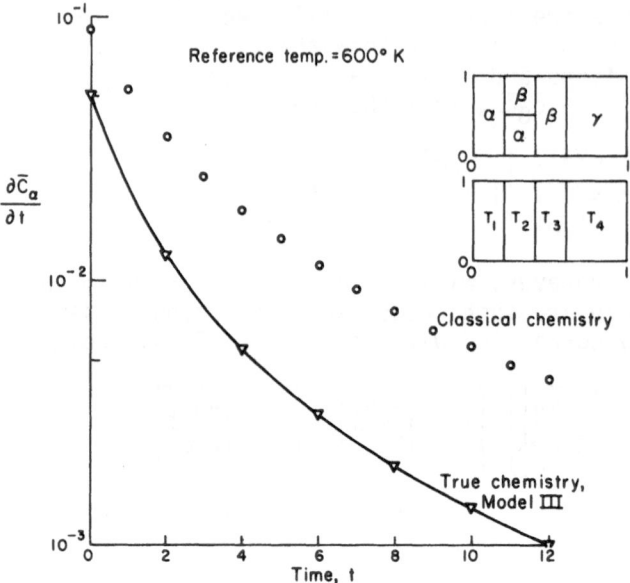

Figure 9.   First Trial Case to Check Correlation Modeling. This is
a four-celled, constant temperature model. Each cell
that goes by at time t = 0 is shown in the upper right-
hand corner of the figure.

Let us consider an example and see how well this simplified
model works. Consider the test case shown in Figure 9. At time
t = 0, the known eddy shown in the upper right-hand corner of the
figure is taken to be the generator of a turbulent reacting flow.
Since I can compute the state of the eddy at any future time, the
true chemistry of this thought experiment is given by the solid
curve for $\partial \bar{c}_\alpha / \partial t$ in the figure. If I had used the classical
expression

$$\frac{\partial \bar{c}_\alpha}{\partial t} = - k_1(\bar{T}) \bar{c}_\alpha \bar{c}_\beta \tag{1}$$

to compute the chemistry in this situation, I would have computed
the results given by the circular data symbols of Figure 9. It is
obvious that a significant error is made because I have not
accounted properly for molecular mixedness. There is no error due
to temperature nonuniformity since T is constant. Given only the
initial values of the mean mass fractions $\bar{c}_\alpha$ and $\bar{c}_\beta$ and the initial
values of $\overline{c'_\alpha c'_\beta}$, $\overline{c'_\alpha c'_\gamma}$, and $\overline{c'_\beta c'_\gamma}$ at time t = 0, the correlation model
just described evaluates a seven-celled model having the same
initial five properties (some cells are quite small). If one
follows the chemistry of this system via a second-order closure
calculation, one obtains the triangular data points shown in

Figure 9. The agreement is seen to be surprisingly good between the modeled chemistry and the true chemistry.

Let us try a slightly more difficult problem. Let us consider that the turbulent pattern generating the basic or true chemistry is that shown in the upper right-hand corner of Figure 10. This flow is identical in molecular mixedness to that shown in Figure 9; however, we now introduce a thermal unmixedness into the problem. Since I can compute the future behavior of this eddy, I can compute the true turbulent reaction rate of this second thought experiment. The result is shown by the solid curve for $\partial \bar{c}_\alpha / \partial t$ in Figure 10. Notice that if I had used classical chemistry to describe the situation, i.e., Eq. (4), I would have been very much in error, as can be seen by comparing the circular data points in Figure 10 with the solid curve for $\partial \bar{c}_\alpha / \partial t$. The error now consists of two parts. First, the degree of molecular mixedness has not been properly taken into account (as we have seen in Figure 9) but further, the average temperature is controlled by the final cell in which there are no reactants. Given only the values of $\bar{c}_\alpha$, $\bar{c}_\beta$, $\overline{c_\alpha' c_\beta'}$, $\overline{c_\alpha' c_\gamma'}$, $\overline{c_\beta' c_\gamma'}$, $\bar{T}$, $\overline{T'^2}$, $\overline{c_\alpha' T'}$, and $\overline{c_\beta' T'}$ at time t = 0 and no other information, the simple modeling we are discussing here was able to predict a seven-celled model having the same nine original characteristics and follow the true chemistry with phenomenal accuracy through a second-order closure calculation in which all the higher-order moments required for closure were obtained from the seven-celled model and the initial conditions on the problem.

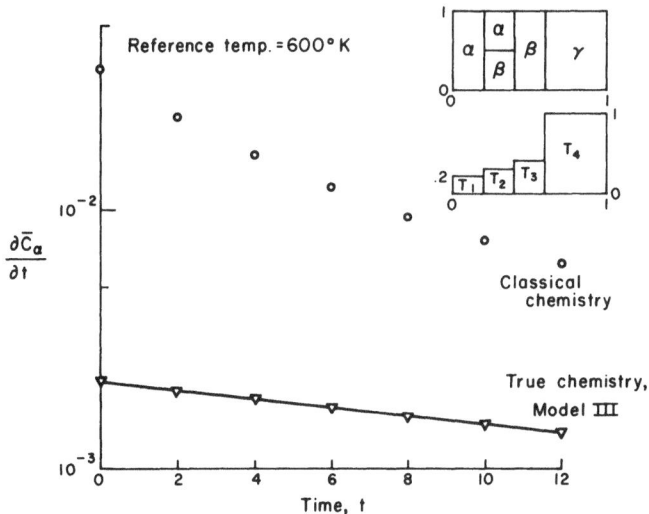

Figure 10. Second Trial Case to Check Correlation Modeling. The temperatures of the four cells are assumed to stay constant even though α reacts with β to form γ at T_2 and remains at T_2 in cell number two.

We have performed many such thought experiments. In general, if the generating flow is something reasonably like what might be expected in a turbulent diffusion flame, the closure model we are proposing here to handle the problem of molecular and thermal mixedness does very well.

There are some rather "pathological" flows for which it cannot do well. These are generally some rather odd two- or three-celled generating flows. Indeed, considering the fact that we have simplified the model rather arbitrarily from a general thirteen to a restrictive nine-parameter model and that this restrictive model is required to have seven cells, one of which is already defined, this is not surprising. I will show an example of such a failure.

Consider the test case shown in Figure 11. The generating model is a three-celled eddy of rather bizarre form. The true chemistry for this case is given by the solid curve in the figure. Classical chemistry, as we would expect, cannot cope with the unmixedness that is apparent in the generating model. This may again be seen by comparing the circular data points for $\partial \bar{c}_\alpha / \partial t$ with the solid curve. The "modeled" chemistry given by the tri-angular data points does not do too badly. It is very difficult for a restrictive seven-celled model to do such a "pathological" flow. "Bizarre," "pathological," whatever description I give of the generating flow, it is nevertheless a test of what the closure model can do - what it can tolerate and what it cannot. In this particular case, the restrictive nine-parameter model we used for the calculations actually came up with a negative cell, i.e., one of the ε 's was slightly negative. It is truly surprising then that the model did as well as it did in computing the flow.

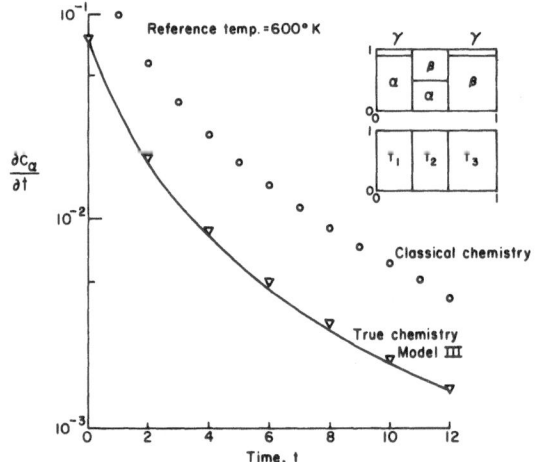

Figure 11.   Third Trial Case to Check Correlation Modeling. The generating flow has three cells and the temperature is constant.

Matters are even worse if we add to this "tough" case a thermal unmixedness, as may be seen by reference to Figure 12.

At the present time, we are working to see how much of the error the modeling makes in handling flows such as that given in Figure 12 can be eliminated by use of the complete thirteen-parameter model. I suspect a lot of it can be. Nevertheless, there will always be flows which can be invented that will "confuse" the model that is being used. This must be so because a model based on all the available independent mean scalars and second-order correlations of scalars just cannot contain enough information to completely describe a turbulent flow. Fortunately, for flows which are of engineering importance, it does appear that closure models can be made from a complete second-order description of a turbulent reacting flow that will be able to cope, for engineering purposes, with the problems of molecular and thermal unmixedness.

No discussion of the modeling of a turbulent diffusion flame can be complete without some mention of the part played by fluctuating heat release rates on the generation of turbulence in such flows. If the shears in the turbulent flame are high and the heat release rates low, then the major contributor to the production of turbulent energy is a term of the form

$$\frac{D(\overline{u'^2} + \overline{v'^2} + \overline{w'^2})}{Dt} \sim 2\overline{u'_j u'_i} \frac{\partial \overline{u}_i}{\partial x_j} \tag{25}$$

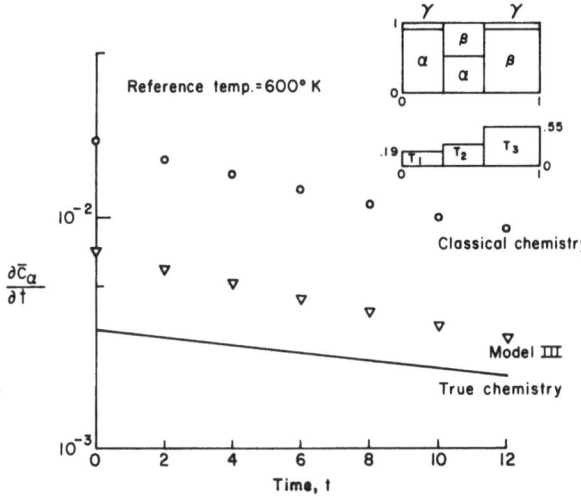

Figure 12. Fourth Trial Case to Check Correlation Modeling. The generating flow at time t = 0 is shown in the upper right-hand corner. It might be recognized that this is the same flow discussed earlier in connection with Figure 4.

If, on the other hand, the shears are very low and the heat release
rates are very large, a large contributor to the production of
turbulent energy is

$$\frac{D(\overline{u'^2} + \overline{v'^2} + \overline{w'^2})}{Dt} \sim \overline{\frac{2p'}{\overline{\rho}} \frac{\partial u_i'}{\partial x_i}} \tag{26}$$

Of course, when the flow is incompressible, $\partial u_i'/\partial x_i$ is zero so
that there is no such production. Even for high Mach number
shear layers, $\partial u_i'/\partial x_i$ is generally not an important term in the
equations of motion and certainly

$$\overline{\frac{p'}{\overline{\rho}} \frac{\partial u_i'}{\partial x_i}} \ll \overline{u_j' u_i'} \frac{\partial \overline{u_i}}{\partial x_j} \tag{27}$$

In order to handle the problem of turbulence generation by chemical
heat release, a method must be found for estimating $\overline{p'(\partial u_i'/\partial x_i)}$. It
is not too difficult to obtain an expression for $\partial u_i'/\partial x_i$ from the
basic equations for ρ', T', and the c_α'. If, in this equation, low
Mach numbers are assumed, the dominant cause of $\partial u_i'/\partial x_i$ is the
fluctuating release of heat in the turbulent flame. If, for such
a local flow, one models the relationship between p' and $\partial u_i'/\partial x_j$
(no averages taken), then one is able to solve for $\partial u_i'/\partial x_i$ from
the equations mentioned above, and an expression for the turbulent
energy production by fluctuating chemical heat release can be
obtained. But this is getting off my main topic and is a problem
at least as thorny as the mixedness problem when it comes to modeling, and it deserves at least one lecture of the length of this one
all by itself.

I would like to make one more remark concerning the modeling
I have suggested for handling the molecular and thermal unmixedness
problems of turbulent flow. What we are really saying is that one
can, somehow, out of all the information available if one uses a
complete second-order closure, construct some sort of "typical"
local structure or eddy. The probability distribution functions
and joint probability distribution functions (if one wishes to
calculate them from this typical eddy) are spikey Dirac function
affairs that probably give some of the general features of the real
distribution functions. They are, probably, all that one can hope
to get out of a second-order closure calculation. They are also
related very closely to the way in which second-order closure
schemes go about modeling such things as the tendency-towards-
isotropy and dissipation. For, in a second-order closure modeling
of, say, dissipation, even though we know there is dissipation
over a complete spectrum of eddy sizes, we say that all the dissipation in the model takes place at the "typical" dissipative eddy

size λ. In the same way, although we know that the tendency-towards-isotropy takes place at all eddy sizes, we say that there is a "typical" eddy for the tendency-towards-isotropy and it has a scale Δ_1. These are Dirac function spectra for these elements of the dynamics of turbulence which are of importance in conventional second-order closure schemes. What I have set before you today are my as yet "unhatched" thoughts on the problem of how to handle the problems of molecular and thermal unmixedness in a manner which I think is consistent with the whole philosophy behind second-order closure schemes for computing turbulence.

I would like to take this opportunity to acknowledge the enormous contribution of my colleague, Dr. Ashok K. Varma, towards making this paper possible and to thank the National Aeronautics and Space Administration for its support of our endeavors under Contract NAS1-12412.

DISCUSSION

LEE: (ARL, Wright-Patterson Air Force Base, Ohio)

Is this such that there are enough equations to get the stochastic region? I am wondering if you have thought of comparing your model with the stochastic model.

DONALDSON:

I am not just sure how to answer that question. In the past we have compared with stochastic models. But we have actually learned more about the problem and what the limitations of modeling are from considering specific, rather limiting, cases. For example, if the reactive rates are very, very large, then the half α/half β cell is exceedingly small, the $\overline{c_\alpha}$, $\overline{c_\beta}$, $\overline{c}$ cell (cell 7) is very, very small, and I have an α cell, a β cell, a γ cell, and a half α/half β cell in which all the reaction is going on at a very fast rate.

In general, rather than look at complicated stochastic models, we have looked at simple deterministic models which would tend to be difficult for the model, and then to try to fix the difficulty which was encountered. As I mentioned in my talk, we can sometimes get negative cells with our present simple model. Can we always avoid this problem? If we can't, for what flows will this undesirable characteristic occur? I believe that for most real physical flows a model can be made according to the principles I have outlined that will have no difficulties.

By the way, I might mention something about reactions that require three species to be present simultaneously. It is my belief that to do such problems properly in turbulent media, one is going to have to go to a third-order closure scheme. With the model I am putting forward here, it would only be possible to make such a calculation in the last, or completely mixed, cell of the "typical" eddy model.

One can, as I mentioned in my talk, always construct the complete "typical" eddy model for any number of species. In this case, there is one pure cell of each species, there are all the possible half-and-half cells, and there is always one cell in which all the species are mixed in some proportion. There are always enough equations to determine such a "typical" eddy, no matter how many species are involved. It is merely a matter of work.

MORGANTHALER: (Bell Aerospace Company, Buffalo, New York)

I think your approach is interesting, but I am not sure I understand it completely. Is it not rather arbitrary what you put in each one of your boxes? You can put all alpha or you can put 90-10 alpha-beta, etc. or whatever you like in a box. Or is your model, as shown in figures, separate boxes containing either all alpha, all beta, all gamma, and finally equal quantities of each?

DONALDSON:

Well, we made the modeling the way we did because, obviously, there are cases where you want the limiting solution to be able to give an α cell and a β cell and no mixing whatsoever. This would be true if you were to put the viscosity, the thermal conductivity, and the species diffusion coefficients identically equal to zero in the equations.

MORGANTHALER:

So you consider such limiting cases to be a boundary condition?

DONALDSON:

I think of these things as limiting cases that the model must be able to handle. For example, if the reaction rates are extremely slow and the diffusion rates extremely high, then the completely mixed cell (cell 7) must be a possibility. Also, the model must collapse in a completely satisfactory way from four, to three, to two, or to one species. As far as the half and half cells in the model are concerned, they could have been chosen at some other numbers. However, if symmetry is considered, I guess half and half is about the best guess.

MORGANTHALER:

And then when you are all done, you get the ten parameters to fit some real data; is that the final closure?

DONALDSON:

I think the final closure is as I have described it in the talk. You have the equations for all the mean values of parameters at a point. You have equations for the second-order correlations that are independent at the point in question. To proceed further, you need models for some higher-order correlations in terms of these quantities. One could try, I suppose, to guess these from some physical insight afforded by experimental data together with a knowledge of the limiting constraints on these correlations that I have mentioned in the talk, but it seems to me to be a rather mind-boggling job. As I struggled with this problem a few years ago, I got the idea that the second-order closure equations were really trying to tell me what a "typical" eddy looked like. There are just enough equations to make this model (or close the system) when there are no large pressure fluctuations. The large pressure fluctuation problem is another matter; it will be a lot more work, but I think it can be done.

O'BRIEN: (State University of New York, Stony Brook, New York)

You seem to have brought the temperature in in such a way that if you know the concentration, you know what the temperature increase would be. That is, if you know how much of α is gone, and how much of β has disappeared, you know how much the temperature has increased. Is that a fair statement?

DONALDSON:

If you take the mean equations and you do the reaction rates properly, you will get out of it how the mean temperature rises. What you have to do, and this is the trick, is identify where it is in a typical eddy that the rise in the mean temperature is accounted for.

O'BRIEN:

Yes, but in the cell model turbulence acts quite differently on the temperature field than on the α or β distributions.

DONALDSON:

Sure it does.

O'BRIEN:

So that the existence or the loss of alpha and beta really, at that point, is not connected with the temperature at that point. At least for fast reaction rates, you have a lot of species, mostly the temperature is not the same.

DONALDSON:

Quite so. But what I am saying is that if I knew all of the variables tracked in a complete second-order closure, I could follow these differences. For example, the $\overline{c'^2}$'s do not go like the $\overline{T'^2}$'s (somewhat like but not exactly) and the $\overline{c_\alpha' T'}$'s and the $\overline{c_\beta' T'}$'s behave very differently than the $\overline{c_\gamma' T'}$'s. Thus, if I follow the values of all these correlations and means, that should tell me what a typical eddy looks like and then allow me to write down an expression for all of the higher-order correlations so as to get a closure. The model I have proposed is just a way of doing this in a very general way consistent with all the constraints that I know of that govern these higher-order correlations.

O'BRIEN:

Have you done any concentration calculations with very rapid reactions?

DONALDSON:

We have, and I should have brought some; they have been given at other meetings. We have computed reactions in an isothermal medium that show how these things go. We are just really getting our program set to do compressible multispecies turbulent mixing with chemistry and heat transfer. We have not done it yet!

O'BRIEN:

It is not the compressibility; it is simply, in fact, a transcendental nonlinearity in the coefficient in the Arrhenius expression, for example, makes things very sensitive to the specific temperature. If your model is such that you somehow assume a kind of local, statistical dependence of the temperature on the concentration at that instant, the small differences in the temperature you get by that approximation are really going to make the model very inaccurate in the wake.

DONALDSON:

I could not agree with you more, and that is why we go around looking for pathological test cases which will cause the model to

fail. In this way we can know the model's limitations. Within a
second-order closure model of itself alone, I see nothing better to
do at present. We must find out from both numerical and real
experimental results what the shortcomings of the model are.

Let us go back to my original statement. The Lord never heard
of any of this, and it really is a matter of how well the method
works on practical problems. In some ways, it is a hideous bending
of the truth. One knows that the probability distribution function
for c_α is not a bunch of Dirac spikes. However, for computing the
higher-order correlations of scalars needed for certain engineering
problems at the present, I don't know anything better to do.

O'BRIEN:

The suggestion that Dopazo and I have here is simply looking
at the solution of just the chemistry, the temperature, that you
can do. It is easy enough to get the exact solution for both the
concentration and the temperature statistically distributed and see
how that compares with the model.

DONALDSON:

But, that is what we are doing.

O'BRIEN:

I mean from the exact solution, not the computer solution
which is modeled here. However, what I understand is in your mind
is that it has been proved many times.

DONALDSON:

Yes, but I can obtain my "true" solution for not just one
given distribution going through again and again but for a completely stochastic process. It turns out that the more cells one
takes, the better the thing works. If you really want to louse the
model up, take only two cells, because then you have a seven-cell
model trying to represent a two-celled flow which is very difficult
for the model to do.

SUMMERFIELD: (Princeton University, Princeton, New Jersey)

You have demonstrated rather impressively the disagreement
that we get between going classical chemistry, as you call it,
which is the product of the mean values, and then the true chemistry
where you take microscopic accounts of each little cell as it goes
by and the reaction thus generated. I suspect you get that big
difference, a factor of ten, and in some cases 10^2, because you use

a rather large activation energy and a big spread of temperature.
Okay, that is fine for demonstration. I think it demonstrates
something and I am reaching a conclusion from this. That is, that
is the way fuels burn. We start with cold reactants and we enter
a reaction zone and then they somehow get reacted very rapidly.
Yet discouragingly, your calculation says, if I follow the true
chemistry, it would be much slower than we would get if we followed
the classical chemistry. Yet, somehow, the classical chemistry
gives an answer in real flames which is of the order of, I think
it is something like, correct. To put it another way, I suspect
that if you burn the fuels and oxidants the way you describe, each
one homogeneously burning, then we would not have a turbulent
flame. I wonder, therefore, whether something is left out. Let
me throw in my suggestion of what might be left out. A cell of
alpha and beta mixed at low temperature in your true chemistry
would slowly react, almost nonreact because it waits until it somehow got into a combination with a high temperature in the statistics
that plays itself out. But there is a micro propagation phenomena
from the adjacent hot product cell into this premixed cell. That is
of course not taken into account, not in this kind of computation.
Have you compared your model with the real world?

DONALDSON:

     Yes, let me answer that. We have taken a very stringent test
case, because those always tend to reveal the real nature of the
problem and model. If the model will not work when it is being
stressed, it really is not a very good model. What really happens
in real flows is that, due to heat transport, the big fluctuations,
the $\overline{T'^2}$, tend to go away, as also do the correlations between c_α'
and c_β', c_α' and T', and c_γ' and T'. These reductions are due to the
conduction terms in the equations for these correlations. When you
consider a real flame, what people have done (you can see it if you
just examine the flames) is that they have always cut the metal
(designed the device) so that the mixedness problem does not exist.
If the problem exists, they do not have a good burner. If so, they
go back and decrease the eddy scales which makes the conduction from
one cell into another better, they slow the overall velocity down
in the region of the flame and, when this is done properly, you find
that mixedness is not a significant problem. When you do chemistry
in the atmosphere, however, and in certain cases where you are not
able to go in with your tin-snippers and build the flame holder the
way you want and you are just stuck with the scales and speeds that
are there, then the mixedness problem can be important. How
important? In most of our computations, it is anywhere from a
factor of three to ten, never more to date - but our experience
is rather limited.

BROADWELL: (TRW Systems, Redondo Beach, California)

It seems to me the crucial problem would be demonstrated by the following example. Forget the temperature fluctuations and have alpha and beta on the two sides of a splitter plate and entering a mixing zone with an infinite rate constant. Now, have you tried to work that problem? The reason I have cited this is because it seems to me that crucial problem is: what is the mixing on the molecular level?

DONALDSON:

That is what this address is about.

BROADWELL:

Right. So, why not forget about the temperature and simply use the rate constant?

DONALDSON:

Because, they are exactly the same thing.

BROADWELL:

Okay, just so we understand that. Suppose we must mix alpha and beta at constant temperature with an infinite rate. Can you work that problem for a mixing layer? And what do you find?

DONALDSON:

Yes, we can work that problem. You find that as far as the model is concerned, when the reaction rate is infinite, the cell in which alpha and beta exist tends to go to zero.

BROADWELL:

No, I mean by "working the problem" the following thing: at a given distance downstream can you say how much gamma you have got?

DONALDSON:

Yes, it is done by computing how much alpha and beta can stay there and it is always diffusion limited. And what happens is, just take the mean one, just this --

BROADWELL:

Well, what I am asking for is, in a given Reynolds number flame, at a certain distance downstream, at a given velocity, how much gamma have you got?

DONALDSON:

Well, I do not know just how to answer that; it depends on what Reynolds number you choose.

BROADWELL:

My question simply is, can you do that problem?

DONALDSON:

Yes, and that particular problem we have written on. We did not do the modeling of the third-order correlations in the way presented here. We estimated what they were and the estimates, although adequate for the problem discussed, could not be extended in any consistent way to handle temperature mixedness. The result, in that particular case, was that the reaction rate was strongly diffusion-limited and the limitation depends on the scale and on the Reynolds number based on the scale of the turbulence.

BROADWELL:

So the answer is, it is Reynolds number dependent.

LIEPMANN: (California Institute of Technology, Pasadena, California)

I am impressed by the long way we have come from the days when we said only "laminar" or "turbulent." It looks to me that what you are trying to do is something that is similar to the discrete velocity model of rarefied gas flows. I do not quite understand what you get out of it. I agree with Broadwell; I am not sure what you can compute, how you fix the length of the "eddy," etc.

DONALDSON:

It is interesting that the length of the eddy does not enter the calculation of the third-order correlation of the scalars.

LIEPMANN:

That is not what I meant. What I meant is that if you stop and think, there are two features: First there is the diffusion at the interface and second there is the geometry of the essentially

nonviscous interface. Somehow you have to come to grips with the
nonviscous model of the interface convolution. That is the crucial
point.

DONALDSON:

   Yes, that is right.

GRANT: (Shell Research Ltd., Chester, England)

   I wanted to bring up a pathological case. In a real flame,
one can have fuel and oxygen in the same cell, say, but with the
problem of flammability limits. It does not mean that a reaction
will take place with fuel and oxygen in the same cell. Does your
model take care of this sort of case?

DONALDSON:

   That would be very difficult to put in. I have not approached
or even tried to approach that problem yet. I do not know whether
I could or could not put it into the model at present.

BRODKEY: (Ohio State University, Columbus, Ohio)

   One case not considered is that of extreme non-stoichiometric
reactions. These seem to be insensitive until you get to the order
of 10 to 1 stoichiometry. At one to one stoichiometry, the problem
is relatively easy; at three to one, there does not seem to be much
effect on predictions. But at ten to one, when you have concentrations of one species ten times the other, the story is different;
have you had any experience testing your model under these conditions - even if it is isothermal?

DONALDSON:

   Just thinking of your question and the one previous, you have
to be able, in order to use this model, to write down, if it is a
second-order reaction, what K depends on. It has to be some
function of T and of the concentration of species involved. If
you can write down that functional form, the model can, in principle,
handle it.

BRONFIN: (United Aircraft Corp., East Hartford, Connecticut)

   As an additional comment to the previous two, what you are
really saying is that the chemistry is extremely limited to one
forward running second order or third order reaction. You are not
handling forward and backward reaction.

DONALDSON:

I can do the backward reactions. I just have not chosen to discuss that problem in this presentation.

BRONFIN:

Well, can you handle kinetics which include trace species like oxygen atoms? In fuels, you know, 10^{-8} or more fraction of oxygen atoms will trigger entire reaction, whereas if that concentration goes to 10^{-12} there will be no reaction ...

DONALDSON:

Only if you can write down the functional form.

BRONFIN:

But how many, in other words, has the model tried k_1, k_2, k_3? We are talking now about multi-species, multi-equations. What has the model tried so far, one forward and backward running reaction?

DONALDSON:

In principle, if you can write them down. It may be that when you do that case, it is very simplified Dirac function model or the thing would break down. Those would be the pathological cases. What we are trying to do in our research at present is to bring it out and see where the kind of modeling I have proposed does break down.

COMPUTATIONAL STUDIES OF TURBULENT FLOWS WITH CHEMICAL REACTION

Roland Borghi

Office National d'Etudes et de Recherches Aérospatiales

92329 CHATILLON (France)

ABSTRACT

Research about turbulent mixing of non-reactive flows has progressed recently, particularly in the field of the computational modeling of turbulence. The first practical purpose of these studies is to acquire a more precise knowledge of the fluxes of turbulent transport of momentum, mass and enthalpy, which are determining factors for the time-mean velocity, mass fraction, temperature.

When the turbulent mixing of chemically reactive flows are considered, the time-mean characteristics are determined also by the reaction rates, and, as the molecular fluxes are not the only data to be taken into account, the molecular reaction rates must be modified by the fluctuations themselves, especially the temperature and the concentration fluctuations.

The analytical basis for an approach of this problem, which is clearly in the line of the methods already proposed for non-reactive flows, is presented. The effects that the fluctuations and their correlations can have on the mean reaction rate are discussed and, alternatively, the influence the chemical reactions can have on the fluctuations. Some numerical calculations are presented and discussed, and particular difficulties and limitations of the method, due to the occurrence of chemical reactions, will be shown.

1. INTRODUCTION

The first figure shows some flows of practical interest in which are occurring simultaneously the phenomena of turbulent mixing and combustion or chemical reactions.

The first case pictured is that of a jet of hot gases in combustion, for instance a jet of ramjet or turbojet with afterburner, that mixes with the ambient air, at rest or coflowing. In such a flow, the combustion which takes place in the homogeneous core is quenched progressively in the mixing zone because of the strong temperature decrease.

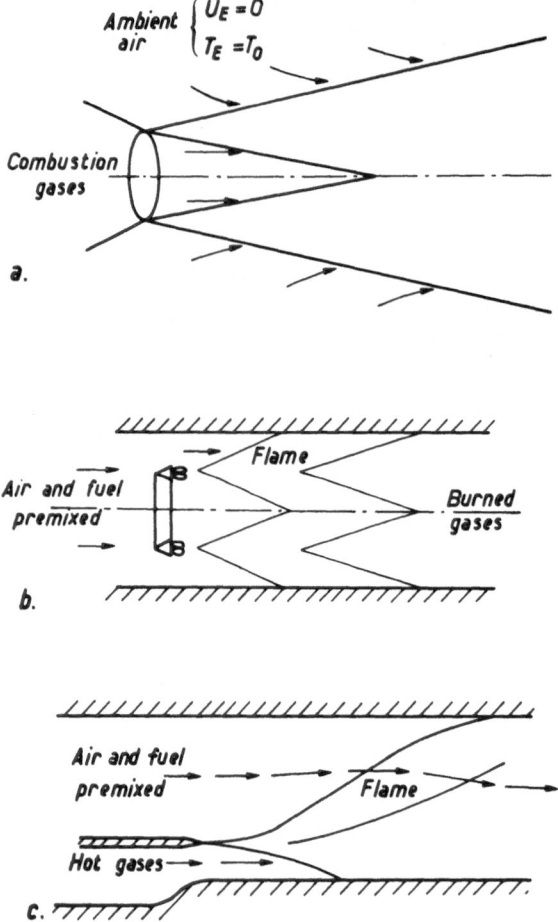

Figure 1a. Chemical Evolutions in Turbojet Plumes.
Figure 1b. Propagation of a Flame Stabilized by Solid Bodies.
Figure 1c. Propagation of a Flame Stabilized by a Pilot Burner.

The two following cases show the development of a flame in
premixed gases of fuel and oxidizer; the flame can be stabilized
either by a solid body or by a pilot burner; in these two cases
the chemical reactions occur simultaneously with the diffusion of
the burned hot gases into the fresh mixture, sustaining the combustion.

The prediction and the computation of the consumption of fuel,
and the other properties of this type of flows is rather difficult,
because the interaction between the turbulence and reaction phenomena is expected to be strong, specially in the more practical
cases. Comparisons between experimental and computational studies
have confirmed this point, many years ago (10, 20, 21), and many
investigators attempted theoretical approaches of the different
aspects of this problem (6, 7, 13, 14, 16, 20, 23). In this paper, a
computational study of the coupling phenomena between turbulence
and chemical reactions is performed, in a simplified but characteristic case; the scope of this study is the following:

- two dimensional (or axisymmetrical) mixing layer flow;
- non-isothermal bimolecular reaction;
- the chemical reaction time and a characteristic turbulence
  time are of the same order.

The analytical basis for the approach is clearly in the line
of the methods already proposed for non-reactive jet flows, namely
the use of second order moments and their balance equations properly closed in order to represent the turbulent fluctuations.

Results of numerical calculations will show some peculiar
properties of the fluctuation correlations and intensities, due to
the occurrence of a chemical reaction, directly or non-directly.
Some difficulties relative to this approach and to the closure
assumptions will appear, due also at first to the combustion, and
will be discussed; because of these difficulties, the practical
problem of the development of a turbulent flame, as shown in the
figure, is always open to theoretical predictions, at this time.

2.   ANALYTICAL STATEMENT OF THE PROBLEM
OF REACTIVE MIXING FLOWS

2.1. Let us consider a mixture in which can occur only a one-
way bimolecular reaction:

$$K + \nu\, 0 \rightarrow P$$

(K = Fuel; 0 = Oxidizer; P = Products; ν = stochiometric coefficient)

In this the volumetric reaction rate is expressed, instantaneously, by a law of the type:

$$\rho\dot{w} = k\ C_K\ C_0\ T^\alpha \exp\left(-\frac{T_A}{T}\right) = k\ \frac{p^2}{R^2 M_0 M_K}\ Y_K\ Y_0\ T^{\alpha-2}\ \exp\left(-\frac{T_A}{T}\right)$$

(C is the concentration in mole number per volume unit, Y the mass
fraction, mass per total mass, p the pressure, T the static temperature, M the molecular mass, $R = R_0/M_m$ the mixture gas constant).

The basic equations for a flow of the three species O, K, P,
are the following balance equations (written with tensorial orthogonal notations):

$$\frac{\partial\rho}{\partial t} + \frac{\partial}{\partial x_\alpha}\ (\rho u_\alpha) = 0$$

$$\frac{\partial\rho u_\beta}{\partial t} + \frac{\partial}{\partial x_\alpha}\ (\rho u_\alpha u_\beta) = -\frac{\partial p}{\partial x_\beta} + \frac{\partial}{\partial x_\alpha}\ (\tau_{\alpha\beta})$$

$$\frac{\partial\rho Y_K}{\partial t} + \frac{\partial}{\partial x_\alpha}\ (\rho u_\alpha Y_K) = \frac{\partial}{\partial x_\alpha}\ \left(J_\alpha^K\right) + r_{SK}\ \rho\dot{w}$$

$$\frac{\partial\rho\phi}{\partial t} + \frac{\partial}{\partial x_\alpha}\ (\rho u_\alpha\phi) = \frac{\partial}{\partial x_\alpha}\ \left(J_\alpha^\phi\right) + 0$$

$$\frac{\partial\rho Z}{\partial t} + \frac{\partial}{\partial x_\alpha}\ (\rho u_\alpha Z) = \frac{\partial}{\partial x_\alpha}\ \left(J_\alpha^Z\right) + \frac{u_\alpha}{Cp}\frac{\partial p}{\partial x_\alpha} + \frac{1}{Cp}\frac{\partial p}{\partial t} + \frac{\tau_{\alpha\beta}}{Cp}\frac{\partial u_\beta}{\partial x_\alpha}$$

In these equations we suppose that the reaction term can be of
the same order as the others; the reaction is not necessarily
infinitely fast. r_S is the massic stochiometric coefficient ($r_{SK} =$
$\nu_K M_K - M_K$, here). The function ϕ is defined by $\phi = Y_K/r_{SK} - Y_0/r_{SO}$;
because the one-way reaction is such that the consumptions of K and
O are proportional (by the factor r_{SO}/r_{SK}), the consumption of ϕ
by the reaction is null.

In the same way, if we define $Z = T + \frac{\Delta Q}{Cp\ r_{SK}}\ Y_K$ (here we assume
an averaged Cp for all species, and a heat ΔQ per unit of mass of
fuel) the equation for Z has also no term provided by the reaction;
the production term is due only to the kinetic energy.

2.2.   In a turbulent flow, the mean values of u_α, ρ, Y_K, ϕ, Z,
T, Y_0 ... are the first interesting quantities; they satisfy a set
of equations found by averaging the previous ones; with the

the assumptions that the flow is steady for the mean values:

$$\frac{\partial}{\partial x_\alpha} \left(\overline{\rho u_\alpha} \right) = 0$$

$$\overline{\rho u_\alpha} \frac{\partial}{\partial x_\alpha} \bar{u}_\beta = -\frac{\partial \bar{p}}{\partial x_\beta} + \frac{\partial}{\partial x_\alpha} \left[\overline{\tau}_{\alpha\beta} - \overline{(\rho u_\alpha)' u'_\beta} \right]$$

$$\overline{\rho u_\alpha} \frac{\partial}{\partial x_\alpha} \bar{y}_k = \frac{\partial}{\partial x_\alpha} \left[\overline{J^k_\alpha} - \overline{(\rho u_\alpha)' y'_k} \right] + r_{sk} \overline{\rho \dot{w}}$$

$$\overline{\rho u_\alpha} \frac{\partial}{\partial x_\alpha} \bar{\phi} = \frac{\partial}{\partial x_\alpha} \left[\overline{J^\phi_\alpha} - \overline{(\rho u_\alpha)' \phi'} \right]$$

$$\overline{\rho u_\alpha} \frac{\partial}{\partial x_\alpha} \bar{Z} = \frac{\partial}{\partial x_\alpha} \left[\overline{J^Z_\alpha} - \overline{(\rho u_\alpha)' Z'} \right] + \bar{\delta}$$

The flux terms appearing in these equations are well known; all the studies for turbulent shear flows or mixing flow for non-reactive gases are focussed on the properties and the calculations of these terms. The source term $\bar{\delta}$ is important in high speed non-isothermal flows.

The reaction source term $\overline{\rho \dot{w}}$ is important in reactive flows. In the same way that the fluctuations must be taken into account in the total transport fluxes, modifying the molecular fluxes, the fluctuations must also be taken into account in the mean turbulent reaction rate $\overline{\rho \dot{w}}$, modifying the molecular expression (but the modification is more complicated, because of the mathematical form of the reaction rate, a function of Y_0, Y_k, T).

2.3. As a simple example of the influence of the fluctuations of Y_0 and Y_k on $\overline{\rho \dot{w}}$, which is proportional to $\overline{Y_0 Y_k}$ if the T fluctuations are neglected, one can see that, if Y_0 and Y_k are fluctuating in phase opposition: $\overline{Y'_0 Y'_k} < 0$ and $\overline{Y_0 Y_k} < \bar{Y}_0 \bar{Y}_k$, and reciprocally $\overline{Y'_0 Y'_k} > 0$ and $\overline{Y_0 Y_k} > \bar{Y}_0 \bar{Y}_k$ if they are fluctuating in phase. On Figure 2, one sees a limiting situation, where $\overline{Y_0 Y_k} = 0$ and $\bar{Y}_0$ and $\bar{Y}_k \neq 0$ ($\overline{Y'_0 Y'_k} = -\bar{Y}_0 \bar{Y}_k$); in this situation, it is evident that the reaction cannot occur, because the oxidizer and the fuel are not at the same place at the same time.

Usually, in the reactive homogeneous non-turbulent flows, it is expected that the variation of Y_0 and Y_k are linearily related, i.e. the function ϕ is a constant; then $Y'_0 = \frac{r_{so}}{r_{sk}} Y'_k$, $\overline{Y'_0 Y'_k} = \frac{r_{so}}{r_{sk}} \overline{Y'^2_k}$ $= \frac{r_{sk}}{r_{so}} \overline{Y'^2_0}$, $\overline{Y'_0 Y'_k} / \sqrt{\overline{Y'^2_k}} \sqrt{\overline{Y'^2_0}} = 1$ and $\overline{Y_0 Y_k}$ is zero only if $\bar{Y}_0$ or $\bar{Y}_k$

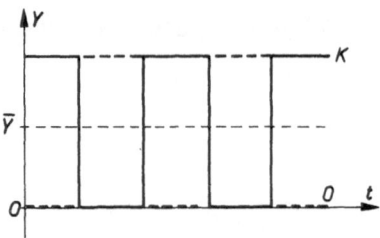

Figure 2.   Extreme Case of Concentration.   Fluctuations Preventing
            Reaction.

is zero; but in turbulent and non homogeneous flows, ϕ is not
necessarily a constant and is also generally a fluctuating quantity;
so situations with a negative value of $\overline{Y_0'Y_k'}$ can arise.   Corrsin (8)
and O'Brien (19), in their first studies on chemical reactions in
homogeneous turbulence, pointed this fact out.

     It is also simple to think that the temperature fluctuation
should be very important for the calculation of $\overline{\rho w}$, because the
exponential factor appearing in the formula.   Also as an example,
Figure 3 shows some numerical results presented by Vulis (22),

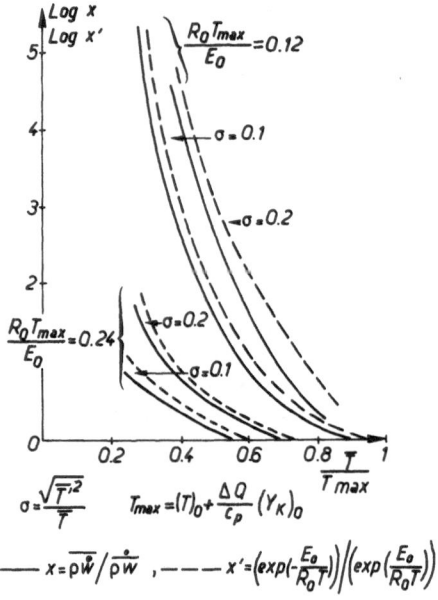

Figure 3.   T Fluctuations Effect on the Chemical Reaction Rate.

assuming a periodic fluctuation of T, (the fluctuations of Y_0 and Y_k being proportional to T fluctuations) and computing $\overline{\rho\dot{w}}$ by:

$$\overline{\rho\dot{w}} = \frac{1}{\tau} \int_0^\tau \rho\dot{w}\left[Y_0(t), Y_k(t), T(t)\right] dt$$

The figure shows that the ratio $\overline{\rho\dot{w}}/\rho\dot{w}(\overline{})$ can be very high, specially if T_{max}/T_A $\left(T_A = \frac{E_0}{R_0}\right)$ and $\overline{T}/T_{max}$ are low, i.e. in the initial zone of the combustion. Reciprocally, if $\overline{T}$ is close to T_{max}, this ratio is lower than 1, but close to 1; the influence of $\overline{T}$ fluctuations is not so important in the final zone of the combustion.

3.   COMPUTATIONAL MODELING OF THE REACTIVE MIXING FLOW

3.1.   The two simple examples above showed how the fluctuations can modify the mean reaction rate, which is a very interesting conclusion from a practical viewpoint. These modifications of $\overline{\rho\dot{w}}$ depend strongly on the fluctuations of T, Y_0, Y_k, in the reaction zone; and these fluctuations are not known but depend on the mixing and the reaction phenomena.

In order to answer the question of knowing why the fluctuations can affect the mean production rate, it is necessary to answer also that of knowing how the reaction can affect the fluctuations.

It is possible to develop an approach which can yield some answers, in the particular case of low fluctuations or low energy reactions. This method is based on the use of the second order moments $\overline{Y_0'^2}$, $\overline{Y_k'^2}$, $\overline{Y_0'Y_k'}$, $\overline{Y_0'T'}$, $\overline{Y_k'T'}$, $\overline{T'^2}$; and approximate expression of $\overline{\rho\dot{w}}$ as a function of these moments is easily found and the balance equations for these moments are handled, after their classical closure.

3.2.   The exact expression of $\overline{\rho\dot{w}}$ as a function of the moments is found (by means of series expansions) in (3) or (5, 23). This expression, valid if $T'/\overline{T} < 1$, can be written in our case:

$$\overline{\rho\dot{w}} = -\frac{k\overline{p}^2}{R^2 M_0 M_k} \overline{Y_0}\,\overline{Y_k}\,\overline{T}^{\alpha-2} \exp\left(-\frac{T_A}{\overline{T}}\right)\left[\Lambda + \frac{\overline{Y_0'Y_k'}}{\overline{Y_0}\,\overline{Y_k}} + P_\Lambda\left(\frac{T_A}{\overline{T}}\right)\left(\frac{\overline{Y_0'T'}}{\overline{Y_0}\,\overline{T}} + \frac{\overline{Y_k'T'}}{\overline{Y_k}\,\overline{T}}\right)\right.$$

$$\left. + P_2\left(\frac{T_A}{\overline{T}}\right)\frac{\overline{T'^2}}{\overline{T}} + P_\Lambda\left(\frac{T_A}{\overline{T}}\right)\left(\frac{\overline{T'^2 Y_0'}}{\overline{T}^2\overline{Y_0}} + \frac{\overline{T'^2 Y_k'}}{\overline{T}^2\overline{Y_k}}\right) + P_3\left(\frac{T_A}{\overline{T}}\right)\frac{\overline{T'^3}}{\overline{T}^3} + \cdots\right]$$

(where the pressure fluctuations are neglected). The P are poly-
nomials with respect to $T_A/\overline{T}$ (and dependent on α). As a first
remark, it can be seen that the T fluctuations intervene by the
mean of the self-correlations $\overline{T'^2}$, $\overline{T'^3}$, ..., but also by the
cross-correlations between T' and Y'_0 and Y'_k; since generally the
polynomial P_Λ is positive, the sign of $\overline{T'Y'_k}$ and $\overline{T'Y'_0}$ (i.e., if T
and Y_k or Y_0 fluctuate in phase opposition or in phase) is important
as well as the sign of $\overline{Y'_0Y'_k}$, in order to see if the fluctuations can
decrease or increase the reaction rate.

Secondly, moments of all orders are involved in this expression, when, in the case of quasi isothermal reaction (neglecting T'),
only the correlation $\overline{Y'_0Y'_k}$ appears. For this reason, particular
closure assumptions will be necessary, in addition to the standard
assumptions for the case of non-reacting flows; in the following
calculations, we shall neglect the moment of orders higher than two,
but this will constitute a limitation for the present method which
we shall examine later.

3.3.   Before the formation of balance equations for the second
order moments, we assume for simplicity the following;

-   kinetic and pressure terms are neglected with respect
    to the reaction term in the energy equation ($\overline{\delta}$ and
    $\delta' \simeq 0$);
-   averaged C_p and M_m (molecular mass) are used;
-   all species have the same single molecular diffusion
    coefficient and the molecular Lewis number is unity;
-   we write the balance equation within the "Boundary
    Layer approximation."

As convention, we express all turbulent fluxes by a diffusion
coefficient (the same for all fluctuating quantity)

$$\overline{(\rho v)' f'} = - D \frac{\partial \overline{f}}{\partial r}$$

The balance equations requested by $\overline{\rho w}$ are relative to:

$$\overline{Y'_0 Y'_k}, \ \overline{Y'_k}, \ \overline{Y'_0}, \ \overline{Y'_0 T'}, \ \overline{Y'_k T'}, \ \overline{T'^2}$$

But these moments are related linearily to:

$$\overline{Y'_0 Y'_k}, \ \overline{Y'^2_k}, \ \overline{\phi'^2}, \ \overline{Z'^2}, \ \overline{Y'_k Z'}, \ \overline{Y'_k \phi'}$$

which verify simpler balance equations.

The formation of balance equations can be performed by a classical method (2), and the different terms obtained are also classical. Formally, $\overline{\phi'^2}$ and $\overline{Z'^2}$ satisfy the same equation, written as:

$$\overline{\rho u}\,\frac{\partial \overline{Z'^2}}{\partial x} + \overline{\rho v}\,\frac{\partial \overline{Z'^2}}{\partial r} = \frac{\Lambda}{r^\delta}\,\frac{\partial}{\partial r}\left[r^\delta(d_m+D)\,\frac{\partial \overline{Z'^2}}{\partial r}\right] + 2D\left(\frac{\partial \overline{Z}}{\partial r}\right)^2$$

$$- 2d_m\,\overline{\frac{\partial Z'}{\partial r}\frac{\partial Z'}{\partial r}}$$

($\delta = 0$ for a two-dimensional flow and 1 for an axisymmetrical flow).

In the left hand side, there are convection terms; in the right hand side, there are: the diffusion term, the production term by the gradients of the mean-value, and the dissipation term.

For $\overline{Y_k'^2}$, the balance equation involves also a reaction term:

$$\overline{\rho u}\,\frac{\partial \overline{Y_k'^2}}{\partial x} + \overline{\rho v}\,\frac{\partial \overline{Y_k'}}{\partial r} = \frac{\Lambda}{r^\delta}\,\frac{\partial}{\partial r}\left[r^\delta(d_m+D)\,\frac{\partial \overline{Y_k'^2}}{\partial r}\right] + 2D\left(\frac{\partial \overline{Y_k}}{\partial r}\right)^2$$

$$- 2d_m\,\overline{\frac{\partial Y_k'}{\partial r}\frac{\partial Y_k'}{\partial r}} - r_{s_k}\,\frac{\overline{kp^2}}{R_0^2}\left[2\overline{Y}_o\overline{Y}_k\overline{T}^{\alpha-2}\,\exp\left(-\frac{T_A}{\overline{T}}\right)\right]$$

$$\left(\frac{P_\Lambda}{\overline{T}}\,\overline{T'Y_k'} + \frac{\overline{Y_k'^2}}{\overline{Y}_k} + \frac{\overline{Y_o'Y_k'}}{\overline{Y}_o} + \cdots\right)$$

This additional reaction term is, for a part, a consumption term, because the decrease of $\overline{Y}_k$ must be accompanied by a decrease of $\overline{Y_k'^2}$, but for another part a production term (if $\overline{T'Y_k'}$ or $\overline{T_o'Y_k'}$ are negative).

The equations for $\overline{Y_o'Y_k'}$, $\overline{Y_k'Z'}$, $\overline{Y_k'\phi'}$ are of the same type:

$$\overline{\rho u}\,\frac{\partial \overline{Y_o'Y_k'}}{\partial x} + \overline{\rho v}\,\frac{\partial \overline{Y_o'Y_k'}}{\partial r} = \frac{\Lambda}{r^\delta}\,\frac{\partial}{\partial r}\left[r^\delta(d_m+D)\,\frac{\partial \overline{Y_o'Y_k'}}{\partial r}\right] + 2D\,\frac{\partial \overline{Y}_o}{\partial r}\frac{\partial \overline{Y}_k}{\partial r}$$

$$- 2dm\,\overline{\frac{\partial Y_o'}{\partial r}\frac{\partial Y_k'}{\partial r}} - \frac{\overline{kp^2}}{R_0^2}\,\overline{Y}_o\overline{Y}_k\overline{T}^{\alpha-2}\exp\left(-\frac{T_A}{\overline{T}}\right)\left[\frac{P_\Lambda}{\overline{T}}\left(r_{s_o}\,\overline{Y_o'T'}\right.\right.$$

$$\left.+ r_{s_k}\,\overline{Y_k'T'}\right) + \frac{r_{s_o}\overline{Y_o'^2}}{\overline{Y}_o} + r_{s_k}\,\frac{\overline{Y_k'^2}}{\overline{Y}_k} + \left(\frac{r_{s_o}}{\overline{Y}_k} + \frac{r_{s_k}}{\overline{Y}_o}\right)\overline{Y_o'Y_k'} + \cdots\right]$$

$$\overline{\rho u}\,\frac{\partial \overline{Y_k'Z'}}{\partial x} + \overline{\rho v}\,\frac{\partial \overline{Y_k'Z'}}{\partial r} = \frac{\Lambda}{r^\delta}\frac{\partial}{\partial r}\left[r^\delta(d_m+D)\,\frac{\partial \overline{Y_k'Z'}}{\partial r}\right] + 2D\,\frac{\partial \overline{Y}_k}{\partial r}\frac{\partial \overline{Z}}{\partial r}$$

$$- 2d_m\,\frac{\overline{\partial Y_k'}}{\partial r}\frac{\partial Z'}{\partial r} - r_{s_k}\frac{k\overline{p}^2}{R_0^2}\,\overline{Y}_0\overline{Y}_k\overline{T}^{\alpha-2}\exp\left(-\frac{T_A}{\overline{T}}\right)\left[\frac{\Delta Q}{C_p r_{s_k}}\left(\frac{P_\Lambda}{\overline{T}}\overline{T'Y_k'}\right.\right.$$

$$\left.+ \frac{\overline{Y_0'Y_k'}}{\overline{Y}_0} + \frac{\overline{Y_k'^2}}{\overline{Y}_k}\right) + \frac{P_\Lambda}{\overline{T}}\overline{T'^2} + \frac{\overline{Y_0'T'}}{\overline{Y}_0} + \frac{\overline{Y_k'T'}}{\overline{Y}_k} + \dots\right]$$

3.4.   The closure of all equations, as well the equations for the mean value, as the new balance equations, needs the knowledge of relations for the turbulent transport fluxes, or for the eddy diffusion coefficient $D = - \overline{(\rho v)'f'}/(\partial \overline{f}/\partial r)$ and for the dissipation terms.

The modeling of the diffusion fluxes or coefficients in a classical problem; a comprehensive review of the techniques used is presented by Mellor and Herring (17); Donaldson (11) and Lumley (15) also performed very useful studies, specially for the dissipation term. In the following calculations, a very simple turbulence model is used, involving:

   a)   a mixing length hypothesis for the eddy viscosity
        coefficient:

$$\varepsilon = \overline{\rho}\,\ell^2\,\frac{\partial \overline{u}}{\partial r}$$

with constant Prandtl and Schmidt number.

The mixing length is assumed to be proportional to the mixing layer thickness.

   b)   an hypothesis of quasi isotropy for the dissipations
        functions; for instance:

$$\frac{\overline{\partial Y_0'}}{\partial r}\frac{\partial Y_k'}{\partial r} = \frac{\Lambda}{\lambda^2}\,\overline{Y_0'Y_k'}$$

where

$$\lambda = \frac{c_1\ell}{c_2 + c_3\,Re_\ell}$$

the Reynolds number of the turbulence $Re_\ell = \ell\sqrt{2k}/\nu$ being computed locally by a balance equation for the kinetic turbulence energy k. The constant appearing in this model has been chosen with the help of Donaldson (11). This model could be of course modified if comparison with experiments should be necessary. The model used seems perhaps representative enough for the present purposes.

3.5. It is necessary at this point to outline the expected limitations of the method; three different types of limitations can be pointed out:

a) The expansion of $\overline{\rho w}$ is only valid for $T'/\overline{T} < 1$; in addition, the series expansion of an exponential is rather poorly fast convergent. Then, only low fluctuations intensity is required. As an example, a simple comparison of the results given by the expansion (truncated after the second term) and the exact expression can be performed, with given fluctuations; for two-valued periodic fluctuations, and for $\rho\dot{w} = T^2\exp(- T_A/T)$, $(T = \overline{T} + T')$, one finds:

for $T_A/\overline{T} = 10$ and $T'/\overline{T} = 0.1$, $\dfrac{|\overline{\rho\dot{w}}_{approximated} - \overline{\rho\dot{w}}_{exact}|}{\overline{\rho\dot{w}}_{exact}} = 1.5\%;$

for $T_A/\overline{T} = 20$ and $T'/\overline{T} = 0.1$, $\dfrac{|\overline{\rho\dot{w}}_{approximated} - \overline{\rho\dot{w}}_{exact}|}{\overline{\rho\dot{w}}_{exact}} = 15\%;$

The approximation becomes poorer if $T_A/\overline{T}$ increases, i.e. for high temperature activation or low temperature cases.

b) The truncature of the expansion eliminates also the third and higher order moments; but temperature and mass fraction are non-negative random variables, and it can be expected that their skewness, for instance, must be non negligible; this case can arise specially in the final stage of the reaction, when the fuel or oxidizer mass fractions are small. As a consequence of this truncature in extreme cases, it can lead to a production of fuel (because of positive erroneous values of $-\overline{\rho\dot{w}} r_{sk}$), and also unacceptable behavior of the fluctuations (for instance negative values of the mean square of some fluctuations). In order to avoid such troubles, it can be convenient to use additional closure assumptions, which would express the third order moments as function of the second order moments and mean values, as proposed by Hilst (12) or Lin and O'Brien (14).

c) Finally, the closure assumptions used here for the flux and dissipation terms are classic; of course, the present method can be erroneous in the cases where non classical closure should be used, specially if compressibility effect must be taken into account, as pointed out by Bray (4).

4.   TURBULENCE PROPERTIES OF A HOT REACTING JET

4.1.   The method presented in the previous Chapter has been numerically expanded (with the integration method of Patankar and Spalding) and applied to the case of a hot jet in combustion, as the first figure shows.

The practical fuel assumed is CO, the oxidizer O_2; the jet has the following characteristics:

$$\overline{T}_0 = 1500°K \qquad\qquad \overline{\mu}_0 = 200 \text{ m/s}$$

$$\overline{Y}_{CO_0} = 3\% \qquad\qquad \overline{p} = 10^5 \text{ Pa.}$$

$$\overline{Y}_{O_{2_0}} = 12\% \qquad\qquad r_0 = 3.33 \text{ cm}$$

Typical reaction data for the CO oxidation have been chosen:

$$T_A = 15000°K \qquad k = 2.5 \cdot 10^6 \quad (SI: \text{ kg mole per kg per second}).$$

The external stream is assumed to be air at rest in the normal conditions:

$$\overline{T} = 300°K, \qquad \overline{p} = 10 \text{ Pa}, \qquad \overline{Y}_{O_2} = 23\%, \qquad \overline{Y}_{CO} = 0$$

The fluctuations intensity in the initial section has been taken into account, as necessary; in all cases however the cross-correlation has been assumed to be zero:

$$\overline{Y'_{O_2} Y'_{CO}} = \overline{Y'_{O_2} T'} = \overline{Y'_{CO} T'} = 0$$

4.2.   The Figures 4 to 7 are related to jet without combustion (with an initial turbulence intensity of 10%). The transverse profiles of temperature (Figure 4) are shaped as usual; the profiles of $\overline{T'^2}$ (Figure 5) show the production of fluctuations within the mixing zone, and also a dissipation in the homogeneous core and in the external stream. The calculations concern the non-developed zone of the jet, and the transverse profiles at different x are, as expected, non-self-similar. The unplotted transverse profiles of $\overline{Y}_k$, $\overline{Y'^2_k}$ are very similar to the profiles of $\overline{T}$ and $\overline{T'^2}$.

The Figures 6 and 7 show the correlation factors between Y'_0 and Y'_k and Y'_k and T'. It can be pointed out that these correlation

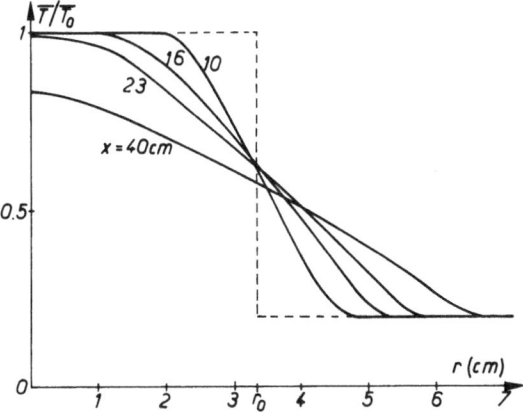

Figure 4.   Hot Jet without Combustion - Temperature Profiles.

factors go very fastly to their extreme values (-1 or +1) within the mixing zone, though their external values are zero, by hypothesis. The reason why $\overline{Y_0^\prime Y_k^\prime}$ is negative, but $\overline{Y_k^\prime T^\prime}$ is positive can be seen in the balance equations: the production turbulent terms are proportional to $\partial\overline{Y}_0/\partial r \cdot \partial\overline{Y}_k/\partial r$ and $\partial\overline{T}/\partial r \cdot \partial\overline{Y}_k/\partial r$, respectively; then, if (in our case) there is more oxidizer in the external flow and more fuel in the internal jet, $\partial\overline{Y}_0/\partial r \cdot \partial\overline{Y}_k/\partial r$ is negative; and when (in our case) the jet is fuel rich and hot, $\partial\overline{T}/\partial r \cdot \partial\overline{Y}_k/\partial r$ is positive. One can also expect that $\overline{T^\prime Y_0^\prime}/\sqrt{\overline{T^{\prime 2}}}\ \sqrt{\overline{Y_0^{\prime 2}}}$ (not plotted) has a shape similar to $\overline{Y_0^\prime Y_k^\prime}/\sqrt{\overline{Y_0^{\prime 2}}}\ \sqrt{\overline{Y_k^{\prime 2}}}$. The existence, within the mixing zone, of a constant value of these correlation factors, seems to be a property similar to the Bradshaw hypothesis; see next page:

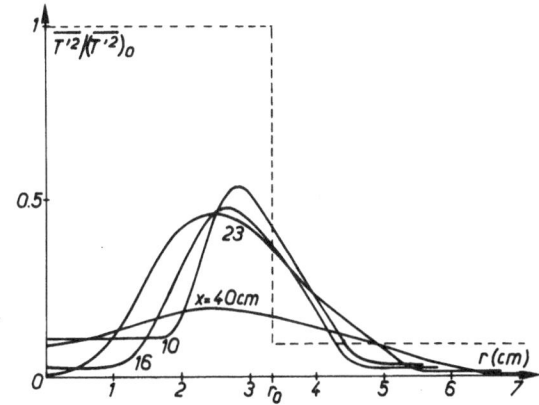

Figure 5.   Hot Jet without Combustion - Temperature Fluctuations
            Profiles.

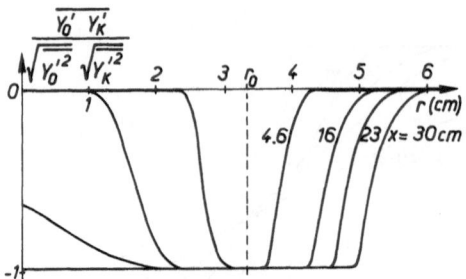

Figure 6.   Hot Jet without Combustion; Correlation between Fuel
            and Oxidizer.

$$\overline{u'v'} = a \ k = a(\overline{u'^2} + \overline{v'^2} + \overline{w'^2})/2$$

(perhaps more general, because the mass balance equations are more
similar than the momentum balance equations). In the case of the
reacting jet, we shall also plot these correlation factors and
examine the resulting modifications.

    4.3. The Figures 8 to 15 concern the case of the reacting jet,
showing the concentration ($\overline{Y_k'^2}$) and the temperature ($\overline{T'^2}$) fluctuations, the mean reaction rate and the transverse profiles of $\overline{Y_k}$ and
$\overline{T}$, and the cross-correlation factors.

    One can first of all remark that the shape of $\overline{Y_k}$ is very dif-
ferent to $\overline{T}$, rather contrary to the non-reacting jet; if the
behavior of the temperature is of the same type in a reacting or
non-reacting jet (the total increase of the temperature level,
with the equivalence ratio assumed here, is almost 200°K) the CO
mass fraction is strongly modified by the consumption; here the

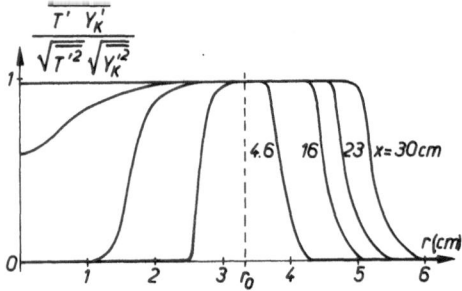

Figure 7.   Hot Jet without Combustion; Correlation between Tempera-
            ture and Fuel.

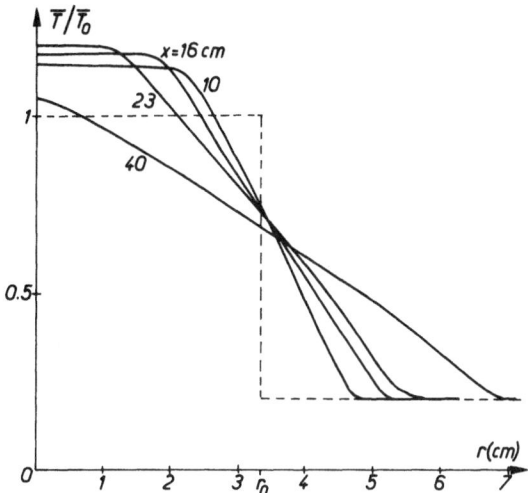

Figure 8. Hot Jet with Combustion: Temperature Profiles.

reaction is such that the CO transverse profiles are depressed on the center line. Secondly one can detect a high level of fluctuations $\overline{Y_k'^2}$ in the homogeneous core (due to the reaction) but a low level in the mixing zone after the core (due also to the reaction which, consuming $\overline{Y_k}$, consumes also $\overline{Y_k'^2}$); on the other hand, the two-peaked shape of the profiles is characteristic; this fact can be explained by the transverse profiles of $\overline{Y_k}$, that exhibit two zones of strong gradient and then bring two production zones of concentration fluctuations on. As in the case of the $\overline{T}$ profiles, the $\overline{T'^2}$ profiles also possess classical shapes.

The correlation factors curves exhibit also a particular behavior; in the zone where the reaction occurs (where $\overline{\rho w}$ is not

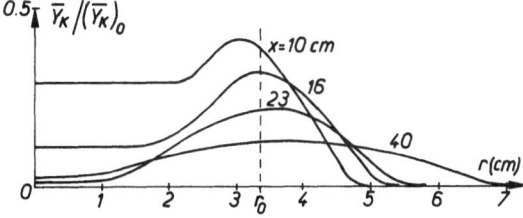

Figure 9. Hot Jet with Combustion: Fuel Concentration Profiles.

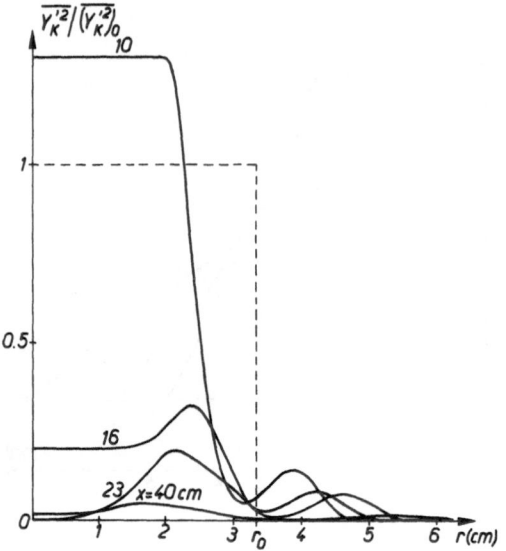

Figure 10. Hot Jet with Combustion: Fuel Fluctuations Profiles.

negligible) the profiles are strongly affected. One sees clearly
that the reaction effect on these correlations is to produce
negative values for $\overline{T'Y_k'}$ and $\overline{T'Y_0'}$, and a value different from -1
for $\overline{Y_0'Y_k'}$; this fact can be physically explained if we consider
that, where the reaction is the predominant phenomenon, an increase
of temperature is followed by a consumption (or decrease) of fuel

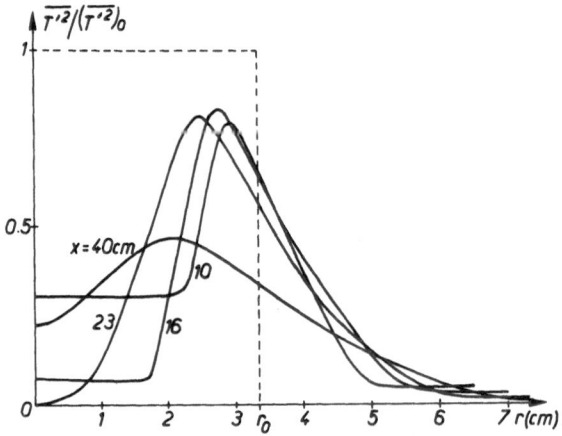

Figure 11. Hot Jet with Combustion: Temperature Fluctuations
            Profiles.

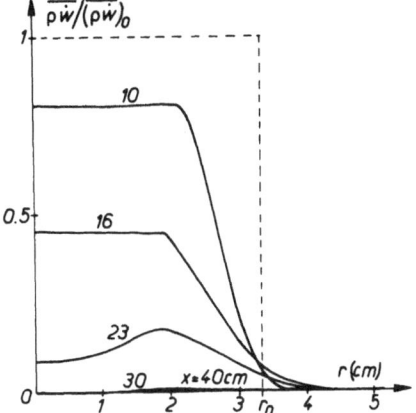

Figure 12. Hot Jet with Combustion: The Mean Reaction Rate.

or oxidizer; in our case of a fuel-lean combustion, the limiting factor will be the fuel consumption, and so $\overline{T'T_k'}/\sqrt{\overline{T'^2}}\sqrt{\overline{Y_k'^2}}$ will be -1 before the other correlation factors and will limit their evolutions.

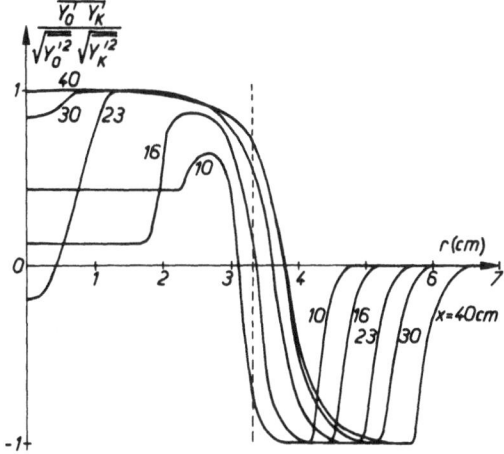

Figure 13. Hot Jet with Combustion: Correlation between Fuel and Oxidizer.

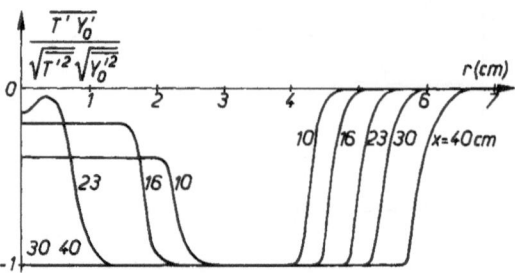

Figure 14.   Hot Jet with Combustion: Correlation between Tempera-
             ture and Oxidizer.

4.4.   The Figures 16 to 19 show some comparisons between
different calculations:

   i)   without combustion,
  ii)   with combustion and 10% initial turbulence intensity,
 iii)   with combustion and no initial turbulence intensity,
  iv)   with combustion and 10% initial turbulence intensity,
      but neglecting the temperature fluctuations,
   v)   with combustion and 10% initial turbulence intensity,
      but neglecting all fluctuations in the mean reaction
      rate.

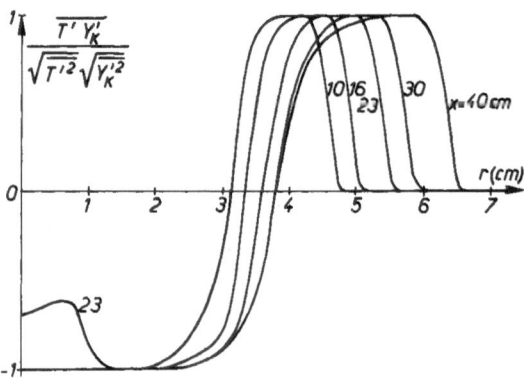

Figure 15.   Hot Jet with Combustion: Correlation between Tempera-
             ture and Fuel.

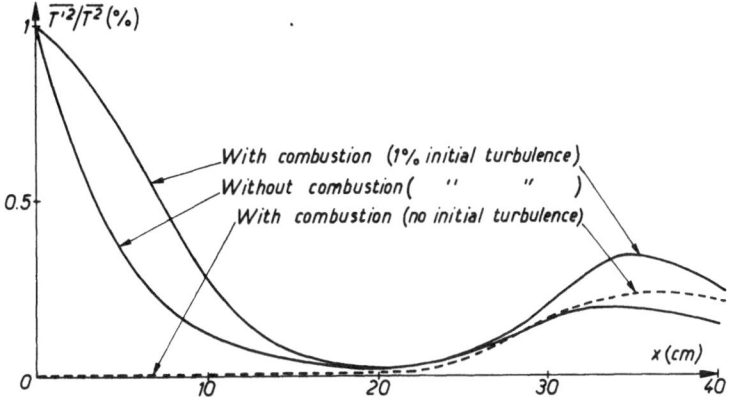

Figure 16.   Temperature Fluctuations on the Center Line.

The Figures 16, 17, 18 compare the levels of T fluctuations; one sees that the combustion effect is to increase the fluctuations with a variable factor, but for which an averaged value is about 2; as pointed out above, the $\overline{T'^2}$ curves with or without combustion are of the same type, but that is not true for the $Y_k'^2$ curves.

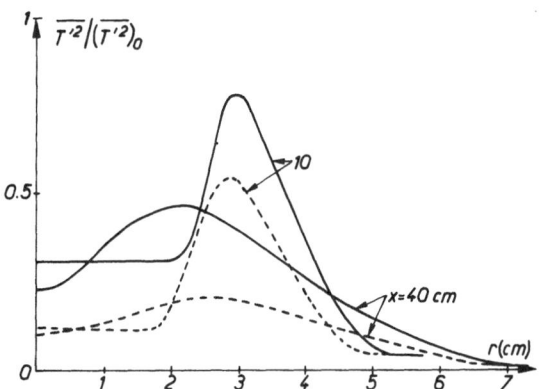

Figure 17.   Temperature Fluctuations Profiles:
————————— with combustion
---------- without combustion

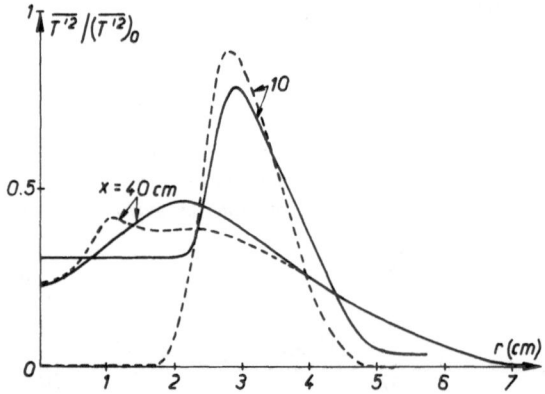

Figure 18.   Temperature Fluctuations Profiles:
              —————— with 10% initial turbulence
              ---------- without initial turbulence

        Finally, some global comparisons are made for the total mass
flow rate of CO destroyed in the plume. It appears that, in our
case, the influence of the turbulence on the reaction is not too
large: in the last computed section (40 cm) an error of about 25%
on the unburned mass flow rate can be produced by neglecting the
turbulence effect. This value is lower (or of the same order) than
the errors which can be made choosing the chemical exponential k
constant within the abundant literature.

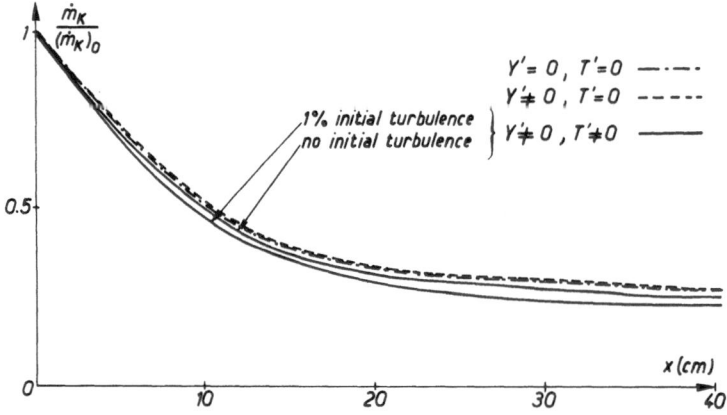

Figure 19.   Hot Jet with Reaction - Global Fuel Consumption.

5. CONCLUSIONS AND FURTHER DEVELOPMENTS

A more critical and more interesting flow pattern is the case of the development of a turbulent flame, stabilized by a solid body or by a pilot burner; such configurations are pictured also on the Figure 1.

Numerous experimental and theoretical studies related to this problem of practical importance have shown that all its physical aspects have not been truly explained (24, 9, 1, 18), e.g. some recent experimental studies of high speed cinematography by shadow-graph techniques (18) at ONERA. Let us hope for the possibility of the application of the method presented here, at least for the case of a low heat release.

Some particular properties pointed out in the present studies (particularly the ± 1 or $- 1$ value for the cross correlations factors) can be used generally; however, one can see that the truncature and the closure of the balance equation will be more difficult, because of the high temperature increase. For this purpose, it may be convenient to use the third order moments, either calculated as a function of the second order moments following Lin (7) or Hilst (19), or, better, by their balance equations.

REFERENCES

1. D. R. Ballal, A. H. Lefebvre - 4th Int. Colloquium on gas dynamics of explosions and reactive systems, San Diego (July 1973).

2. N. J. Beran - Statistical Continuum Theories - Chap. 3, Monographs in statistical Physics, vol. 9, Interscience publishers, 1968, N. Y.

3. R. Borghi - 2nd Symp. IUTAM-IUGG on Turbulent Diffusion in Environmental Pollution (Charlottesville, Va, 1973) - to be published in; Advances in geophysics, vol. 18, Ed. F. N. Frenkiel, R. E. Mann, Academic Press, N. Y.

4. K. N. C. Bray - A. A. S. U. Report no. 330, University of Southampton, (October 1973).

5. K. N. C. Bray - AGARD Specialists meeting on "Analytical and numerical methods for investigation of flow fields with chemical reactions, especially related to combustion." (Liège 1-2 April, 1974).

6. W. B. Bush, F. E. Fendell - Tech. Rep. TRW-7-PU, Project SQUID (July 1973).

7. P. Chung - The physics of Fluid, vol. 13, no. 5, pages 1153-1166 (May 1970).

8. S. Corrsin - The physics of Fluid, vol. 1, no. 1 (Jan.-Feb. 1958).

9. B. S. Cushing, et al. 11th Symp. (Int) on Combustion, p. 817 (1967).

10. G. Damköhler - NACA - TM 1112, (April 1947).

11. C. Donaldson, R. D. Sullivan, M. Rosenbaum - AIAA Journal, vol. 10, no. 2, (Feb. 1972).

12. G. R. Hilst - AIAA paper no. 73, 101 (1973).

13. P. A. Libby - Combustion Science and Technology, vol. 6, pages 23-28 (1972).

14. C. H. Lin, E. E. O'Brien - Astronautica Acta, vol. 17, pages 771-781 (1971).

15. J. L. Lumley and B. Khajeh-Nouri - 2nd Symp. IUTAM-IUGG on Turbulent Diffusion in Environmental Pollution (Charlottesville, Va, 1973).

16. H. B. Mason, D. B. Spalding - Report HTS/73/11, Imperial College of Science and Technology (February 1973).

17. G. L. Mellor, H. J. Herring - AIAA Journal, vol. 11, no. 5 (May 1973).

18. P. Moreau - La Recherche Aérospatiale no. 1974-3, p. 125-135.

19. E. E. O'Brien - The Physics of Fluid, vol. 11, no. 9, pages 1883-1888 (1968).

20. D. B. Spalding - 13th Symposium (Int.) on Combustion, pages 649-657 (1971).

21. A. V. Talantov, V. M. Emolaev, V. K. Zotin, E. A. Petrov - Combustion, explosion and shock waves, vol. 5, no. 1 (January 1969).

22. L. A. Vulis - Explosion, Combustion and Shock Waves (Russian issue) Tome 8 no. 1 (1972).

23. F. A. Williams - Tech. Report 4CSD-4-PU-Project SQUID
     (July 1973).

24. F. M. Wright, E. E. Zukoski - 8th Symp. (Int.) on Combustion,
     page 933 (1962).

DISCUSSION

SPALDING: (Imperial College, London, England)

I take this as a very fine achievement, that which you have
shown us. Now, consider the equation for $\overline{Y_0' Y_k'}$ which you introduced
earlier. There is something very remarkable which you have introduced into these equations. The first one is for the root mean
square fluctuation or the square of the fluctuation of a single
concentration. So also is the second. These are for passive
scalars of the kind that we have been talking about.

Now when we come farther down, you have a remarkable equation
there for $\overline{Y_0' Y_k'}$ which has enabled you to proceed. The term to which
I draw your attention is the second on the right-hand side, the one
beginning +2D. Now, you introduced this very calmly by analogy
with the earlier equation. It would seem to me that it has remarkable implications, not all of which I can see, and I would really
like to hear your justification for that particular presumption.
Now, someone pointed out earlier how the Schmidt number is very
significant. Let us suppose that we were dealing with a case where
0 and K simply did not mix. Then, no matter what turbulence you
have, you could never get a finite concentration from the two quantities at the same point. Yet, that is the meaning of the model
represented by your equations. Can you show from experiments or
from reasonable conclusions from that presumption that such a presumption is justifiable?

BORGHI:

The adoption of that term for integration is a result of my
proposed single turbulent diffusion coefficient applied to the
oxidizer and fuel, and it is indeed an assumption. At this time
I have no experimental data for this. But this does not appear to
me to lead to unrealistic conclusions.

BRAY: (Southampton University, Southampton, England)

I noticed that you used the incompressible form of the turbulence kinetic energy equation. I believe that additional terms in
that equation could be important, in particular, the term representing the divergence of the mean velocity. I wonder if you are able
to put such terms in.

BORGHI:

Not at present.

DONALDSON: (ARAP, Princeton, New Jersey)

I think this approach is very interesting. We have proceeded along the same route, but my whole struggle for the last couple of years has been to include the third-order correlations (the material that I presented to you earlier) and also to not linearize the expression for the reaction rates.

Now, you can draw some interesting conclusions from what you have done, but you have to be very careful interpreting a large part of your solution since you find that the product $\overline{Y_0'Y_k'}$ is the negative of $\overline{Y_0}\overline{Y_k}$. This is tantamount to saying that K and O do not exist at the same place at the same time.

BORGHI:

$\overline{Y_k'Y_0'}$ is negative but not equal to minus $\overline{Y_k}\overline{Y_0}$. $\overline{Y_0'Y_k'}$ is equal to $-\sqrt{\overline{Y_0'^2}}\,\sqrt{\overline{Y_k'^2}}$ in large parts of the calculation domain.

DONALDSON:

Then I misunderstood the numbers which you had; that is my problem.

Again, if you are trying to calculate fast reactions, it is just impossible to neglect the third-order correlations.

BORGHI:

Yes, I agree. The third order correlations are important in many practical cases if the turbulence intensity is large.

O'BRIEN: (State University of New York, Stony Brook, New York)

I am curious as to the appropriateness of the linearization of the temperature term. To what extent can you show that the answer is a posteriori consistent with having linearized a non-linear term.

BORGHI:

The justification of the consequences of linearization is not known a priori but is based on the result of the computations; for instance, if we find the turbulence intensity to be low in the calculated results, we can deduce that the term is not wrong. For high turbulence intensity on the other hand, there is no guaranteed accuracy in this model and the calculated values can diverge. An example of the errors introduced by the approximation is shown in the paper for $T'/T = 0.1$.

DONALDSON:

Bradshaw has just said to me, perhaps there was still some question about your choice of the modeling of the production term. And, I believe that what you did is correct to within the eddy viscosity approximation.

BRADSHAW: (Imperial College, London, England)

Perhaps, what Professor Spalding is attacking is the use of the eddy diffusivity concept, which is, in effect, used in simplifying the term in question.

SPALDING:

What I am saying has nothing to do with the eddy diffusivity-concept.

DONALDSON:

There still seem to be questions in some minds as to the correctness of the representation of the generation term for $\overline{Y_k'Y_0'}$. I ask myself, how is it that the two components actually get together? First they must be folded together and they must get together on the molecular scale in order to undergo reaction. Well, let me just see if I understand what has been done for the generation or folding together term. The term actually consists of two terms: $-\overline{(\rho'v')Y_0'}\dfrac{\partial \overline{Y}_k}{\partial r}$ and $-\overline{(\rho'v')Y_k'}\dfrac{\partial \overline{Y}_0}{\partial r}$. What the author has done is to replace $-\overline{(\rho'v')Y_0'}$ by $D(\partial \overline{Y}_0/\partial r)$ and $-\overline{(\rho'v')Y_k'}$ by $D(\partial \overline{Y}_k/\partial r)$ and add the two expressions to obtain $2D\dfrac{\partial \overline{Y}_0}{\partial r}\dfrac{\partial \overline{Y}_k}{\partial r}$. It seems to me that this is legitimate within the limits of an eddy diffusivity concept.

SPALDING:

I still question the result.

BORGHI:

In fact, I think that it is possible to have an equation for $\overline{(\rho v')Y_0'}$ and another equation for $\overline{(\rho v')Y_k'}$ and in those two equations there are reaction terms, and with these reaction terms, the difference between the diffusivity of $\overline{Y}_0$ and the diffusivity of $\overline{Y}_k$ appears. Thus the single diffusivity notion can be avoided. But the turbulence model used here in the computation was to test the influence of the reaction and not the influence of the turbulence model.

RECENT ADVANCES IN THEORETICAL DESCRIPTIONS OF TURBULENT DIFFUSION FLAMES

F. A. Williams

Department of Applied Mechanics and Engineering Sciences

University of California, San Diego, La Jolla, California

ABSTRACT

The theoretical analysis of the structure of a laminar diffusion flame in a shear flow is reviewed. Emphasis is placed on the "flame stretch" or critical Damköhler number required for extinction, in order to delineate the regime in which a turbulent diffusion flame can correctly be described as an ensemble of laminar diffusion flames. It is concluded that under many conditions of practical interest such a description is applicable. Potential benefits of exploiting this fact are then reviewed.

1. INTRODUCTION

The most fundamental distinction in the field of combustion is that between premixed and diffusion flames. In the former case reactants are intimately mixed prior to initiation of the flow process during which reaction occurs, and in the latter mixing of fuel and oxidizer occur during the flow process of interest. Although this added mixing phenomenon may appear to make diffusion flames more difficult to analyze, in laminar flows the opposite has been found true. Chemical kinetics play an essential role in the structure and propagation of premixed flames, but the major aspects of laminar diffusion-flame structure can be described without addressing kinetics. Specifically, as recognized by Burke and Schumann,[1] reaction zones may be approximated as thin sheets separating broad diffusion zones, whose structures may be calculated independently of reaction-zone structure by use of coupling functions.[2] Does this simplification extend to turbulent diffusion

flames? If so, then progress can be made in theoretical analyses
of turbulent diffusion flames. The present review considers the
question of conditions under which turbulent diffusion flames will
be composed of a statistical ensemble of laminar diffusion flames.

Before beginning, it is of interest to give a qualitative
description of a turbulent diffusion flame. A schematic diagram
appears in Figure 1. At the lip of the fuel duct, usually there
exists a laminar mixing layer containing a laminar flame sheet.
Laminar instability causes the mixing layer to become turbulent,
and farther downstream there is a highly turbulent reacting jet.
As the turbulence begins, the flame sheets are stretched, convoluted and rapidly whipped about, forming a turbulent flame brush
which appears on the average to fill the entire jet with flame.
Whether the brush is indeed composed of a nearly homogeneous,
premixed reacting mixture, or of a collection of rapidly moving
and highly sheared small laminar diffusion flamelets, is the subject of the present review. Experimentally it is often observed
that near the end of the brush blobs of flame continually break
away and burn as they flow downstream. This indicates that if the
diffusion flame sheets persist, they do not always remain as intact
surfaces, but instead develop holes, and parts eventually become
completely disconnected. Nevertheless, even under these severe
conditions, it is possible for the reaction zone to be so thin that
it remains locally planar and consumes reactants in a laminar
fashion.

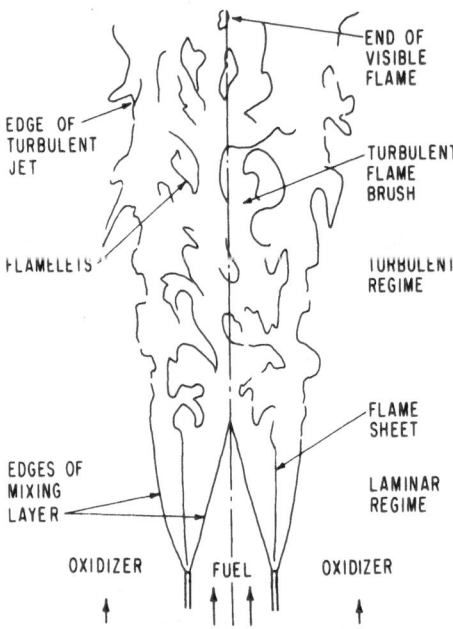

Figure 1. Schematic Illustration of a Turbulent Diffusion Flame

2. FLAME STRETCH

To analyze a laminar diffusion flame in a shear flow, select a point on the flame sheet, identify the normal to the reaction plane at that point, and adopt a coordinate system translating along the sheet in the laboratory frame, such that the component of velocity of the fluid parallel to the sheet in the translating frame is zero at the selected point. In addition, allow the coordinate system to translate normal to the sheet in the laboratory frame at a velocity such that the selected point on the reaction plane remains at the origin. For the results obtained from subsequent development it is immaterial whether the moving frame is inertial, its origin coinciding in position and velocity with the selected point possibly only at a selected time, or non-inertial, such that the coincidence is maintained for the entire lifetime of the reaction plane at the selected point. Since in the following analysis area changes of the flame sheet will be found to be of critical importance, in the moving frame the component of the vorticity vector normal to the reaction plane is inconsequential; if the non-inertial option is chosen, the moving system may be defined such that the normal component of vorticity vanishes identically at the selected point. On the other hand, the vorticity component parallel to the flame sheet can be important if by virtue of continuity it causes changes in flame area. To be definite, it seems to be desirable to adopt a (non-inertial) rotating frame in which the reaction plane remains normal to the ξ axis. Thus, assume that the ξ coordinate points from the fuel side to the oxidizer side of the flame, with the selected point on the reaction plane always located at the origin. With respect to this frame, the fluid may have velocity components both normal and tangential to the reaction plane, but the tangential components vanish at the origin. The local flow field is illustrated schematically in Figure 2.

If u, v and w are the ξ, η and ζ components of velocity in the non-inertial cartesian system just defined, then on the

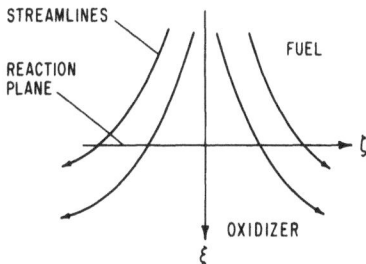

Figure 2.   The Local Flow Field for a Laminar Diffusion Flame in a Shear Flow

reaction plane near the origin $v = a\eta + c\zeta$ and $w = c\eta + b\zeta$, where a, b and c are functions of time only, and the vanishing of the normal component of vorticity has been employed. At any selected instant of time, it is possible by means of a simple rotation about the ξ axis to orient the η and ζ axes so that $c = 0$, thereby aligning these two coordinates with the principal axes of the intersection of the rate-of-strain quadratic with the reaction plane. In this final system, $v = a\eta$, $w = b\zeta$, and the instantaneous time rate of increase of the local area A of the flame sheet is $dA/dt = A(a + b)$. The flame stretch γ is defined as the fractional time rate of increase of the flame-sheet area, viz.,

$$\gamma \equiv d \ln A/dt = a + b . \tag{1}$$

The flame stretch defined here is of central importance to the structure and dynamics of a laminar flame in a shear flow. This importance has been demonstrated clearly for premixed flames by Klimov,[3] who recognized in unpublished work that the same ideas are applicable to diffusion flames. Estimates suggest that in turbulent flows flame stretch usually will be more important than flame curvature, which cannot be treated on the basis of a planar flame-sheet approximation. Corrections for curvature are of the order of the ratio of a characteristic flame thickness to a characteristic radius of curvature. This ratio typically is smaller than the product of flame stretch with residence time, which characterizes the order of magnitude of the influence of flame stretch. There are special conditions, e.g. near burnout of a flamelet, under which curvature may be important.

It is easy to visualize some of the ways in which stretch influences flame structure. First observe that the sign of the flame stretch can be either positive or negative; negative stretch may be termed flame compression. From continuity concepts it is clear that to produce positive stretch there must be a tendency for flow toward the reaction plane to occur at large positive and negative values of ξ. On the other hand, compression is associated with flow away from the reaction plane at large ξ. These convective effects influence flame structure differently. Consider the usual case in which fuel and oxidizer both are cooler than the reaction plane. The influence of the convective effects on the temperature profile is illustrated in Figure 3. The convective influence of stretch tends to steepen the temperature profile, making the region of heated gas thinner. Conversely, compression of the sheet smooths the profiles, thickening the heated region. It is also apparent that the convective effects tend to increase the reactant flux into the reaction plane in the presence of stretch, and to decrease the reactant flux under compression.

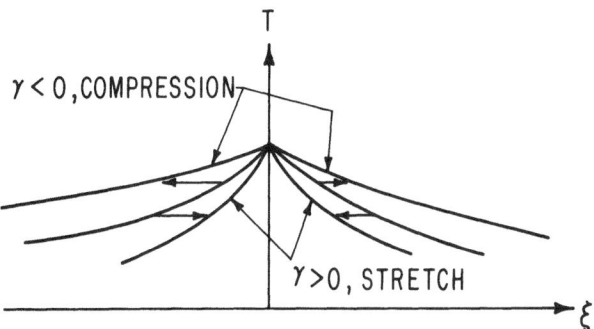

Figure 3. Illustration of Influence of Stretch and Compression on Temperature Profiles

A simplified interpretation of flame stretch may be obtained by neglecting the substantial derivative of density $D\rho/Dt$ at the origin. In this case, the continuity equation requires that $u = u_0 - \gamma\xi$ near the origin, where u_0 is independent of ξ. It is then seen that γ may be viewed as a particular component of the strain-rate tensor, the diagonal component for velocity and distance normal to the reaction plane. Although this interpretation emphasizes that γ is essentially a strain rate or velocity gradient normal to the flame, the definition given in Eq. (1) is more general and therefore more useful for accurate analyses.[3]

3. APPLICABILITY OF STRETCH IN TURBULENT FLAMES

The interpretation of γ as a strain rate helps to clarify the role of flame stretch in turbulent flows. Typical strain rates in turbulent flows are of the order of the ratio of the kinematic viscosity ν to the square of the Kolmogorov scale η. Thus, values of γ of the order of ν/η^2 are expected to be encountered. Of course, there will be a distribution of flame stretch or of the various components of the strain-rate tensor of flame throughout a turbulent flow. Improved statistics on strain rates in turbulence would be helpful in defining better the conditions under which turbulent diffusion flames are composed of ensembles of laminar diffusion flames.

Use of the approximation $\gamma = \nu/\eta^2$ allows rough estimates to be made of conditions for the occurrence of appreciable stretch prior to reactant depletion in turbulent diffusion flames composed of laminar flamelets. Consider a fuel blob of some characteristic size δ. The diffusion flame around the blob consumes it essentially by a laminar diffusion process. In gases laminar Schmidt and

Prandtl numbers are both of order unity, and therefore the effective laminar diffusion coefficient is approximately ν, so that the
burning time of the blob will be roughly δ^2/ν. Thus, burnout will
occur prior to stretch if $\gamma\delta^2/\nu \ll 1$, and substantial stretch
prior to burnout if $\gamma\delta^2/\nu \gg 1$. The estimate of γ then shows that
blobs small compared with the Kolmogorov scale will be consumed
before being stretched, while blobs large compared with the
Kolmogorov scale will be stretched and distorted considerably
prior to consumption. Thus, the eddy size at which spectral transfer is terminated by viscous dissipation, approximately coincides
with the eddy size below which turbulent straining fails to modify
combustion of a fuel eddy. This result probably is to be expected
in advance from the similarity between diffusion and viscous dissipation in gases. However, in view of the fact that initial sizes
of fuel eddies in turbulent diffusion flames generally are very
much larger than the Kolmogorov scale, the result emphasizes the
great importance of flame stretch in turbulent flames.

4. STRUCTURE OF LAMINAR DIFFUSION FLAMES IN SHEAR FLOWS

The equation of conservation of energy for a one-dimensional
laminar diffusion flame in the presence of flame stretch helps to
reveal further characteristics of laminar flamelets in turbulent
flows. Let $Y \equiv (T - T_0)/(T_{af} - T_0)$, where T is temperature, T_0 is
the initial oxidizer temperature and T_{af} is the adiabatic flame
temperature for a stoichiometric mixture of fuel and oxidizer.
Under the assumption that conditions are one dimensional but
possibly transient, energy conservation may be shown[3] to be expres-
sible as

$$\frac{\partial Y}{\partial t} - \gamma\xi \frac{\partial Y}{\partial \xi} + u_0 \frac{\partial Y}{\partial \xi} - \nu \frac{\partial^2 Y}{\partial \xi^2} = \frac{1}{\tau} F(Y) . \qquad (2)$$

Here τ is a characteristic chemical time and $F(Y)$ a nondimensional
rate of chemical heat release, τ being defined so that the peak
value of $F(Y)$ is unity. The assumptions of negligible radiant
energy loss, of a one-step reaction and of a Lewis number of unity
have been introduced so that through coupling functions F is expres-
sible as a function of Y alone. These approximations often are
reasonably good[2] and have been introduced here for the purpose of
simplifying the presentation by removing the necessity of writing
species conservation equations. It may be noted that suitably
defined nondimensional fuel and oxidizer concentrations also obey
Eq. (2). Also for purposes of simplification in presentation, the
thermal conductivity has been taken constant and the Prandtl number
unity; without these approximations the factor ν is modified and
moves inside one of the two ξ derivatives, a coordinate transformation being needed to bring the factor outside as it appears in Eq.

(2). Unless $F(Y)$ is approximated as a delta function, the reaction plane becomes a region of finite thickness, and the origin, at which the ξ component of velocity is $u_0(t)$, must be defined to lie somewhere within this reaction region, e.g. at the point where $F(Y)$ has its maximum.

A good approximation for $F(Y)$ may be obtained by assuming an Arrhenius rate function. It is found that

$$F(Y) = C(Y - Y_1)^m \, Y^n \, \exp[- \beta Y/(1 - \alpha Y)] \,, \tag{3}$$

where C is a constant, n is the order of the overall reaction with respect to fuel, m is the order of the overall reaction with respect to oxidizer, $Y_1 = (T_1 - T_0)/(T_{af} - T_0)$ with T_1 being the initial temperature of the fuel, $\alpha = (T_{af} - T_0)/T_{af}$, and β a non-dimensional activation energy. The nondimensional activation energy is related to the overall activation energy E and the universal gas constant R by the formula $\beta = E(T_{af} - T_0)/R \, T_{af}^2$. The function $F(Y)$ is sharply peaked near the maximum value of Y because β is large; typically $\beta \approx 10$.

The transient term in Eq. (2) accounts for energy accumulation. There are two convective terms, the first associated with the strain rate and the second with u_0. Diffusion of heat is accounted for by the last term on the left-hand side and chemical heat release by the right-hand side.

Certain important aspects of the nature of the solution to Eq. (2) were not discussed earlier. First, observe that $\sqrt{\tau\nu}$ defines a characteristic length associated with reaction and diffusion. Although reaction times vary over wide ranges of values, typically $\tau \sim 10^{-6}$ sec, and $\nu \sim 1$ cm^2/sec, so that this characteristic length is very small, of the order 10^{-3} cm. It defines the characteristic thickness of the reactive-diffusive zone, the narrow region (analyzed by Fendell, Liñán and others, see Ref. 2) within which the reaction rate is important. Within this zone, which earlier was approximated as the reaction plane, all other terms in Eq. (2) are negligibly small; for example, when the thickness of the reactive-diffusive zone is used to nondimensionalize ξ, the second convective term contains the small parameter $\sqrt{u_0^2 \, \tau/\nu}$.

At distances large compared with $\sqrt{\tau\nu}$, an appropriate characteristic length is the representative dimension δ of a fuel or oxidizer blob, and the right-hand side of Eq. (2) is negligibly small, there being a balance among diffusion, convection and transient accumulation of energy. The important convective term is the stretch term for the prevalent case in which $\gamma\delta^2 \gg \nu$. There is a significant question as to whether there exists a steady-state solution in this outer region. Obviously a steady state requires a balance between convection and diffusion. Physically, the balance can exist only if convection and diffusion of heat occur in opposite directions. Since heat is diffusing away from the reaction plane, the convective velocity must be directed toward the reaction plane at large ξ if a steady state is to exist. Thus, γ must be positive; stretch is necessary for a steady state. Mathematically, in Eq. (2) $\partial^2 Y/\partial \xi^2$ is positive, and $\xi \partial Y/\partial \xi$ is negative, so γ must be positive. For zero stretch, or in the presence of compression, a steady state is impossible.

In turbulent flows, both steady and unsteady conditions are to be expected for laminar diffusion flamelets. However, since stretch statistically predominates over compression in turbulence, a tendency to favor locally steady state flamelets is expected. This tendency is accentuated by the dynamics of flame motion. From Eq. (2) it is estimated that in a region of positive stretch, the characteristic transient time for the diffusion flame to achieve a steady state is of order $1/\gamma$, which is short compared with the burning time of a blob that is large compared with a Kolmogorov scale. The picture that emerges is one in which flamelets encountering regions of compression rapidly move to seek out regions of stretch, where they prefer to exist in a steady state.

An additional observation following from Eq. (2) is that the preferred steady flamelets possess structures and reactant consumption rates that are independent of the sizes of fuel and oxidizer blobs, provided that $\gamma \delta^2 \gg \nu$. In this case, comparison of the stretch and diffusion terms defines a characteristic length of $\sqrt{\nu/\gamma}$ (i.e., of the order of the Kolmogorov scale) for the thickness of the convective-diffusive zone. With the typical value $\gamma \sim 10^3$ sec^{-1}, this thickness, of order 3×10^{-2} cm, is large compared with the reaction-zone thickness but small compared with representative blob sizes. Statistical properties of the laminar diffusion-flame structure then depend only on statistical properties of the portions of the strain-rate tensor which affect flame stretch. Direct correlations between strain rates and rates of reactant consumption may be developed.

5. EXTINCTION

What can go wrong with the general situation sketched above? Why won't all turbulent diffusion flames be composed principally of ensembles of steady laminar diffusion flames in shear flows? The main limitation appears to be the phenomenon of diffusion-flame extinction. Steady-state solutions to Eq. (2) have been explored by a number of investigators.[4-8] An important parameter which arises is a Damköhler similarity group, $D \equiv (\gamma \tau)^{-1}$, the ratio of a characteristic flow time (reciprocal of flame stretch) to a characteristic reaction time, or alternatively the ratio of a reaction rate to a strain rate. The calculated dependence of the non-dimensional maximum temperature on D is illustrated schematically in Figure 4, which has been reviewed in Ref. 2. The key result is the S shape with an unstable intermediate branch and with critical Damköhler numbers D_i and D_e for ignition and extinction, respectively. As stretch increases, D decreases, and therefore associated with D_e is a critical value of stretch above which the flame is extinguished. Physically, if γ is too large, the residence time of a fluid element within the reaction zone is too small compared with the chemical time for reaction to occur.

Experimental verification of the existence of the extinction point on the S curve has been achieved. Kent designed an apparatus for establishing a flat, stretched diffusion flame in the laminar stagnation-point boundary layer above a liquid fuel.[9] The flame is made flat by directing a uniform flow of air or other oxidizing gas mixtures downward onto the surface of the liquid. Figure 5 is a photograph of the thin, blue, stretched flame produced with heptane as fuel and an oxygen-nitrogen mixture as the oxidizer. The flame stretch in the experiment is approximately U/ℓ, where U is the velocity of the oxidizing gas leaving the duct, and ℓ is the spacing between the outlet of the duct and the liquid surface. By increasing U in the experiment, the value of γ was increased, and it was observed that at a critical value of U sudden extinction of the flame occurred. The extinction was too rapid for its dynamical aspects to be measured. However, flame temperatures were measured with a fine coated thermocouple as extinction was approached. A representative graph is shown in Figure 6, where the small decrease in flame temperature prior to extinction, as predicted theoretically, is seen to occur. In Figure 6 the Damköhler number is roughly $D = \ell/(U\tau)$, which decreases as U increases, the maximum value of U corresponding to the extinction Damköhler number D_e.

The value of the extinction Damköhler number is most important, since the diffusion flame clearly cannot exist if $\gamma > (D_e\tau)^{-1}$.

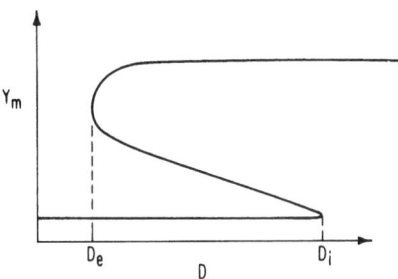

Figure 4.   Dependence of Nondimensional Maximum Temperature on Damköhler Number

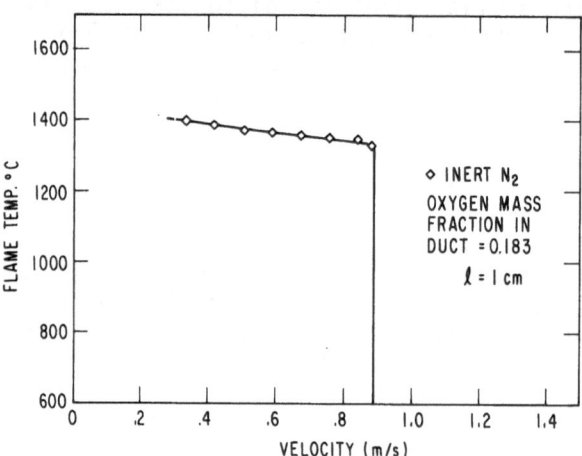

Figure 6.   Measured Dependence of Flame Temperature on Flame Stretch

Figure 5. Photograph of Stretched Laminar Diffusion Flame

Asymptotic analysis, treating β as a large parameter, produces good analytical expressions for D_e,[8] which can be used to extract overall reaction-rate data from measured conditions of diffusion-flame extinction.[9] The idea of measuring reaction rates essentially by increasing γ until extinction of a laminar diffusion flame is achieved, has been known for a number of years[10] and has been termed the measurement of flame strength. Thus, laminar flame-strength experiments turn out to be directly relevant to turbulent diffusion-flame structure.

The numerical value of D_e determines whether the stretched flamelets will be found for a given fuel-oxidizer mixture under a given strain rate. This value may vary appreciably from one set of conditions to another. In some cases, its order of magnitude is available experimentally from the measurements cited above. Figure 7 shows extinction conditions measured for various oxidizing gas mixtures, with ℓ of the order of 1 cm. It is seen that the critical flame stretch for extinction is of the order of 10^2 sec^{-1} in these cases, which correspond to rather highly diluted mixtures with relatively long reaction times. For these mixtures in turbulent flows with Kolmogorov scales below roughly 10^{-1} cm, the ensemble of stretched laminar diffusion flames will not exist. On the other hand, in richer mixtures the flame strength will be much higher (since the reaction time is much shorter at the higher flame temperatures), and the critical value of γ for extinction is likely to be larger than the values encountered in the turbulent diffusion flame. Thus, it is anticipated that turbulent diffusion flames are composed of stretched laminar diffusion flamelets only for sufficiently reactive fuel-oxidizer combinations.

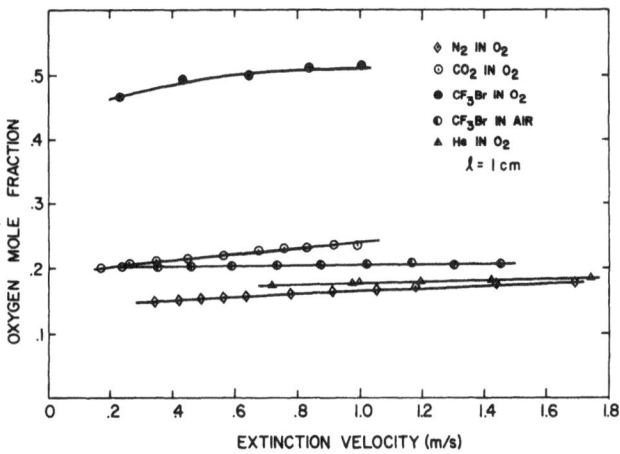

Figure 7. Measured Extinction Conditions for Heptane with Various Oxidizing Gases

What happens when the range of values of γ in the turbulent
flow largely exceeds the critical extinction value? The answer to
this question is unknown; the subject needs more study. Although
a single laminar flamelet is extinguished above the critical γ, it
seems highly unlikely that the turbulent flame will be extinguished.
Mixing of reactants will continue (even at an enhanced rate at the
higher γ) and eventually the fuel-oxidizer mixture is likely to
encounter a hot spot, ignite and burn, now in a principally premixed fashion. Thus, the structure of turbulent diffusion flames
at high strain rates may resemble more closely the structure of
premixed turbulent flames.

6. MACROSTRUCTURE OF TURBULENT DIFFUSION FLAMES COMPOSED OF STRETCHED LAMINAR DIFFUSION FLAMELETS

If the strain rates are small enough for stretched laminar
diffusion flames to comprise the turbulent diffusion flame, then a
number of macroscopic characteristics of the turbulent diffusion
flame can be established on a firm basis. Thus, it can be demonstrated[11-13] that through use of coupling functions, fields of
fuel, oxidizer and product concentrations, as well as temperature,
can be related to the concentration field for a normalized inert,
on a space-time resolved basis, under conditions that often are
reasonably realistic. It follows that measurement of sufficiently
detailed statistical properties of an inert added to either fuel
or oxidizer, enables corresponding statistical properties of the
fields of interest to be computed. The importance of this result
stems from the extremely great difficulty in being confident of
a priori predictions of mixing in turbulent shear flows. Although
the correspondence strictly requires the inert to be measured in
the hot flow, it is possible that cold-flow measurements may provide a reasonable correspondence. The need for further studies to
test this possibility, through consideration of influences of
local rapid heat releases on turbulent velocity fields, thus is
underscored.

An untried approach to calculation of volumetric production
rates of nitric oxide in turbulent diffusion flames, based on
deductions drawn by use of the coupling functions, is also worth
pursuing.[13] Production occurs only in the vicinity of the laminar
diffusion flamelets, and the laminar production can be related to
the local flame structure for an assumed kinetic scheme. Use of
the probability distribution function for the normalized inert
concentration then provides an explicit expression for the average
local production rate. An asymptotic expansion for production
kinetics having high activation energies reduces this expression
to a formula involving the local probability density function for
the inert, evaluated at the inert concentration which corresponds

to the flame location. Thus, by measuring the local probability density function for inerts, local average production rates of nitric oxide can be computed. Total production rates may then be obtained by integrating the local average rates over the flame.

The methods just discussed are designed to circumvent the necessity of describing theoretically the turbulent macrostructure of diffusion flames. They use certain measured statistical properties, principally probability densities, to calculate other properties that are of interest. While the methods work well for certain quantities, such as production of nitric oxide, they become difficult to apply for other qualities, such as volumetric rates of heat release.[13] Thus interest remains in the difficult problem of modeling turbulent reacting shear flows.

ACKNOWLEDGEMENT

Preparation of this presentation has been supported in part by the Air Force Office of Scientific Research, Office of Aerospace Research, U. S. Air Force under Grant AFOSR-72-2333.

REFERENCES

1. S. D. Burke and T. E. W. Schumann, Ind. Eng. Chem. 20, 998 (1928).

2. F. A. Williams, "Theory of Combustion in Laminar Flows," Annual Reviews of Fluid Mechanics 3, 171 (1971).

3. A. M. Klimov, Zhur. Prikl. Mekh. i Tekhn. Fiz. 3, 49 (1963).

4. Y. B. Zel'dovich, Zhur. Tekh. Fiz. 19, 1199 (1949).

5. D. B. Spalding, Fuel 33, 255 (1954).

6. D. B. Spalding, ARS J. 31, 763 (1961).

7. F. E. Fendell, J. Fluid Mech. 21, 291 (1965).

8. A. Liñán, "The Asymptotic Structure of Counterflow Diffusion Flames for Large Activation Energies," to appear in Acta Astronautica.

9. J. H. Kent and F. A. Williams, "Extinction of Laminar Diffusion Flames for Liquid Fuels," to appear in Fifteenth Symposium (International) on Combustion.

10.   A. E. Potter, Jr., S. Heimel and J. N. Butler, "Apparent
      Flame Strength, a Measure of Maximum Reaction Rate in Diffu-
      sion Flames," Eighth Symposium (International) on Combustion,
      Williams and Wilkins Co., Baltimore (1962), p. 1027.

11.   W. R. Hawthorne, D. S. Weddell and H. C. Hottel, "Mixing and
      Combustion in Turbulent Gas Jets," Third Symposium on Combus-
      tion, Flame and Explosion Phenomena, Williams and Wilkins Co.,
      Baltimore (1949), p. 266.

12.   H. L. Toor, A.I.Ch.E. Journal 8, 70 (1962).

13.   F. A. Williams, "A Review of Some Theoretical Considerations
      of Turbulent Flame Structure," to appear in proceedings of the
      April 1974 meeting of the Propulsion and Energetics Panel of
      AGARD, entitled Analytical and Numerical Methods for Investi-
      gation of Flow Fields with Chemical Reactions Especially
      Related to Combustion.

DISCUSSION

LIEPMANN: (California Institute of Technology, Pasadena)

How is your model related to the Marble-Carrier-Fendell[1] model
in regard to some of these ideas on flame stretch?

WILLIAMS:

I have seen a SQUID report by Carrier, Fendell and Marble.[1]
What they have done in that report is to work out a number of
different flow models. Some time-dependent flows may have also
been included, but I am not sure. As a collection of model calculations, it has some very useful elements. But, as I understand
it, they have not addressed the question of extinction in that
paper. They assume a structure and describe some of the flame
zones. But, I think it is important also to address the question
of extinction. Is that fair?

FENDELL: (TRW Systems, Redondo Beach, California)

I think as much is said on extinction in that paper as you
said in this address, with perhaps a few exceptions. But, it was
interesting listening to it.

---

[1]   Carrier, G. F., F. E. Fendell and F. E. Marble, "The Effect of
      Strain on Diffusion Flames," PROJECT SQUID Technical Report
      TRW-5-PU, October 1973, to appear, SIAM J. App. Math.

WILLIAMS:

You covered extinction as well?

FENDELL:

About as far as perturbation theory permitted.

SWITHENBANK: (Sheffield University, Sheffield, England)

I am interested in your comments on the following thoughts that struck me as a consequence of your paper. Most of the combustion systems that we deal with in practice use liquid or solid fuels and hence are two-phase flow systems. In particular, oil evaporates and burns as a diffusion flame in many instances. Sjögren mentioned this in a paper[2] in the last Combustion Symposium. He suggested that when the flame lifted off the drop by the velocity difference between the drop and the air, then one got a premixed flame, whereas, when the flame was stabilized on the drop, one got the diffusion flame, corresponding to the two cases of the blue flame and the yellow flame. Now, if one considers a gas turbine combustor, the droplet is typically a hundred microns or less. The size of the dissipation eddy is comparable to the size of the diffusion eddy which turns out to correspond to an eddy of about a hundred microns, something, if anything, slightly bigger than the drop. This poses an interesting problem of what happens when the droplet is larger or smaller than the shear flow that one is dealing with. These are some practical thoughts which I feel are relevant to real problems.

WILLIAMS:

What happens when the blob is smaller than the shear flow, depends on whether it has liquid in it or not. If the droplets have vaporized, then the sizes you mention are small compared with the Kolmogorov scale, and the small fuel pockets simply burn as individual blobs in the diffusion process. When the pockets become very small, the residence time decreases and extinguishment of very small pockets occurs. But, these small pockets comprise a very small part of the amount of fuel, and the effect therefore probably is negligible. If the pockets contain liquid fuel surrounded by diffusion flames, then extinction of these diffusion flames at small sizes can leave behind appreciable amounts of fuel and contribute to combustion inefficiency.

---

[2] Sjörgan, A. (1972) "Soot formation by combustion of an atomized liquid fuel," Fourteenth International Symposium on Combustion, Combustion Institute, p. 919.

SWITHENBANK:

Do you feel that could give us the blue flame rather than the yellow flame?

WILLIAMS:

It is not true that diffusion flames are always yellow and premixed always blue. The flame in Figure 5 is a diffusion flame, and it is blue. Whether the flame is blue or yellow is mainly a question of residence time. If the residence time is long enough in the somewhat heated, fuel-rich zone, there is time enough for fuel pyrolysis, and carbon is formed, giving rise to a yellow flame.

For turbulent diffusion flames at the exits of the pipes in turbulent flow, at high flow rates there is a lifted flame which is typically blue, whereas at low flow rates there is an attached flame which is typically yellow. There have been arguments that a lifted flame is probably a premixed flame and therefore blue, and the attached flame is a diffusion flame and therefore yellow. But I think that is an oversimplification. It is true that the lifted flame does have a higher fuel velocity, and it has a shorter fuel residence time, and so there will be less pyrolysis, but that does not necessarily mean that it is going to be premixed. It may, in fact, be premixed because the strain rate is higher under those conditions, and the laminar diffusion flamelets may be extinguished, but I do not think you can tell if it is premixed or not by the color. For example, there are yellow premixed flames. It has been said, for instance, that oxygen can catalyze fuel pyrolysis and so, by some premixing you can enhance the rate of production of carbon to produce yellow flames, according to chemists.

O'BRIEN: (State University of New York, Stony Brook, NY)

I am interested in the idea of characterizing the strain of the flame by the mean strain rate. Is that really the rate of generation of flame area in a turbulent flow?

WILLIAMS:

It is related in an order-of-magnitude sense.

O'BRIEN:

I am not sure if it is that. I do not know whether it is an appropriate measure or not. It seems to be too simple for the description of the rate at which material surface is generated. I do not really know too much about this, and I thought somebody might.

BRODKEY: (Ohio State University, Columbus, Ohio)

Is not that parallel to Batchelor's analysis for the high Schmidt number case? The spectra of the scalar extends well beyond that for the velocity and it is the same strain field which causes the spreading and transfer through the spectral region. I just looked up some numbers and, in our experiment, for between 30 and 150, we settled for a hundred for our values of the estimated strain.

O'BRIEN:

But the idea here, anyway, is that you are talking about the generation of surface area which enhances the reaction, is that correct?

WILLIAMS:

Actually, what comes into the one-dimensional conservation equation is the dependence of flow velocity on distance normal to the flame front. That velocity gradient in the flow direction is the key thing. Although you cannot have such a velocity gradient without generating the flame area, the area generation is not directly in the analysis.

O'BRIEN:

What happens to the mean velocity in all of this?

WILLIAMS:

To lowest order, it always comes out that when you do an analysis, say, for large activation energy, calculating the structure of either a premixed or a diffusion flame, in lowest order convection is negligible in the reaction zone. To lowest order, the temperature effect is so strong that the concentration of reactants is very small in the reaction zone. So convection disappears in lowest order, and there is a diffusion-reaction balance. The mean velocity is present, but its influence is negligible.

RHODES: (ARO, Inc., Arnold Air Force Station, Tennessee)

What about problems where you have very high speed flames, such as in after-burning rocket plumes, where you may have a characteristic chemical time which may be of the same order as the mixing time so that it is hard to conceive of a laminar flame, even a big, fat thick one, in this type of flow. Will you comment on this?

WILLIAMS:

Maybe it is a highly strained little thin one in that type of
flow! I would make the following observation. The situation
depends on pressure and temperature and on whether the flow is
premixed. In a plume there could still be some reaction as burning gases are expelled. But, if you are talking about, say, a
fuel-rich jet reacting with the surrounding air, then the main
question is whether the pressure and temperature levels in the two
streams are high enough to produce a Damköhler number in the reaction zone which is higher than the extinction value. If the jet
is hot, then--unless we are at very high altitude, where the pressure is low, so that you get low reaction rates--there ought to be
this ensemble of laminar diffusion flames. If the jet has expanded
greatly and cooled, then the diffusion flamelets may be extinguished, and a flame of a more premixed character may develop.

RHODES:

Let us take a hot, fuel-rich plume at a high altitude, particularly one with ignition delay, where you have an ignition time
which might be long compared with the flow time. It just is difficult for me to conceive of this thing developing into a series of
laminar reaction fronts. In many cases, you may have a standoff
distance of a few feet. This could be 5 to 10 shear layer thicknesses for an ignition delay distance. Here you have a very
complicated mixture of partially premixed and diffusion flames.

WILLIAMS:

That is right, but you seem to be taking extreme conditions.
With ignition occurring a few feet away, conditions must be on the
verge of totally frozen flow. At least, locally, I suppose there
would be premixed flames, since you did have mixing prior to the
onset of the reaction. However, these could burn away, leaving
predominantly diffusion-type flamelets in the downstream region.

O'BRIEN:

Probably those local regions, once they do start to burn, they
do burn out very quickly, if conditions are such that, once the
thing is established, the reaction rate is high. And I will still
say the effects are that you have obtained local premixed and
farther downstream you should go back to this ensemble with laminar
diffusion flames.

WILLIAMS:

What do you say?

RHODES:

It is still hard to conceive of laminar flames in that type of flow.

MORKOVIN: (Illinois Institute of Technology, Chicago, Illinois)

Now are your scales such that we can disregard the interaction between members of the ensemble?

WILLIAMS:

Well, yes and no. Locally, so far as the structure of the reaction zone is concerned, I think that interaction is negligible. It is unlikely that the width of reaction zone itself will be as large as the Kolmogorov scale. These are very narrow zones.

MORKOVIN:

But are there not larger diffusion regions around the reaction zones which might interfere with each other?

WILLIAMS:

That is right. And, in that sense, they surely interact. The surface is a highly convoluted surface, and the rate of consumption of one part of the blob may be affected by the fact that some other part of the blob is some distance away from it. The equation that I was looking at was purely a local equation. If you want actually to go ahead and describe the structure of the turbulence, then interactions among members of the ensemble may become important.

MORKOVIN:

Suppose you lead nice oncoming turbulence into a stagnation region, quite a common occurrence, burning and heating at a stagnation point. Now can you find a sufficiently sharp distinction between the cases where you can decide whether or not to take the interaction into account?

WILLIAMS:

I do not see how to answer that question, even experimentally. If the turbulence has a very low intensity and only wrinkles the laminar flame insignificantly, then interactions are negligible. At higher intensities, distortions are greater and it seems likely that interactions would become important. I have not addressed interactions in my presentation because I don't know how.

LIBBY: University of California, La Jolla, California)

Listening to this discussion raises the question of whether
or not you believe, at least in the case of fast chemistry--which
O'Brien is talking about--where one can use experimental data on
passive scalars to learn at least something about chemistry
behavior. If one does, then, of course, we should be able to see
to what extent a surface of constant concentration or constant
temperature is multiply connected. To do so is not an easy experiment because there is always some doubt as to whether or not the
surface is wrinkled twice between the probes. But we are trying
with multiple probes to see to what extent there is a contour of
concentration--one percent let us say--outside of a point where
there is already one percent inside. My guess is that that will
occur very rarely.

TURBULENT MIXING IN SYSTEMS WITH SIMPLE REACTIONS

Edward E. O'Brien

Department of Mechanics, State University of New York

Stony Brook, New York 11790

1. INTRODUCTION

The additional complexities that chemically reacting species bring to turbulently mixed scalar fields may be of several kinds. If the reaction is strongly exothermic, the thermal field which is generated may distort the turbulence and modify the mixing process by introducing compressibility effects and buoyancy forces. This general problem has not been thoroughly investigated and in this paper we will not consider it further except to refer to a thesis[1] which summarizes the literature, formulates the dynamical interactions and deduces some results in the final period of decay. We also neglect any significant role the reaction plays in altering the local mass density or the molecular transport coefficients of the flow. Here we will treat only dynamically passive reactive species.

There is in many important reacting flow systems significant uncertainty about the chemical kinetic schemes which might properly describe the phenomena at a molecular level. We are in no position to deal authoritatively with these complex schemes and, where we must confront a specific chemistry in these remarks, we have adopted the usage "simple" reaction as a synonym for a one-step, irreversible reaction of the kind

$$A + nB \rightarrow mP + \text{heat (moderate)}, \qquad (1)$$

where n and m are stoichiometric coefficients. Further, Fick's law is assumed to describe the relation between the molecular diffusion velocity and the instantaneous concentration field of each specie. While this reaction has received a great deal of attention recently,[2-8] the extent to which the methods and results obtained from its investigation can be generalized to situations with more complex chemical kinetics remains virtually unstudied, although some general remarks about stochastic mass production rates will be made in Section 2.

Clearly, from (1), we would like to be able to solve simultaneously the Navier-Stokes equations (N-S), the overall mass conservation equation, the three species mass balance equations and the equation of thermal energy generation and transport. The equation of state and appropriate initial and boundary conditions should also be prescribed. These are stochastically coupled equations. If dynamic passivity of the reaction is assumed the (N-S) equations are self-determined, and the fluid can be considered incompressible.

Recognizing that neither the stochastic incompressible N-S equations nor the equations of simple turbulent mixing of a non-reacting species have been satisfactorily solved and that the reaction itself introduces new non-linearities it becomes important to establish reasonable research goals as replacements for the complete solution referred to in the previous paragraph. Our attitudes have been: first to try to understand the novel features of the statistical mechanics of the reaction itself in a turbulence which has a prescribed stochastic description either in terms of moments or probability density functions; second to consider as progress any reduction of the reacting problem, in whole or in part, to a problem of simple mixing. This latter is based mostly on the notion that experimental investigations of simple mixing are usually simpler than those involving reacting species.[6,9] As we shall see later we are not so sure that this is true of theoretical work where the most intransigent terms continue to be turbulent convection and molecular diffusion.

2. MOMENT FORMULATION

2.1. Introduction

A typical species mass conservation equation associated with a "simple" reaction may be written[10]

$$\frac{\partial \Gamma_A}{\partial t} + \underset{\sim}{u} \cdot \nabla \Gamma_A = D_A \nabla^2 \Gamma_A - C(T) \Gamma_A \Gamma_B , \tag{2}$$

where Γ_A is the mass fraction of species A,
Γ_B is the mass fraction of species B divided by the stochiometric mass ratio,

D_A is the molecular diffusivity of A,
u is the turbulent velocity,
and the rate constant $C(T)$ is taken to be of Arrhenius type, i.e.

$$C(T) = \frac{nA\bar{\rho}}{W_B} \exp\left(-\frac{T_a}{T}\right).$$

T_a is the activation temperature $T_a = \frac{E_a}{R^0}$.
W_B is the molecular mass of B.
A is the preexponential factor.
$\bar{\rho}$ is the mean density.

There are two special cases which have received some attention in the turbulence literature

(a) If Γ_B is everywhere available in considerable excess and the flow is approximately isothermal the mass production term $C(T)\Gamma_A\Gamma_B$ may be effectively written as $C\Gamma_A$ where C is a constant reaction rate. The most interesting result for this linear, nonspectrally selective reaction concerns its intensity spectrum in the inertial subrange, which may be quite strongly distorted by the reaction.[11]

(b) If species A and B share the same identity or if the reaction is everywhere perfectly mixed then the mass production term may be written as $C(T)\Gamma_A^2$, where most research on this equation has, in fact, taken $C(T)$ as a constant. Two noteworthy results in homogeneous turbulence are the apparent approximate statistical independence of the reaction and the mixing[12] for decaying reactions and the existence of an inertial subrange behavior which differs significantly from that associated with a linear reaction.[13]

2.2. Some Properties of the Mass Production Term in Equation 2

A. When T is of order T_a somewhere in the field of reactants the coupling between temperature and species concentration promises to be very complicated and ad hoc moment formulations may be unreliable for the discussion of such phenomena. As we shall see in section 3 when C is everywhere large enough the natural formulation of the problem is in terms of probability density functions.

B. Γ_A is a non-negative quantity which by Liapunov's inequality for absolute moments must satisfy a hierarchy of moment inequalities of the kind, namely

$$\overline{\Gamma_A^2} \geq \overline{\Gamma_A}^2$$

$$\overline{\Gamma_A^3} \geq \overline{\Gamma_A^2}^2 \ \overline{\Gamma_A}^{-1}$$

etc.

A generalization to n positive random variables has recently been reported.[16,17]

C. There are no spatial derivatives in the mass production term. This suggests that turbulence, with its penchant for transferring energy into progressively finer structures and for distorting larger scalar blobs into more and more elongated and wrinkled shapes, has little <u>direct</u> influence on the progress of the reaction but mostly acts <u>indirectly</u> in conjunction with molecular diffusion to enhance the reaction by increasing the apparent diffusivity of each species. In the limit of no molecular diffusion in statistically homogeneous scalar fields one has the strong result that all single point statistical properties are entirely independent of the turbulence.[18,19] It also suggests that spectral transport of scalar intensity due to the reaction term is not strongly coupled to the spectral flux of turbulent energy,[11,13] although it clearly must depend on the level of scalar intensity in each wave number band.

D. Experimentally it is known that diffusion control for unpremixed reactants must apply when C is very large.[20]

Thus, from (2), $\overline{\Gamma_A \Gamma_B} = O(1/C)$ as $C \to \infty$

i.e. $\overline{\gamma_A \gamma_B} = -\overline{\Gamma_A}\,\overline{\Gamma_B} + O(1/C)$ as $C \to \infty$

In fact one can show theoretically under certain diffusion-controlled circumstances

$$\frac{\overline{\gamma_A^2}}{\overline{\Gamma_A}^2} \geq (\pi - 1)$$

In conjunction with property (B) this implies, for example, $\overline{\gamma^3} > 0$ and therefore the probability density function of concentration must be far from Gaussian, a result now strongly supported by experimental observations.[6,9]

2.3. A Strategy for Solution of the Moment Equations

For isothermal homogeneous flow and concentration fields a technique of solution for the reactant species statistics has been presented previously.[17,22] This has now been extended to computations of the product statistics and to formulation of closed equations for reversible reactions.[16] A typical closure is the relation $\overline{Y_A Y_B Y_P} = [\overline{\Gamma_A}^{-1}\overline{Y_A Y_B} + \frac{1}{2}(\overline{\Gamma_A \Gamma_B \Gamma_C})^{-1}\overline{Y_B Y_P}(\overline{\Gamma_A \Gamma_B} + \overline{Y_A Y_B}) + K\overline{\Gamma_B}]\overline{Y_A Y_P} +$ [A↔B], where K is not an arbitrary constant but is determined by the initial conditions. [A↔B] means repeat the previous term interchanging subscripts A and B. Here we will give only a brief outline of the method, which has not yet been exploited in inhomogeneous flows.

In equation (2) three effects are represented: turbulent convection, molecular diffusion and chemical reactions. One attempts to model with moment closures all three of these phenomena, taken pairwise, by satisfying all known invariances and asymptotic conditions associated with each of the pairs. The first pair are turbulence and molecular diffusion, a well-studied partnership for which numerous approximation schemes are available, including some of the most sophisticated from modern turbulence theory.[23] The second pair involves turbulence and chemical reaction without diffusion. The rather meagre constraints that dominate this pair's behavior arise from the general nature of the convective term in turbulent transport and from the previously mentioned feature that turbulent convection in a homogeneous system plays no role in determining single point scalar moment evolution. Since one can generally obtain exact solutions for stochastically distributed reactants[18] this later feature is crucial in establishing necessary conditions for the accuracy of proposed closure approximations. The third pair involves molecular diffusion and reaction, which are amenable to numerical calculation and for which there also exists theoretical results in the diffusion-controlled limit[21] (very rapid reaction in final period turbulence). Both of these features can be used to guide the reactive field closure approximations already mentioned above.

The hypothesis that accurate pairwise modeling of the phenomena will reproduce adequately the behavior of the entire system will remain just that until experimental checking in a homogeneous system is possible. Perhaps complete numerical simulation at moderate Reynolds numbers is closer to being a reality.[24] By our method of construction all known asymptotic conditions are already incorporated into the problem in such a way as to be produced under the appropriate asymptotic circumstances, so that they cannot be used to verify its accuracy. For example all of the asymptotic results on two species diffusion-controlled behavior will be automatically produced[25] (approximately) at very rapid reaction rates if the numerical computation can be achieved.

In the case of simple turbulent mixing it has long been established that one cannot treat convection and diffusion as

statistically independent events.[26]   The reasons are well known
and intimately related to the spectral selectivity of the diffusion
term.   It is our hunch that the nature of the reactive terms, discussed in 2.2, may be such that they can be partially decoupled
from complete turbulent mixing in the manner described above.

It is not at all clear that a similar approach can be useful
when the flow is nonisothermal and the reaction strongly temperature dependent.   In fact, in this situation it has seemed to us
more promising at present to abandon a moment formulation in favor
of a direct confrontation with probability density functions.

3.   PROBABILITY DENSITY FUNCTION FORMULATION

3.1.   Introduction

It has become apparent in recent years that the more productive
approach to diffusion-controlled turbulent mixing of reactants is in
terms of probability density functions rather than moments.[2,3,5,27]
This limit, when it exists everywhere in the flow field, when the
reactant species are dynamically passive and, when the coefficients
of scalar diffusion are effectively equal,[3] produces a profound
simplification of the equations referred to in the introduction.

Without any assumptions on reaction rate the mass conservation
equations associated with equation (1) become

$$L\chi = 0 \tag{3a}$$

$$L\psi = 0 \tag{3b}$$

$$L\Gamma_A = -C(T)\Gamma_A(\Gamma_A - \chi) \tag{3c}$$

where $L = \left\{\dfrac{\partial}{\partial t} + \underset{\sim}{u}\cdot\nabla - D\nabla^2\right\}$ the simple mixing operator,

$$\chi = \Gamma_A - \Gamma_B$$

and $\psi = \Gamma_A + \Gamma_P$.

With minor assumptions that appear reasonable for low speed flows[10]
it is possible to incorporate the thermal energy equation into the
above set in the form

$$L\phi = 0, \tag{3d}$$

where $\phi = \Gamma_A + T$, T being a dimensionless temperature, $\dfrac{C_p T_{real}}{Q}$.

3.2. The Very Rapid Reaction

As $C(T) \to \infty$, $\Gamma_A(\Gamma_A - \chi) = O(1/C)$;

hence either $\Gamma_A \to 0$ i.e. $\Gamma_B \to -\chi$

  or $\Gamma_A \to \chi$ i.e. $\Gamma_B \to 0$

$$(4)$$

Although this result is of no help in solving (3c) since its right hand side cannot be zero in the limit $C \to \infty$ (as one can easily see by taking the mean of (3c) in a homogeneous flow), the direct correspondences between Γ_A and positive χ and between Γ_B and negative χ permit a solution of the probability density function (p.d.f.) $P[\Gamma_B; \underset{\sim}{x}, t]$ or $P[\Gamma_B; \underset{\sim}{x}, t]$ when $P[\chi; \underset{\sim}{x}, t]$ is known. That is, equation (3c) can be bypassed. It also turns out that $\bar{\Gamma}_p(x, t)$ and $\bar{T}(x, t)$ can be determined when $P[\chi; \underset{\sim}{x}, t]$ is known, but to obtain $P[\bar{\Gamma}_p; \underset{\sim}{x}, t]$ (or $P[T; \underset{\sim}{x}, t]$) one needs the full joint p.d.f. $P[\chi, \psi; \underset{\sim}{x}, t]$ (or $P[\chi, \phi; \underset{\sim}{x}, t]$).[28]

Obtaining $P[\chi; \underset{\sim}{x}, t]$ is an objective of simple mixing research and there exists both theoretical approximations[1,29,38] and, especially valuable for real shear flows at the moment, experimental techniques for measuring the p.d.f. of some nonreacting scalar fields.[30] The details of the computation outlined above have been carried out for one transverse plane of a three-dimensional jet of heated air[30] to produce prediction of low order moments of the reactants when a jet of species B is injected into a field of species A with which it undergoes a very rapid reaction. The mean product and thermal fields are also predicted.

Other laboratories have reported measurements of nonreacting scalar p.d.f.'s in free and bounded turbulent shear flows.[31-34] Some of them have been converted by the above method into chemical species distributions in the same flow when the reaction is very rapid. Libby has mentioned schemes for the measurement of $P[\chi, \psi; \underset{\sim}{x}, t]$ using temperature and helium as jointly distributed scalars.[27] This kind of data is necessary for obtaining complete information about the product species.

From a physical point of view the mathematical model of diffusion-controlled reactions invokes the notion of infinitely thin reaction zones separating regions of the fluid in which either A is present or B is present but not both. The limit $C \to \infty$ is the requirement that the time scale of reaction be very much smaller than any time scale of the turbulent mixing field. A reasonable amount of experimental information exists to support this view but there is a fundamental difficulty in probing the interior of the reaction zones whose dimensions are at least as small as the smallest probes available.[17] Since chemical effects are located

in these zones, experimental pursuit of their physics and chemistry
is vital to further understanding of the phenomena. At its simplest level it appears to be a difficult problem in the statistical
geometry of random surfaces[35] with seemingly little connection to
the behavior of intermittent regions of high energy dissipation
which are of current interest in turbulence.[36] It appears to us
unlikely that measurement of product species or temperature can
adequately define these zones since, once produced, such scalars
undergo regular turbulent mixing--the reaction zones, per se, do
not.[17] A direct measurement of pH seems more likely to be useful.

3.3. Reactions of Moderate Rate

When the time scale of reaction is not everywhere very much
smaller than any other characteristic time in the problem, equation (3c) cannot be bypassed by the device described in section
3.1. Nevertheless (3c) is the only equation in the system which
retains the mass production term and therefore the reaction rate.
The other three quantities χ, ψ and ϕ remain as simple mixing
statistically coupled through sharing the same random velocity
field as a coefficient in the operator L and possibly through
correlated boundary or initial conditions.

Closed functional formulations of the problem have been presented[37] using both Hopf's[38] and Lewis-Kraichnan's[39] formalisms.
The equations are complicated and so far explicit solutions have
been obtained for only two homogeneous situations: binary mixing
in final period turbulence[1,39] and autoignition of a turbulent
mixture.[40] In the latter case one focuses on the probability
density function of temperature in a regime where the Arrhenius
reaction rate is strongly temperature dependent in the limit of
large activation energies. As we mentioned in Section 2 such a
situation is extremely awkward in a moment formulation. In the
p.d.f. formulation, however, the difficulties arising from the
transcendental nonlinearities of the reactive term are transferred
into a partial differential equation for the p.d.f. for which the
coefficients now become strongly dependent on an independent variable. The problem then becomes one of numerical integration[1]
rather than statistics.

In fact, by deriving an equation for the single point joint
p.d.f. $P[\Gamma_A, \chi, \psi, \phi; \underset{\sim}{x}, t]$ where the variables satisfy (3a, b, c and d)
we can show that the above simplification always occurs for the
mass production term thanks, again, to the absence of spatial
operators in that term. We find the equation for the single point
joint p.d.f. becomes the following.

$$\frac{\partial P[\Gamma_A, X, \psi, \phi; \underset{\sim}{x}, t]}{\partial t} + \frac{\partial}{\partial \Gamma_A} C(T) (\Gamma_A^2 - \Gamma_A X) P[\Gamma_A, X, \psi, \phi; \underset{\sim}{x}, t]$$

$$= -D \lim_{\underset{\sim}{x}' \to \underset{\sim}{x}} \nabla_{\underset{\sim}{x}'}^2 \frac{\partial}{\partial \Gamma_A} \int \Gamma_A' P[\Gamma_A, \Gamma_A', X, \psi, \phi; \underset{\sim}{x}, \underset{\sim}{x}', t] d\Gamma_A'$$

$$- D \lim_{\underset{\sim}{x}' \to \underset{\sim}{x}} \nabla_{\underset{\sim}{x}'}^2 \frac{\partial}{\partial X} \int X' P[\Gamma_A, X, X', \psi, \phi; \underset{\sim}{x}, \underset{\sim}{x}', t] dX'$$

$$- D \lim_{\underset{\sim}{x}' \to \underset{\sim}{x}} \nabla_{\underset{\sim}{x}'}^2 \frac{\partial}{\partial \psi} \int \psi' P[\Gamma_A, X, \psi, \psi', \phi; \underset{\sim}{x}, \underset{\sim}{x}', t] d\psi'$$

$$- D \lim_{\underset{\sim}{x}' \to \underset{\sim}{x}} \nabla_{\underset{\sim}{x}'}^2 \frac{\partial}{\partial \phi} \int \phi' P[\Gamma_A, X, \psi, \phi, \phi', \underset{\sim}{x}, \underset{\sim}{x}', t] d\phi'$$

$$- \int d\underset{\sim}{u} \; \underset{\sim}{u} \cdot \nabla_{\underset{\sim}{x}} P[\Gamma_A, X, \psi, \phi, \underset{\sim}{u}; \underset{\sim}{x}, t] \tag{5}$$

While equation (5) is unclosed in all four diffusion terms and the turbulent convection term and therefore can only be solved by approximation techniques, the point to be made is that the mass production term is closed and adds no complication in this formalism. We note as well that the difficulties of solving simple mixing by means of a p.d.f. formulation give rise to exactly the same closure problems as appear here. From the theoretical point of view mixing with chemical reaction is no more difficult than mixing without it if one eschews moments in favor of p.d.f.'s.

Existing p.d.f. closure techniques come mostly from studies of turbulence dynamics[41],[42] and statistical mechanics[43] and in general they appear to be less developed than their moment closure counterparts. At the simplest level the diffusion terms can be approximated with at least proper qualitative behavior[40] by expressing two point joint p.d.f.'s in terms of conditionally Gaussian joint p.d.f.'s. For homogeneous fields this can be shown to be roughly equivalent to approximating $\overline{\gamma \nabla^2 \gamma}$ by $\overline{\gamma^2}/\lambda^2$ where the behavior of λ is postulated. The convective nonclosure, represented by the appearance of $\underset{\sim}{u}$ in the p.d.f. in the last term of equation (5), is generally more difficult although it disappears altogether in statistically homogeneous systems. In shear flows Lundgren's approximation[29] for the pressure term in the single point p.d.f. equation for velocity in a shear flow may be relevant. He introduced the notion of a relaxation time specified in terms of the local structure of the shear flow in analogy to the Krook model in kinetic theory. Cluster expansions which have been employed in analyzing turbulence dynamics[41],[42] may also be useful in the treatment of reacting scalar fields.

3.4. Conclusions

In the pursuit of reasonable ways to approach problems of reacting turbulent flows one soon discovers a paucity of experimental data against which to test assumptions and intuitive physical images of the phenomena. For example, the idea has been expressed[10] that temperature and species concentration are probably only very weakly correlated in homogeneous, exothermic, two species very rapid reactions in turbulence and, if that is true, considerable simplification of some of the problems posed here becomes possible. Similarly, the nature of buoyancy forces in a rapidly reacting flow is mostly determined by the geometry of the reaction zones, about which we know practically nothing. Finally the theoretical apparatus necessary to bring reactive flow studies into a reasonable framework seems to be already in existence. One needs to apply approximations in simple well-defined contexts in order to build up experience with their strengths and weaknesses and to modify them accordingly. In particular any progress made in the study of probability density functionals in either turbulence dynamics or simple mixing can be immediately absorbed into the description of reactive flows. For long range progress in understanding the subject this approach appears to hold considerable promise.

ACKNOWLEDGMENTS

The work reported here has been supported by the National Science Foundation under Grant K040738.

REFERENCES

1. Dopazo, C. Ph.D. Thesis, State University of New York at Stony Brook, July 1973.

2. Toor, H. L., A.I.Ch.E.J. 8, 70-78 (1962).

3. O'Brien, E. E., Phys. Fluids 14, 1326 (1971).

4. Libby, P. A., Comb. Sci. and Tech. 6, 23-28 (1972).

5. Bush, W. B. and F. E. Fendell, 4th Int. Coll. on Gasdynamics of Explosions and Reactive Systems, La Jolla, July 1973.

6. Gibson, C. H., Lyon, R. R. and Hirschsohn, I., A.I.A.A.J. 8, 1859-1863 (1970).

7. Chung, P. M., Phys. Fluids 15, 1735-1746 (1972).

8. Spalding, D. B., 13th Symp. on Combustion, Combustion Institute, Pittsburgh, p. 649 (1971).

9. Gibson, C. H. and Libby, P. A., Comb. Sci. and Tech. 6, (1972).

10. Dopazo, C. and O'Brien, E. E., Phys. Fluids 16, 2075 (1973).

11. Corrsin, S., J. Fluid Mech. 11, 407 (1961).

12. O'Brien, E., Phys. Fluids 12, 1999 (1969).

13. Corrsin, S., Phys. Fluids 7, 1156-1159 (1964).

14. Lumley, J. L., J. Atm. Sci. 21, No. 1 (1964).

15. Phillips, O. M., Int. Coll. on Fine-Scale Structure of Atmosphere, Moscow (1965).

16. Lin, C. H., Ph.D. Thesis, State University of New York at Stony Brook (1974).

17. O'Brien, E. E., Advances in Geophysics 18, Academic Press (1974).

18. O'Brien, E. E., Phys. Fluids 11 (1968).

19. Hill, J. C., Phys. Fluids 13, 1394 (1970).

20. Vassilatos, G. and Toor, H. L., A.I.Ch.E.J. 11, 666-673 (1965).

21. O'Brien, E. E., Phys. Fluids 14, 1804 (1971).

22. Lin, C. H. and O'Brien, E. E., Astronautica Acta 17, 771-781 (1972).

23. Herring, J. R. and Kraichnan, R. H., Lecture Notes in Physics, Vol. 12, Statistical Models and Turbulence, Berlin, Springer (1972).

24. Orszag, S. A. and Paterson, G. S., Lecture Notes in Physics, Vol. 12, Statistical Models and Turbulence, Berlin, Springer (1972).

25. O'Brien, E. E. and Lin, C. H., Phys. Fluids 15, 931 (1972).

26. Saffman, P. G., J. Fluid Mech. 8, 273 (1960).

27. Libby, P. A., AGARD Meeting, Liege (1974).

28. Lin, C. H. and O'Brien, E. E., J. Fluid Mech., to be published (1974).

29. Lundgren, Phys. Fluids 12, 485 (1969).

30. Tutu, N. K. and Chevray, R., Bull. Am. Phys. Soc., Series 11, 18, 11 (1973).

31. Stanford, R. and Libby, P. A., Phys. Fluids, to be published (1974).

32. La Rue, J. and Libby, P. A., Phys. Fluids, to be published (1974).

33. Feidler, H., Advances in Geophysics 18, Academic Press (1974).

34. Zaric, Z., C. R. Acad. Sc. Paris, Serie A, 275, 459 (1972).

35. Corrsin, S., Lecture Notes in Physics, Vol. 12, Statistical Models and Turbulence, Berlin, Springer (1972).

36. Kuo, A., Y-S. and Corrsin, S., J. Fluid Mech. 50, 285 (1971).

37. Dopazo, C. and O'Brien, E. E., submitted Phys. Fluids (1974).

38. Hopf, E., J. Ratl. Mech. Anal. 1, 87 (1952).

39. Lewis, R. M. and Kraichnan, R. H., Comm. Pure Appl. Mathm. 15, 399 (1962).

40. Dopazo, C. and O'Brien, E. E., 4th Int. Coll. on Gasdynamics of Explosions and Reactive Systems, La Jolla, July 1973.

41. Lundgren, T. S., Lecture Notes in Physics, Vol. 12, Statistical Models and Turbulence, Berlin, Springer (1972).

42. Fox, R. L., Phys. Fluids 16, 957 (1973).

43. Stell, G. Cluster Expansions in The Equilibrium Theory of Classical Fluids, Eds. H. L. Frisch and J. L. Lebowitz, Benjamin, N. Y. (1964).

DISCUSSION

HILL: (Iowa State University)

In the last slide you showed an equation for the marginal distribution of the variable T that was to be approximated as conditionally Gaussian. Just for my clarification, what is T?

O'BRIEN:

The T should have been better written as $T = \phi - \Gamma_A$ as in eqn. (3d). The distributions of all the scalar fields might be approximated as conditionally Gaussian.

HILL:

Then, essentially, the two point conditional distribution of temperature is considered to be Gaussian?

O'BRIEN:

Yes. If I have only two reactants and a product, as an example, one would have the expressions developed here (eqn. (5)). If you did the full problem you would have to close the diffusion terms for all scalars, temperatures and concentrations, and you would have to make some assumption like conditionally Gaussian, on every one of them.

HILL:

I was at first confused by your use of the term conditional distribution, since in my earlier work (Ref. 19) the velocity field was the given variable in the conditional distribution of concentration. Your particular application to the conditional expectation of the scalar is quite novel. Aside from it being a mathematical device, what sort of physical implications does the use of a conditionally Gaussian distribution make?

O'BRIEN:

Well, I am not sure. It has all the properties one needs, qualitatively. We have been studying the probability density functions for some time now and one's intuitive feeling about them is often wrong. For example, I am used to thinking of them as something that diffuses, but they really do not do that. As I said, there are properties of the conditionally Gaussian approximation to the diffusion term which look good. Neighboring temperatures go up together. If the temperature is higher than the mean at one point, the expected temperature at a neighboring point is

also likely to be higher than the mean. And it has the right
limit as the separation distance between the two points increases
greatly. That is, it reduces to the expected temperature for a
single point. There are other properties. If the separation
vector goes to zero the conditional expected value becomes the
regular expected value as it should.

HILL:

Do you think it would be possible to express the closure in
terms of random variables which themselves satisfy the nonnegativ-
ity conditions? For example, if a log-normal distribution fell
out of your approximation, the Liapunov inequalities would
automatically be satisfied.

O'BRIEN:

You mean to ask whether this approximation satisfies the
constraint that all concentrations are never negative? We have
done some specific computations--which we talked[1] about at La Jolla
a year ago--on auto-ignition in which you have a rapidly increasing temperature competing with diffusion of that temperature, to
determine whether you get ignition locally. You can solve separately the temperature probability distribution equation without diffusion and the same probability density equation without reaction.
You get exact solutions to them which of course behave correctly.
I do not know whether Dopazo showed that if you put the two
together you also can never introduce a negative concentration,
did you?

DOPAZO: (SUNY, Department of Mechanics, Stony Brook, NY)

No, we did not. I want to make two remarks on the positiveness of the p.d.f. and on the positiveness of the concentration
fields. With respect to the second one we have not done any work.
I think a simple study of the characteristics in the concentration-
time plane (or space) should easily show that if the initial range
of probable concentrations is in the positive concentration region
it can never move to the negative concentration region. In connection with the first remark, we only considered temperature
fields. The negative p.d.f. which developed there in the tentative
stages of the computations were due to instabilities of the numerical schemes. But, except for that, all the p.d.f.'s obtained were
positive and behaved nicely except very close to the thermal
runaway time.

---

[1] Dopazo, C. and O'Brien, E. E. Fourth Int. Coll. on Gas Dynamics
     of Explosions and Reactive Systems, La Jolla, July 1973.

O'BRIEN:

It is very messy, it has to be done by numerical approximations.

There is one piece of information I wanted to give, but did
not. It is that if you use this diffusion equation, the p.d.f.
approximation for the diffusion that we talked about, the conditionally Gaussian one, if that is the only significant phenomenon
occurring, it turns out that you can prove exactly that the mth-
moment of the concentration decays in a way which is invariant to
the distribution shape. It depends only on the initial m^{th} moment
of the distribution. But whether you have a half Gaussian or
whether you have a truncated Gaussian--no matter what it is--the
m^{th} moment depends only on the initial m^{th} moment, not on the rest
of the shape. It therefore seems hopeful to me when you apply the
mixing approximations which work with almost Gaussian distributions
and talk about low order moment behavior, it probably will not be
seriously compromised by having to deal with the unusual sort of
distribution functions you get for reactive species undergoing
rapid reactions.

WILLIAMS: (UCSD, La Jolla)

You said something about decay dominating over spectral
transfer in an isothermal system. I have tried to describe a
pre-mixed turbulent flame by making an expansion for low intensity,
and the theory breaks down at a much lower intensity of turbulence
than I would have thought. The reason it breaks down is because
of onset of spectral transfers due to the chemistry which come in
even before spectral transfers due to the fluid mechanics in this
particular problem. You might expect that, I suppose, because of
the very strong temperature dependence of the reaction rate. My
comment, therefore, is that in not all flows will decay dominate
spectral transfer.

O'BRIEN:

You know that the intensity is just another moment and that
is why we have left the moments to use the p.d.f.'s when there are
strong thermal terms and I would make no such statement about the
spectra when you have significant combustion. By spectral transfer,
I simply mean the analogous thing to the spectral transfer due to
the turbulence field, that is the amount that goes out from one
part of the spectra and into another as distinct from the amount
that just goes out. My claim is that for isothermal reactions of
arbitrary speed of even two species the amount that just gets
dumped out is very much greater than the amount that gets transferred to other regions of the spectrum. This is true only of the

reactive nonlinearity. The terms with velocity in the moments are all just spectral transfer.

WOLFSON: (Air Force Office of Scientific Research, Arlington, VA)

I noticed that at the beginning of your discussion you simplified your equations so that you introduced, I assume, low Reynolds number approximation in these flows. However, the simplification that you introduced was for relatively slow moving systems. The question I have, is, have you continued your evaluation of analysis towards rarefied gas flows and also towards compressible flows.

O'BRIEN:

Yes, in a very limited sense. Dopazo wrote a thesis[2] a part of which has to do with extending the Kovasznay method based on the modes of interaction in compressible flows, the density mode, the vorticity mode, and the acoustic mode, and to that he added the concentration mode. When he did that, he got this absolutely massive looking system which can be, at least looks to be, penetrable in a direct way only in the limit of final period of decay of the turbulence. So, I would not claim to be able to say anything yet about the compressible flow situation. But, I do not think this analogy is going to work. Because, I do not think you can have measurements of simple scalar fields in compressible flows, and expect to get information out of that for reactive compressible flows, since the density distributions will simply be generated locally in a different way when there is a reaction.

LEE: (Wright-Patterson Air Force Base)

I have a question on your last equation. It appears that the assumption you made is like Krook's approximation! The way that you have chosen the relaxation parameter, does it guarantee you a steady-state p.d.f.?

O'BRIEN:

You mean, without the chemistry? It is not really a Krook's approximation. Even if you have statistically homogeneous fields so that the velocity term does not occur explicitly the only steady state is the asymptotic one in which the p.d.f.'s are all delta functions and there are no fluctuations left. The diffusion terms guarantee that and our conditionally Gaussian approximation reproduces this behavior.

---

[2]   Dopazo, D. Ph.D. Thesis, State University of New York at Stony Brook, July 1973.

TURBULENT MIXING STUDIES IN A CHEMICAL REACTOR

Robert S. Brodkey

Ohio State University

Columbus, Ohio

A great deal of work has been done on mixing from a chemical engineering viewpoint, starting, I believe, in 1953 with Dankwerts' (1953) work; he first defined mixing in terms of intensity of segregation. This is a useful measure which would be unity for two unmixed streams and zero for totally mixed. Nothing was done with this until Wilhelm at Princeton started in the late '50's to do experimental work, and Toor (1962) started the theoretical work on second order chemical kinetics of isothermal dilute systems. There is now information on both liquids and gases that can be used for modeling checks; this is mainly the work of Toor and his student [see Mao and Toor (1971)].

One of the most challenging problems facing the reactor designer is that of predicting conversion as a function of position and time in a chemical reactor. The goal is i) to predict the turbulent field for a given geometry, ii) to predict the mixing from a knowledge of the turbulent field, and finally, iii) to predict the course of turbulently controlled or affected chemical reactions from knowledge of the mixing field. There are two extremes in the process: first, for very slow reactions, when chemical rate is the controlling factor, the turbulent mixing is unimportant. Second, for very rapid reactions, the chemical rate is unimportant, and the conversion is controlled by the turbulent mixing. The former can be approached using residence-time-distribution and kinetic information. The latter is controlled by the turbulent mixing; but the main question is the problem that lies between where both the kinetics and mixing must be considered.

To help answer this, data are needed. The kinetics have been measured by Toor and his co-workers, and we have measured the

mixing and turbulence in the same system. The experimental facil-
ity is a mixing module where each reactant is injected separately
through a multi-tube array. The design is such that one has a
gross uniformity, but totally unmixed. The concentration is uniform across the front, and there is no decay of the mean concentration axially. The only thing that decays is the concentration
fluctuations. The conversion in very rapid reactions is directly
related to this decay and thus, once the mixing is predicted, so
is the chemical conversion. The comparison is excellent.

A theoretical equation for predicting a second-order, irreversible reaction was developed by assuming the velocity and concentration fields are not correlated. Assuming mixing is not affected by
kinetics and neglecting the diffusion term, the final equation was
integrated by a fourth-order Runge-Kutta process. The directly
calculated results agreed extremely well with the experimental data
cited earlier.

One final word, selectivity of a chemical reaction is of key
importance in reactor design. No one to my knowledge has attempted
to incorporate this into the problem as yet.

REFERENCES

1. Dankwerts, P. V. (1953) Appl. Sci. Research A3 279.

2. Toor, H. L. (1962) A.I.Ch.E.J. 8 70.

3. Mao, K. W. and H. L. Toor (1971) Ind. Eng. Chem. Fund. 10
    192.

COMMENTS ON UNMIXEDNESS

James M. Eggers

NASA Langley Research Center

Hampton, Virginia

I would like to present some reacting flow data and the results of computations, both of which illustrate the unmixedness phenomenon. By unmixedness, I mean the condition where eddies of fuel and oxidant have engulfed each other but where the molecular mixing necessary for complete reaction has not taken place. The data and computations, taken from Reference 1, are relevant here in that they illustrate the application of the concentration fluctuation equation previously discussed in the paper by Dr. Spalding. The unmixedness concept is also incorporated in the second-order reacting flow modeling reported on here by Dr. Donaldson.

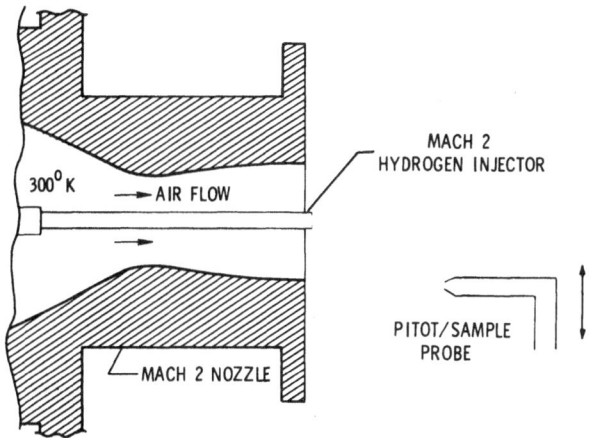

Figure 1. Coaxial Mixing and Combustion Apparatus

The flow geometry corresponding to the data to be discussed
is shown in Figure 1, and consisted of a coaxial Mach 2 hydrogen
jet surrounded by a parallel cold Mach 2 air stream. Combustion
of the hydrogen jet was piloted by reaction in a low velocity
annular oxygen flow between the hydrogen and air flows.

Gas sample measurements taken downstream in the region near
the flame revealed the presence of unreacted hydrogen and oxygen
in the same sample. As the probe had a blunt tip, simple cylindrical entrance and sample velocities were low, sample residence
times were such that the unreacted gases cannot be explained in
terms of chemical kinetic effects.

Calculations were made using the Spalding et al program,
CHARNAL, as documented in reference 2. A simple step function
variation with time for instantaneous concentration was employed
in which total hydrogen concentration has the mean value plus the
root-mean-square fluctuation for one half the time and the mean
value minus the root-mean-square fluctuation for the other half.
The fluctuating concentration model concept utilized is illustrated in Figure 2. When the mean concentration is near stochiometric, both unburned fuel and oxidizer can be present on the
average.

Application of this model produced a reasonable representation
of the fraction of reaction completed of the data, as shown in
Figure 3. However, the complexity of the initial conditions for
these data precluded any conclusions as to the general validity of
the modeling concepts used in the calculations. The ability to
calculate unmixedness is considered a necessary feature for reacting flow models, particularly for near field applications.

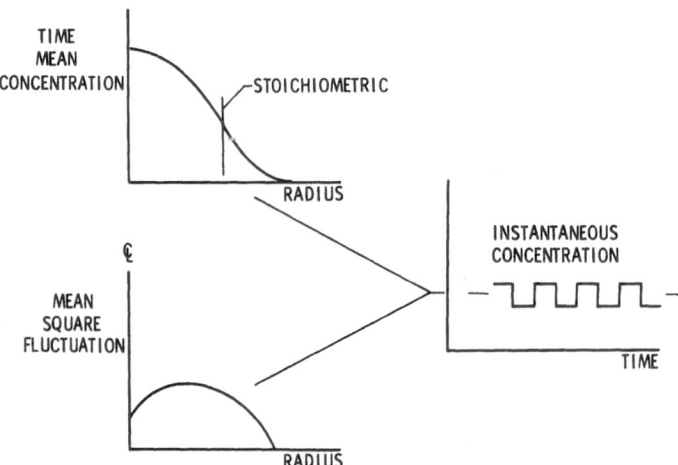

Figure 2. Fluctuating Concentration Model

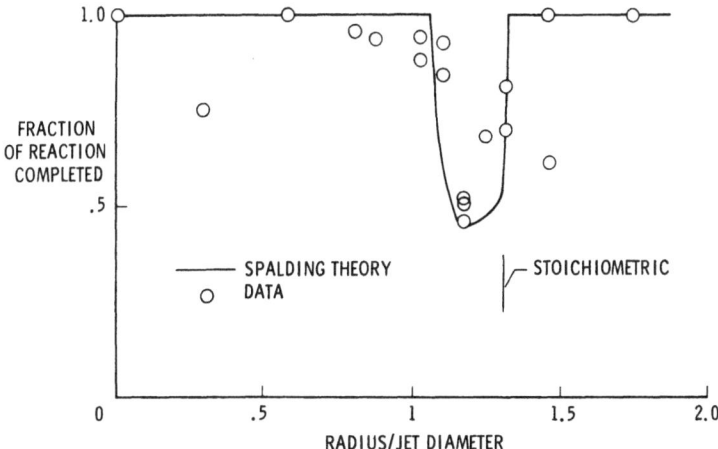

Figure 3.   Coaxial "Unmixedness" Data and Calculation

In conclusion, the validity of turbulence models can only be determined when predictions made with the model are compared to good experimental data. The biggest obstacle to the development and, or, refinement of these models is the absence of sufficient good quality experimental data, together with our inadequate understanding as to what are the necessary parameters required to define a particular turbulent flow. Particularly for reacting flow systems, our computational ability seems to have far out paced our acquisition of data to evaluate the computational techniques. Thus, the need for good quality reacting flow data is readily apparent.

REFERENCES

1.   Evans, John S., and Anderson, Griffin Y.: Supersonic Mixing and Combustion in Parallel Injection Flow Fields. Presented at the AGARD Propulsion and Energetics Panel Meeting (Liege, Belgium) April 1974.

2.   Spalding, D. B., Launder, B. E., Morse, A. P., and Maples, G.: Combustion of Hydrogen-Air Jets in Local Chemical Equilibrium. Report 73-1, Fluid Mechanics and Thermal Systems, Inc., Waverly, Alabama, November 1973. (Available as NASA CR-2407.)

VELOCITY AND PRESSURE CHARACTERIZATION OF COAXIAL JETS*

J. H. Morgenthaler

Bell Aerospace Division of Textron

Buffalo, New York 14240

Dr. L. Moon, Bell Aerospace Company, has made a series of mean velocity and stagnation pressure measurements from which local free-stream static pressure was determined for a coaxial air jet; I believe these data will interest you. We feel the results demonstrate that even for the simple coaxial injector geometry frequently used in H_2/O_2 rockets, conventional modeling techniques do not apply.

The geometry investigated is shown on Figure 1 on which select mean velocity profiles also are presented. Additional profiles, as well as the measurements of turbulence intensity and a detailed discussion are presented in Refs. 1 and 2. An actual Bell H_2/O_2 coaxial injector, previously used in rocket engine performance tests was used in this investigation to guarantee its relevance. The mean velocity profiles plotted in Figure 1 show a dramatic decrease in the velocity of the central jet immediately downstream of the injection station. That is, the velocity on the centerline decreases from 250 ft/sec at z = 0.010 in. to 100 ft/sec at z = 0.50 in., despite the fact that the annular flow has an initial velocity of over 700 ft/sec, which is 2.8 times that of the central jet.

This behavior is caused by significant axial and radial pressure gradients that exist in the recirculation region immediately downstream of the relatively thick (0.050 in.) splitter plate, as

---

* Work supported by NASA-Lewis Research Center, under Contract NAS3-16798, Dr. R. J. Priem, Project Manager.

MEAN VELOCITY U, FT/SEC

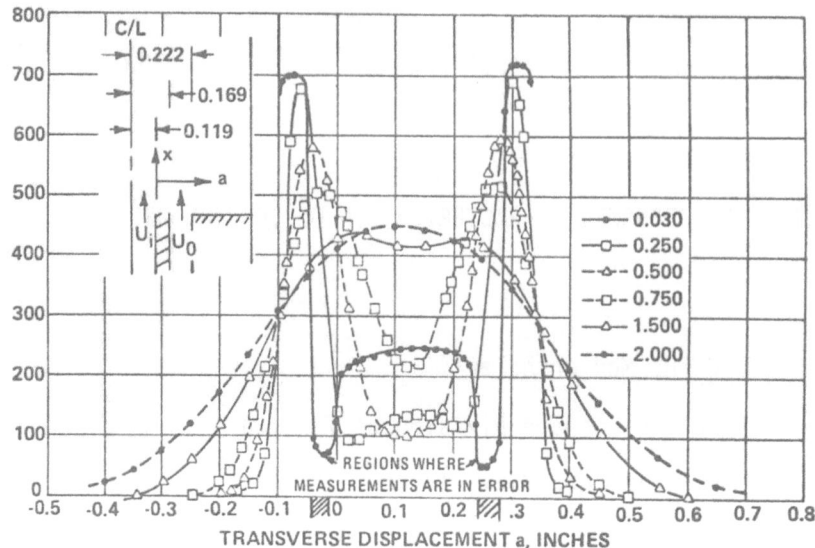

TRANSVERSE DISPLACEMENT a, INCHES

Figure 1.   Mean Velocity Profiles for Air-Air Injection into Ambient
            Atmosphere, obtained for a Single Element of Bell Aero-
            space 19 Element Coaxial Injector

well as viscous momentum transfer to this region from the central
jet. Static pressure profiles computed from detailed pitot probe
and hot wire anemometer measurements, Refs. 1 and 2, are presented
in Figures 2 and 3. We believe, the high-velocity outer jet creates
this reduction in static pressure through an aspiration effect,
which causes the central jet to spread, thereby reducing its velocity. Since the density of the entire flow field is practically
constant, the central jet velocity and area should be very nearly
inversely proportional to each other at axial stations near the
exit plane (before significant entrainment has occurred).

     Clearly, conventional modeling approaches, which use the
standard shear layer equations, and rely on an eddy viscosity or
turbulence kinetic energy model to predict the momentum transport,
cannot predict this simple practical flow because a recirculation
region exists downstream of the splitter plate across which transverse pressure gradients occur, and this effect is omitted in conventional analyses. We believe these results illustrate graphically
the importance of making careful measurements in flows of practical

STATIC PRESSURE
PSIA

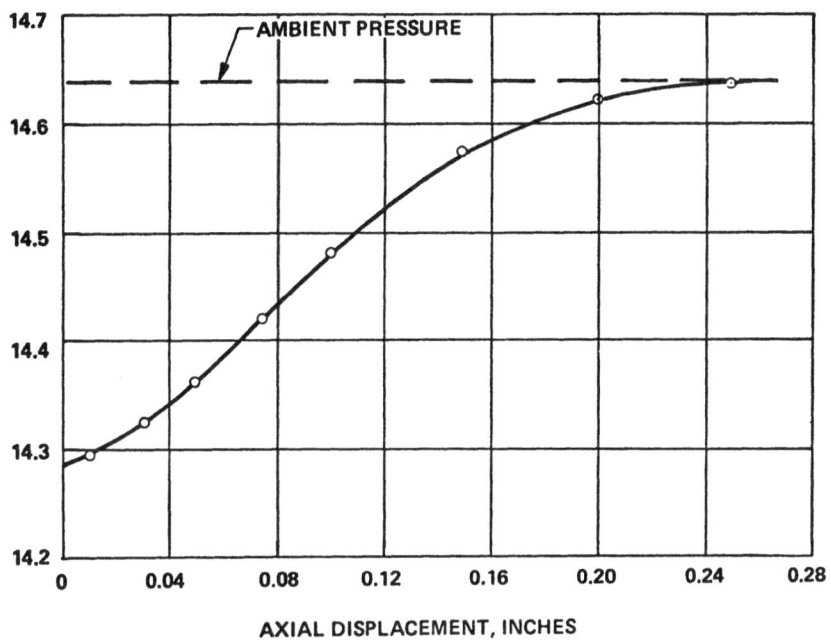

Figure 2. Centerline Static Pressure Versus Axial Displacement

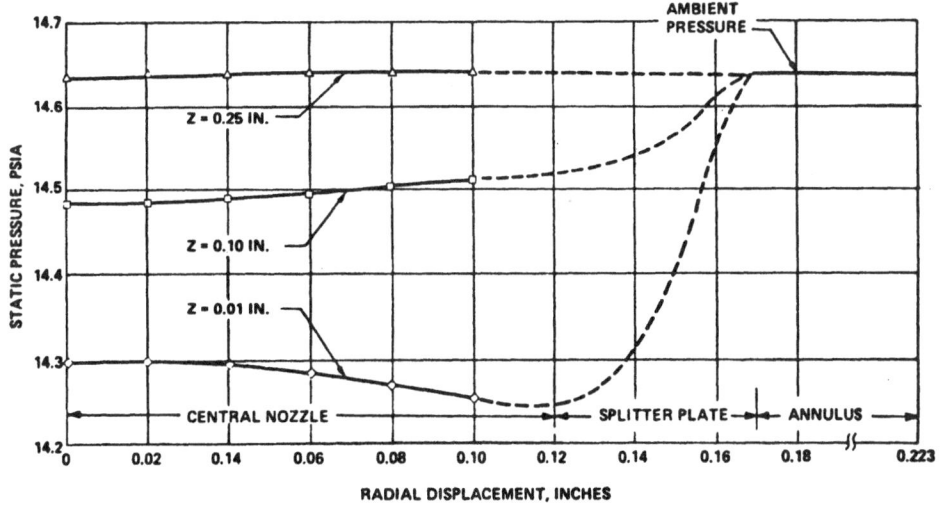

Figure 3

interest <u>prior</u> to attempting predictions utilizing state-of-the-art modeling techniques.

REFERENCES

1. Moon, L. F., "Experimental Investigation of Turbulent Velocity and Static Pressure in the Developing Region of Coaxial Jets," Eleventh JANNAF Combustion Meeting, Pasadena, California, September 1974.

2. Morgenthaler, J. H., Moon, L. F., and Stepien, W. R., "Developing a Gas Rocket Performance Prediction Technique," NAS3-16798, Final Report in Preparation.

A PROBABILITY DISTRIBUTION FUNCTION FOR TURBULENT FLOWS

Robert P. Rhodes

ARO, Inc.

Arnold Air Force Station, Tennessee 37389

NOMENCLATURE

C	Mass fraction of a species
P(x)	Probability density function for the variable x
T	Temperature
U	Velocity
σ	Standard deviation

Subscripts

1	Highest mean value in profile
2	Lowest mean value in profile
m	Centerline

Superscript

-	Average

The research reported in this paper was conducted at the Arnold Engineering Development Center, Air Force Systems Command, Arnold Air Force Station, Tennessee. Research results were obtained by ARO, Inc., contract operator at AEDC. Major financial support was provided by the Air Force Office of Scientific Research under Program Element 61102F, Project 9711; Dr. B. T. Wolfson was project monitor. Further reproduction is authorized to satisfy the needs of the U. S. Government.

There is current interest in being able to account for the presence of fluctuations when calculating velocity, temperature, and composition distributions in turbulent flow fields with the effect of energetic chemical reactions included. Examples of this type of flow occur in combustors and afterburning rocket plumes. There are two major approaches to solving this problem. The first is evaluation of the higher moments of the equations of motion, the energy equation, and the species continuity equations.[1] This approach requires information which is currently not available experimentally. Methods using the concept of probability density functions (PDF's) for the velocity and the scalar properties of the flow field have been described in Refs. 2, 3, and 4. In a turbulent flow each property varies with time at any given point in the flow field, and this variation can be represented using a PDF. The integral of a PDF between two values of a parameter in the field represents the fraction of time the parameter has a value between these two limits. Some attempt has been made to calculate PDF's a priori[2,3] however, these methods require some rather drastic assumptions and are limited to rather simple cases. An alternate approach is to use PDF's which are derived from available experimental data.[4]

If a flow variable is non-dimensionalized so that the resulting parameter has a value of zero at one edge of a mixing layer and one at the other, the PDF must have non-zero values only between these limits. As was pointed out,[5] a beta function fits this requirement. The beta function was used by Richardson[6] for the PDF of a parameter in a reacted fuel oxidizer system.

In this paper it is indeed shown that a beta function is a good representation of the PDF of a passive scalar. The beta function written as a PDF is:[7]

$$P(x)_X = (x)^{\alpha-1} (1-x)^{\gamma-1} / \int_0^1 (x)^{\alpha-1}(1-x)^{\gamma-1}dx$$

where

$$\alpha = \bar{x} (\bar{x} (1-\bar{x})\sigma^2 - 1)$$

$$\gamma = (1-\bar{x}) (\bar{x} (1-\bar{x})/\sigma^2 - 1)$$

$$\bar{x} = \text{mean value of } x$$

$$\sigma = \text{standard deviation of } x$$

Until recently, there has been little or no data for PDF's for any turbulent flow parameter. In 1970, Spencer[8] reported experimental data on the PDF of axial velocity in a planar two-stream mixing layer with a velocity ratio of 0.61. Spencer normalized

these data by the standard deviation and shifted the mean to zero.
In Fig. 1a, showing data taken near the high speed edge of the
layer, the beta function goes to infinity where the velocity corresponds to the mean value of the high speed stream while the data
show velocities higher than free stream value. This would indicate
the high speed free stream is not steady or that velocity fluctuations are induced with an extreme greater than the nominal maximum
velocity. Similar results for the low speed edge of the layer are
shown in Fig. 1b. Even in this case, where the parameter considered
can go beyond the limits 0 and 1, the beta function fits reasonably
well over a large portion of the data. The beta function fits data
quite well in the central portion of the shear layer (Fig. 1c).

The agreement between the beta function and PDF's for passive
scalars is better than for velocity. A PDF for temperature in the
wake of a heated cylinder[9] is shown in Fig. 2. These data were
also non-dimensionalized by the standard deviation and shifted so
the mean is at zero. Because no temperatures lower than the free
stream value were observed, the data and the beta function both
have a lower limit of -1.1σ.

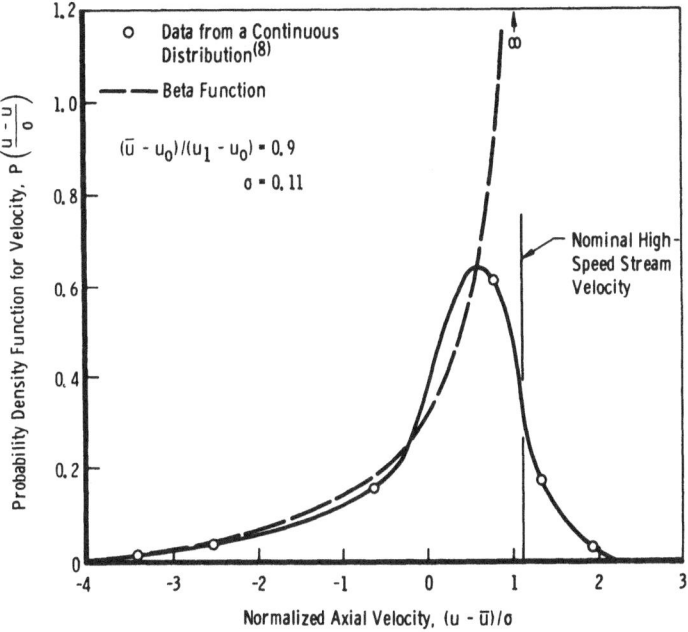

Figure 1a

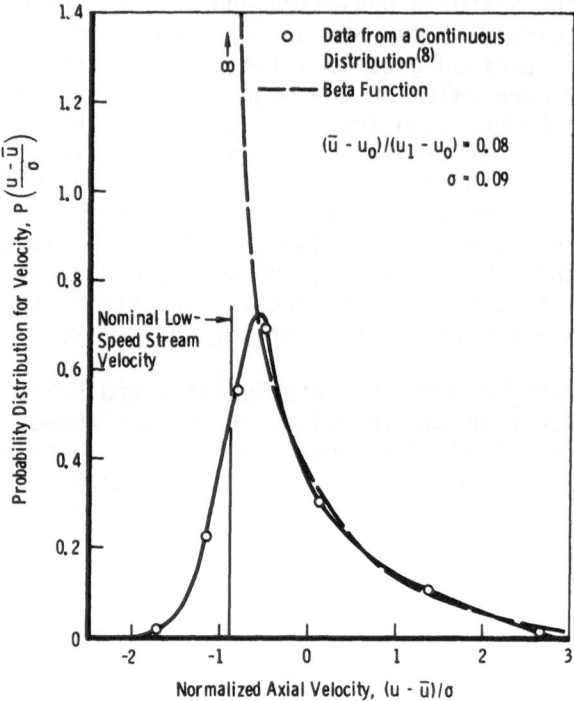

Figure 1b

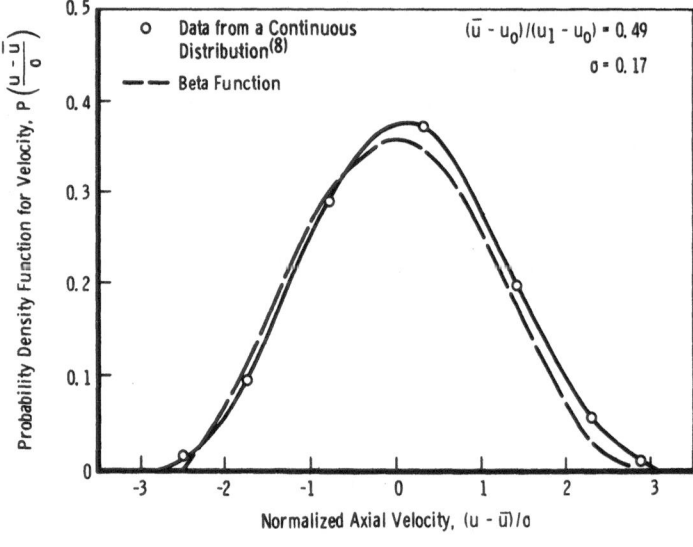

Figure 1c

Figure 1.   Comparison of the Beta Function with Probability Distribution Functions for Axial Velocity

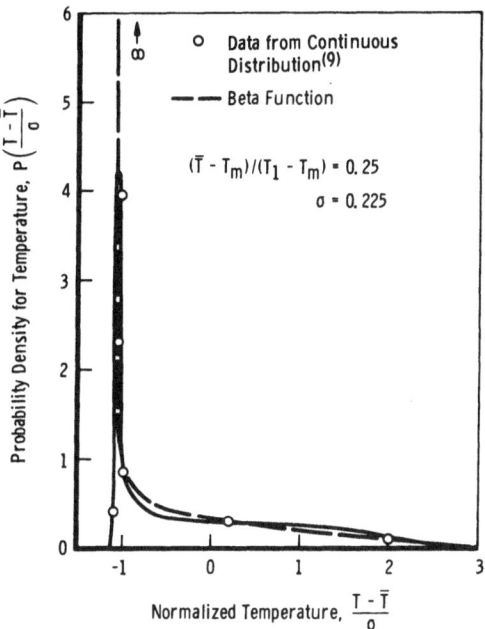

Figure 2. Comparison of the Beta Function with a Probability Distribution Function for Temperature

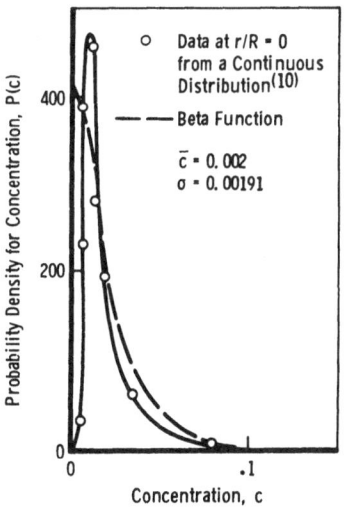

Figure 3a

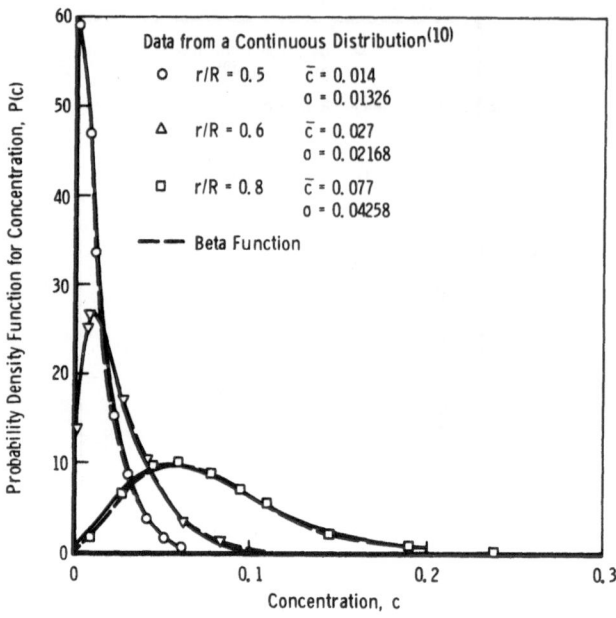

Figure 3b

Figure 3. Comparison of the Beta Function with Probability Distribution Functions for Concentration

        Data on the PDF for helium mass fraction in air at the exit
of a pipe in which the helium was injected from the wall upstream
of the exit were presented in Ref. 10. These are compared with the
beta function in Fig. 3a and 3b for four radial locations at the
pipe exit. In this case, the parameter plotted varies between zero
(pure air) and one (pure helium) with no physically possible way
that any concentration can exist outside this range. The beta
function is a very good representation of this data.

        In conclusion, it has been shown that a beta function is a
good representation for the probability density function of a
passive scalar and is a reasonably adequate representation of the
probability density function for the axial velocity.

                            REFERENCES

1.   Donaldson, C. duP. "A Progress Report on an Attempt to Con-
     struct an Invariant Model of Shear Flows," AGARD CP-93 Turbu-
     lent Shear Flows, Jan. 1972.

2. O'Brien, E. E. "Turbulent Mixing of Two Rapidly Reacting Chemical Species," Physics of Fluids, 14, 1326-1331 (1971).

3. Chung, P. M. "Diffusion Flame in Homologous Turbulent Shear Flows," Physics of Fluids, 15, 1735-1747 (1972).

4. Rhodes, R. P., Harsha, P. T., and Peters, C. E., "Turbulent Kinetic Energy Analysis of Hydrogen-Air Diffusion Flows," Acta Astronautica, 1, 443-470 (1974).

5. Rhodes, R. P., Harsha, P. T., "Putting the Turbulence in Turbulent Flow," AIAA Paper 72-68.

6. Richardson, J. M., Howard, H. C., and Smith, R. W., "The Relation Between Sampling Tube Measurements and Concentration Fluctuations in a Turbulent Gas Jet," Fourth Symposium on Combustion, Williams and Wilkins, Inc., Baltimore, 1953.

7. Abromowitz, M. and Stegin, I. A., Eds., Handbook of Mathematical Functions, Dover Publications, New York, 1965.

8. Spencer, B. W., "Statistical Investigation of Turbulent Velocity and Pressure Fields in a Two-Stream Mixing Layer," Ph.D. Thesis, University of Illinois (1970).

9. LaRue, J. and Libby, P. "Temperature Fluctuations in the Plane Turbulent Wake," Physics of Fluids, to be published.

10. Stanford, R. A. and Libby, P. "Further Applications of Hot Wire Anemometry to Turbulence Measurements in Helium Air Mixtures," Project Squid Technical Report USCD 5PU, January 1974.

MIXING IN "COMPLEX" TURBULENT FLOWS

P. Bradshaw

Department of Aeronautics, Imperial College

London SW7 2BY, England

ABSTRACT

This paper is a discussion, with special reference to mixing, on turbulent flows whose boundary conditions are more complicated than those of simple plane shear layers. Recent work at Imperial College will be used to illustrate (i) the unexpected behaviour of some complex flows and the crucial experiments needed to explore them, (ii) the modern measurement techniques which are especially useful in complex flows with mixing, (iii) the general needs of calculation methods for such flows. I hope that the paper will provide a bridge between the preceding papers on calculations and the following papers on experiments, and also help to remind people that simple shear layers are rather rare in practice.

1. INTRODUCTION

We define a "complex" flow as any flow other than a "simple" shear layer, i.e. a layer with a monotonic velocity profile and nearly-parallel, nearly-straight streamlines. Thus complex flows include nearly-parallel interacting pairs of shear layers, such as jets, wakes and duct flows, whose complexity is often disguised by symmetry but is present nevertheless: complex flows also include monotonic shear layers whose streamlines are longitudinally curved, laterally divergent or otherwise distorted. Interaction of a pair of shear layers does not seem to change the turbulence structure greatly: indeed in some cases a simple "super position" assumption gives a good representation although the process is nearer "time-sharing" than "superposition". Unfortunately distortion of almost

any kind produces a surprisingly large effect on a shear layer:
crudely speaking, the changes in turbulence structure are an order
of magnitude larger than expected from the explicit extra terms
which the distortion adds to, say, the Reynolds stress transport
equation. Therefore although nearly all flows of practical
interest are fairly thin shear layers for most of their lives
(that is, the mean rate-of-strain field is not too far from a
simple shear) the small departures from a simple shear produce
large effects on the turbulence structure. These effects appear
both in the Reynolds stresses and in the turbulent heat or mass
flux rates: there are few data on the latter, but measurements
in buoyant flows, which have some affinities with curved flows,
show that changes in heat or mass transfer are generally of the
same order of those in momentum transfer (though turbulent Prandtl
and Schmidt numbers do not remain accurately constant).

2. INTERACTING SHEAR LAYERS

Dean (1974 a, b) has made extensive measurements in a developing two-dimensional duct flow, with one of the boundary layers
slightly heated so that the fluid could still be traced after the
two boundary layers had merged. The object was to see how the
conditionally-sampled statistics of the "hot" fluid differed from
those of an isolated boundary layer. Figure 1 shows the way in
which the temperature-intermittency profile spreads across the
duct, and Figure 2 shows typical conditionally-sampled intensity

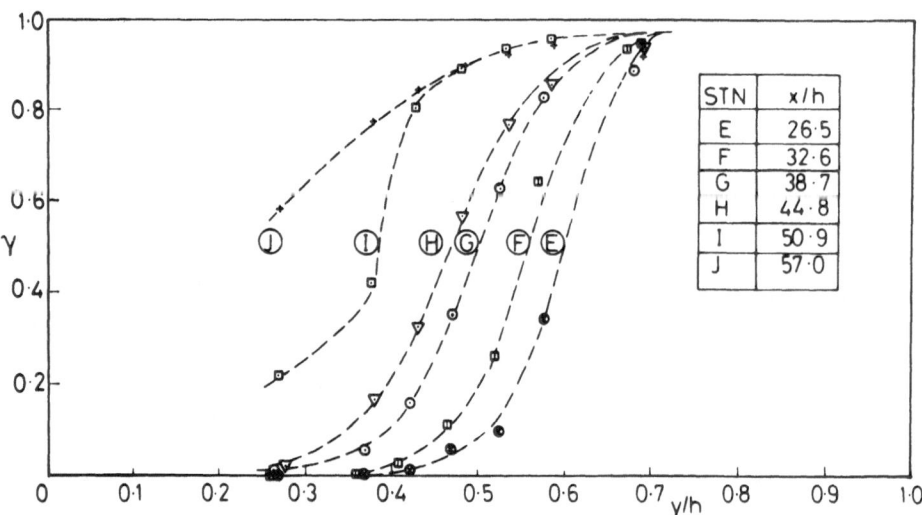

Figure 1.   Temperature-Intermittency Profiles in Entry of Duct with
One Side Heated (Dean)

profiles. The quantities plotted in the latter figure are γ times
the average within the hot fluid and $(1-\gamma)$ times the average within
the cold fluid. In both cases fluctuations are measured with
respect to the unconditional mean velocity. With these definitions
the "hot" and "cold" quantities plotted add up to the conventionally-
averaged quantity, which simplifies interpretation. The profiles
should of course be reflections of each other, a good check on
experimental accuracy. Figure 3 shows typical shear stress profiles: here the "hot" and "cold" contributions are related by
reflection and inversion. The region of overlap is smaller for
the shear stress than for the intensities, but is quite small for
both. The curves in Figure 3 are calculations by the method of
Bradshaw, Dean and McEligot (1973) in which the "hot" and "cold"
turbulence fields are assumed not to influence each other except
via the mean velocity profile. We originally called this "superposition" but Dean's results show that "time-sharing" is a more
accurate description. On either interpretation, one expects the
turbulence structure of each layer to be like that of a boundary
layer so that boundary-layer data can be used in the calculation
method. Our mean-flow predictions are encouraging; Figure 3 and
Dean's investigations of structural parameters add detailed
corroboration. In highly-turbulent jet flows there are significant
structural changes due to interaction, but superposition or time-
sharing is still a fair first approximation: it has some affinities with Prof. Spalding's concept of turbulent transport as a two-
way process (see these Proceedings). I think the explicit treatment

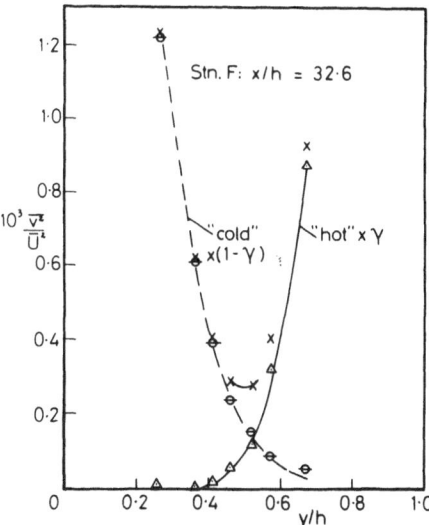

Figure 2.   Conditional Averages of Normal-Component Intensity in
            Half-Heated Duct

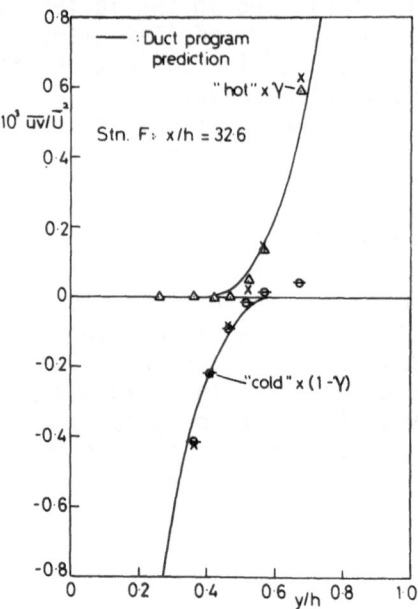

Figure 3.   Conditional Averages of Shear Stress in Half-Heated
            Duct

of shear-layer interactions may be not merely a help but a necessity in understanding asymmetrical jet or duct flows, especially
in predicting heat transfer. As yet I cannot see how to extend the
interaction treatment to axisymmetric (or nearly axisymmetric)
flows where one might say that 360 shear layers are interacting.

3.   DETERMINATION OF INTERMITTENCY

We found little difficulty in obtaining consistent values of
temperature intermittency by digital processing of resistance-
thermometer outputs in the duct flow. This was partly because the
temperature signal is one-sided (no zero crossings within the "hot"
or turbulent region) but mainly because, like most users of analogue intermittency meters, we arbitrarily smoothed the signal -
in our case, by using an uncompensated 1 micron wire as a resistance thermometer giving a response up to about 3 kHz or a streamwise wavelength of 0.2 duct heights. Only when we started looking
at velocity intermittency in a boundary layer, with a much better
frequency response, did we realize that the traditional smoothing
technique completely misses the fine structure of the irrotational/
turbulent interface which is so obvious in smoke pictures. Figure
4 shows the output from a computer program which declares the flow

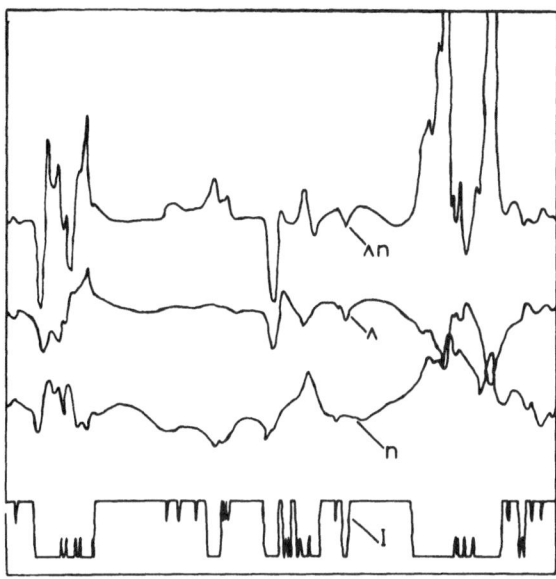

Figure 4. Intermittency in Constant-Pressure Boundary Layer
(Murlis): Common Time Axis for all Traces

to be turbulent when either $|\partial uv/\partial t|$ or $|\partial^2 uv/\partial t^2|$ exceed threshold
values chosen manually by inspection of plots like that shown in
the figure: no smoothing is employed. The top trace is the usual
zero-one intermittency function, but with short-duration excursions
shown as small-amplitude spikes. The maximum duration counted as
"short" was chosen so as to force agreement with Klebanoff's (1955)
intermittency measurements: it corresponds to a streamwise distance of 0.03δ. The difference between the "wholesale" intermittency (rejecting short bursts) and the unsmoothed "retail" intermittency is significant. The difference in average burst length
would be even larger: the probability distribution of burst length
rises with decreasing burst length down to the limit of digital
resolution. We feel that the problem is somewhat more serious than
Prof. Kovasznay indicated when he compared the interface to the
seashore and implied that a generally-acceptable and non-critical
definition could be found. To continue the analogy with, say, the
east coast of the USA, the irregularities vary in scale from the
Gulf of Mexico through Chesapeake Bay and the Mississippi to the
smaller rivers and inlets, all of which may have significance to
somebody. Particularly in fast-reacting flows, where the process
depends rather critically on the surface area of the fuel-oxidant
interface, we may have to measure or predict quite complicated
statistics of the interface. The difficulties that arise are

discussed in more detail by Bradshaw & Murlis (1974). Additional
difficulties arise, particularly in flows at low Reynolds number,
when one tries to determine intermittency by using a tracer (dye
or smoke) whose Schmidt number is far larger than unity: the
tracer will stay on the inside of the viscous superlayer, which
may be quite thick, and the effect is probably to exaggerate the
indentations of the interface.

4. FLOW VISUALIZATION IN MIXING LAYERS

Another (unrelated) case in which careful interpretation of
flow-visualization results is required is the mixing layer, studied
by Roshko (see these Proceedings) and others. Figure 1(c) of
Bradshaw, Ferriss and Johnson (1964) is a schlieren picture of a
circular jet (i.e. an axisymmetric mixing layer) with a small
density difference between the jet and the surrounding fluid. The
structure which shows up, downstream of the transition region, is
almost as pronounced as that in Brown and Roshko's (1971) photographs, but the correlation and spectrum measurements published by
Bradshaw et al. (and satisfactorily repeated by Weber 1974) indicated a conventionally three-dimensional structure. Very recently
we set up a high-aspect-ratio two-dimensional mixing layer and
found, by injecting smoke, that the transitional disturbances did
not always remain two-dimensional through successive stages of
vortex pairing, as appears to have been the case in Brown and
Roshko's work and in the work of Winant and Browand (1974) at very
low Reynolds number. In our case two adjacent vortices often
became significantly non-parallel so that when pairing occurred
the vortices merged into a double helix and breakdown to three-
dimensionality ensued. Sectional views, taken with a plane of
light in the xy plane, did not show the rather smooth interface
apparent in Brown and Roshko's spanwise-integrated shadowgraph
pictures. Further details of the study are in course of publication by Chandrsuda and Bradshaw: our present feeling is that low
Reynolds numbers, the peculiarities of flow-visualization techniques, and possibly the untypical nature of low-aspect-ratio
flows, may have combined to give, via several experiments, a
misleading impression that mixing layers are a different species
of turbulence.

5. SHEAR LAYERS WITH EXTRA RATES OF STRAIN

If a shear layer experiences an extra rate of strain (velocity
gradient), e say, in addition to the simple shear $\partial U/\partial y$ which drives
the layer, the extra terms that appear in equations such as the
Reynolds-stress transport equations are of order $e/(\partial U/\partial y)$ times
the existing terms. However for the cases $e = \partial y/\partial x$, $e = \partial w/\partial z$ and
$e = -\text{div } \underset{\sim}{U}$ (i.e. longitudinal curvature, lateral divergence and

bulk compression in compressible flow) the existing turbulence
terms in the transport equations change by a factor of order

$$F = 1 + 10 \; e/(\partial U/\partial y) \tag{1}$$

and so do the apparent eddy viscosity or mixing length. The factor 10 is not, alas, universal: it varies from case to case. The
effect of $e = \partial U/\partial x$ (or $- \partial V/\partial y$?) seems to be rather smaller but
still large enough to be a possible explanation of the difference
between the turbulence structure of jets and of wakes. The thin-
shear-layer approximation requires $e/(\partial U/\partial y) \ll 1$ but only if
$10e/(\partial U/\partial y) \ll 1$ will the effects of e on the turbulence structure
be negligible. Many flows of practical interest have such large
values of e that the thin-shear-layer approximation is violated:
however if $e/(\partial U/\partial y)$ is larger than about 0.1 the flow is likely
to be dominated by pressure gradients rather than Reynolds-stress
gradients. A particularly difficult class of flows is where a
short region of large extra strain produces very large changes in
the turbulence structure which then affect the growth of the shear
layer for a long distance downstream.

This need to classify turbulent flows according to the size
of the extra strain rate has led us to an approximation which is
intermediate between the thin-shear-layer approximation and the
complete equations of motion, and which is applicable to laminar
or turbulent flow. We call it the "fairly-thin-shear-layer" (FTSL)
approximation: it consists simply of approximating, as well as one
can, the terms which are neglected in the classical thin-shear-
layer (TSL) approximation. Obviously, many people have used this
approach in fragmentary ways in the past, but finite-difference
solutions and, in particular, turbulence models based on approximations (closures) of the transport equations have now become
sufficiently popular for FTSL ideas to be more generally usable.
Indeed the FTSL approximation naturally merges with the way in
which small terms are approximated in turbulence models. The
mathematical behaviour of the FTSL equations depends on the nature
of the approximations used: sometimes the approximations will be
defined only within a computer program, as in the approximation of
$\partial p/\partial y$ by an "upstream" value of U^2/R. In general however the
equations will be elliptic unless $\partial p/\partial y = 0$. Terms which are
neglected in the TSL approximation are of order $e/(\partial U/\partial y)$ times
the retained terms (consider $e = \partial U/\partial x$, for example) so that we
require $e/(\partial U/\partial y) \ll 1$, say $e/(\partial U/\partial y) < 0.01$ to get a TSL accuracy
of the order of magnitude of 1 percent. If we can approximate
these terms to 10 percent accuracy, the overall accuracy of the
FTSL approximation will be 1 percent if $e/(\partial U/\partial y) < 0.1$. For
larger extra strain rates the full Navier-Stokes equations are
needed but we saw above that such flows are likely to be dominated
by pressure gradients rather than stress gradients. Therefore the
region of application of the FTSL approximation is

$$0.01 < e/(\partial U/\partial y) < 0.1$$

These are of course very rough figures. The key to using the FTSL
approximation is the identification of the direction of the shear
layer, which is then used as a (curvilinear) coordinate axis. Like
the TSL equations, equations using the FTSL approximation will be
non-invariant with respect to rotation of coordinates, which permits, even if it does not encourage, the use of non-invariant
turbulence models. Even in a highly-curved shear layer like that
shown in Figure 5, the shear-layer direction is easy to define.

  Figure 5 shows a mixing layer between still air and a blower-
tunnel stream which is deflected through 90 deg. (Castro 1973).
The curvature is a minimum at about 30 cm from the origin of the
layer and the turbulent intensity and shear stress (referred to
axes along and normal to the local shear-layer direction) reach
minima a little further downstream (Figure 6) in response to the
stabilizing effect of the curvature. The amazing feature is that
as the curvature decreases, the intensity, and most of the other
turbulence properties, overshoot the plane-layer values before
finally decreasing. Nominally all quantities should asymptote
back to the plane-layer values but in our rig the mixing layer
spreads to the wall before this process is complete. Figure 7
shows the turbulent energy balance on the (curvilinear) centre
line. The key feature is that although all terms decrease, the

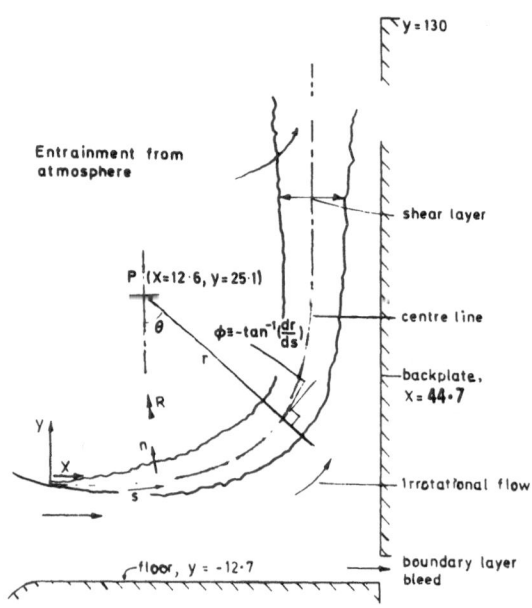

Figure 5.   Curved Mixing-Layer Rig (Castro)

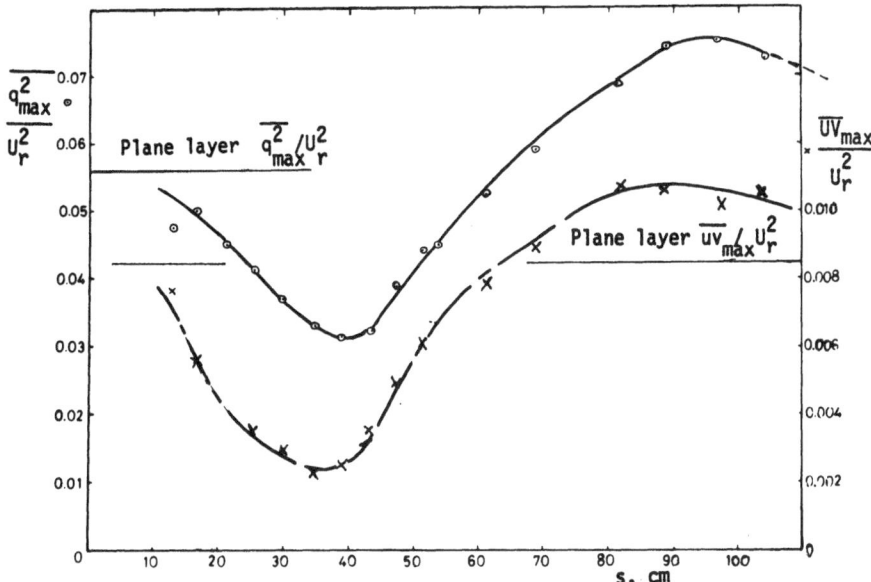

Figure 6. Intensity and Shear Stress in Curved Mixing Layer

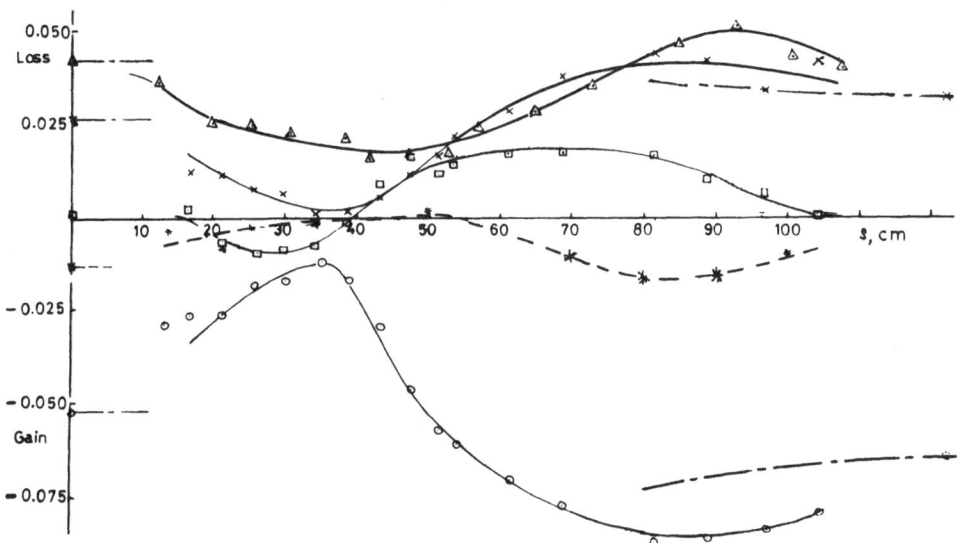

Figure 7. Turbulent Energy Balance on Centre Line of Curved Mixing Layer. ⊚, Production; ▲, Dissipation; •, Advection; x, Diffusion.

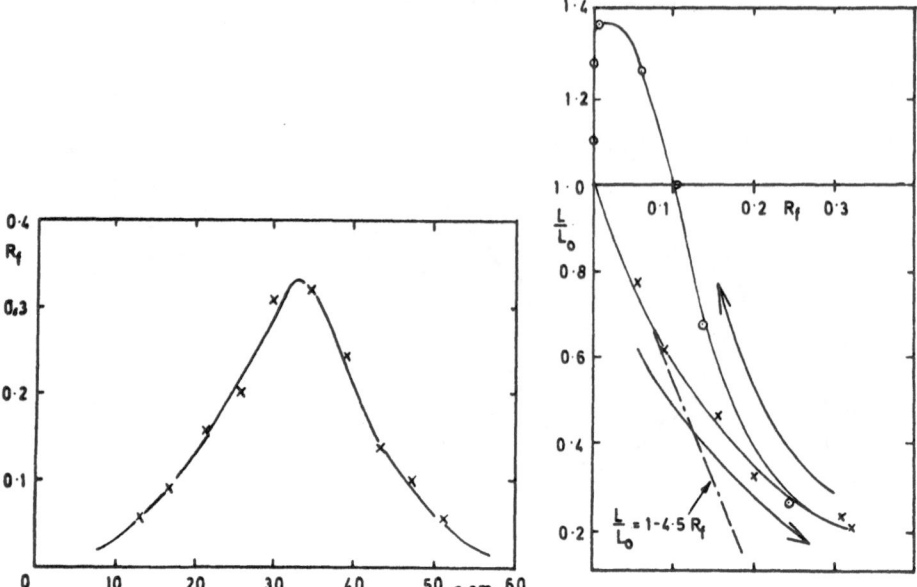

Figure 8. Variation of Dissipation Length Parameter with "Richard-
           son Number" on Centre Line of Curved Mixing Layer:
           L_0 is Value in Plane Mixing Layer of Same Thickness

diffusion term (turbulent transport of turbulent energy) decreases
most of all, evidently because the large eddies are most susceptible
to curvature. The result is that the production recovers much
more rapidly than the diffusion, and the newly-produced energy is
trapped near the centre line instead of being diffused away to the
edges. This is a simple but satisfactory explanation: however
its incorporation into calculation methods is non-trivial. Figure
8 shows the variation of the dissipation length parameter
$L \equiv (\overline{uv})^{3/2}/\varepsilon$ with a curvature parameter ("Richardson number") R_f,
roughly equal to $-2e/(\partial U/\partial y)$. The dotted line shows the variation
predicted if (1) is applied to L (a more refined procedure than
applying it to the mixing length, and one which works well for a
variety of mildly-curved flows). The full line shows the actual
variation, predictable neither by (1) nor by the extension of (1)
(Bradshaw 1973) in which a lagged, rather than local, value of e
is used. Further details of this experiment, and of the use of the
FTSL approximation, are given by Castro and Bradshaw (1974).

     Figure 9 shows that the overshoot in intensity found in
Castro's work is not an isolated phenomenon. Here we have a
boundary layer deflected 30 deg. round a convex bend (measurements
by Young, unpublished). As in So and Mellor's (1973) experiment,
the shear stress in the outer layer is greatly reduced by the
curvature, but in the plane flow further downstream the shear

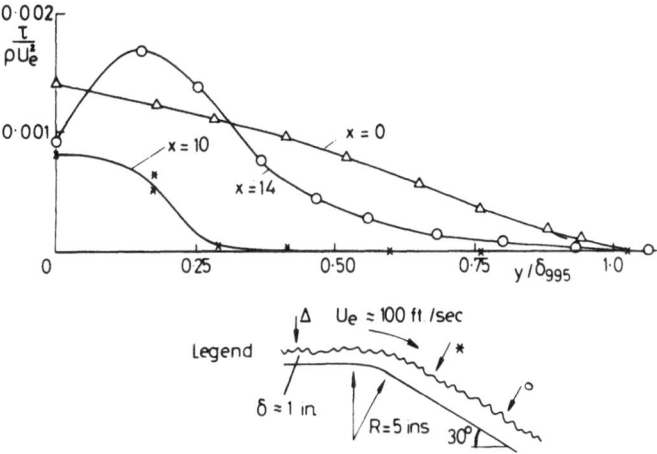

Figure 9.   Shear Stress Profiles in Boundary Layer on 30 Deg.
Convex Bend (Young)

stress recovers, to exceed its initial value in the region of
y/δ = 0.2:   this behaviour is <u>not</u> found in calculations which
simulate the pressure gradient without the curvature. We have also
made measurements on the concave side, where quasi-steady
longitudinal vortices are found. Meroney (1974) has measured the
flow in a more gentle bend in the same rig (δ/R ~ 0.01 - 0.02) and
found good agreement with predictions using the F-factor (1).

Another case in which application of the lagged F-factor to
the dissipation length parameter gives good results is the supersonic boundary layer in pressure gradient (e = -div $\underline{U}$). Without
it, calculations by (for instance) the method of Bradshaw & Ferriss
(1971) disagree badly with experiments in strong pressure gradients
although this method predicts constant-pressure boundary layers up
to M $\simeq$ 5 quite well. Great improvements result (Bradshaw 1974)
from incorporating the lagged F-factor, with an empirical constant
of 10, as in (1). There are of course not enough turbulence
measurements to check the details of what happens, but clearly bulk
compression is another addition to the family of influential extra
strain rates. Green, Weeks and Brooman (1973) have used (1),
unlagged and with an empirical constant of 7, to predict the simultaneous effects of longitudinal curvature, lateral divergence and
bulk compression on the waisted body of Winter, Rotta and Smith
(1968).

6.   LENGTH-SCALE TRANSPORT EQUATIONS FOR TURBULENCE MODELS

In any turbulence model that uses one velocity scale and one
length-scale to describe the turbulence, a length-scale transport

equation can be recovered, whether the original transport equation refers to length scale, dissipation rate, or energy-length product. Supposing for simplicity that the scales are $(\overline{uv})^{\frac{1}{2}}$ and L, we get, for a simple shear layer with no extra strain effects,

$$\frac{DL}{Dt} = c \cdot L \frac{\partial U}{\partial y} - \frac{\partial}{\partial y} (V_L L) + f \cdot (\overline{uv})^{\frac{1}{2}} \tag{2}$$

Here c and f are constants and V_L is a turbulent transport velocity. If two velocity scales, such as $\overline{uv}$ and $\overline{q^2}$, were available, each term could include a function of the slowly-varying parameter $\overline{uv}/\overline{q^2}$, which would not make much difference to the results. In most of the published methods I have analyzed, c is small. The second term of (2) is not small, but can be eliminated by integrating (2) with respect to y, when we get

$$\frac{d\hat{L}}{dx} \propto \frac{(\widehat{\overline{uv}})^{\frac{1}{2}}}{\hat{U}} \tag{3}$$

where $\hat{L}$, $\widehat{\overline{uv}}$ and $\hat{U}$ all represent some sort of average across the width of the shear layer. This is the only dimensionally-correct result for a model based on $\overline{uv}$ and L with the above non-critical simplifications. Now experiments on families of jets, wakes and boundary layers suggest that a much better approximation than (3) is

$$\frac{d\hat{L}}{dx} \propto \frac{\widehat{\overline{uv}}}{\hat{U}^2} \tag{4}$$

For instance if we accept that $\hat{L}$ is proportional to δ in a simple boundary layer the linear correlation between entrainment rate and maximum shear stress shown in Figure 2 of Bradshaw, Ferriss and Atwell (1967) agrees with (4). Unfortunately (4) implies that the last term in (2) depends on the mean velocity, and whatever may be one's views about the necessity of rotational invariance one must certainly require translational invariance in any major terms of a turbulence model intended for general application. (The correlation used by Bradshaw et al. was intended to represent a minor term in a small range of flows.) I do not see any simple way out of the difficulty, except rather speculative steps like assuming that f in (2) is strongly dependent on $(\partial V/\partial y)/(\partial U/\partial y)$ which is in turn proportional - other things being equal - to $\overline{uv}/U^2$ or the experimentally-observed dL/dx. In this way one can insert a legal version of (4), but there is no direct experimental justification at all for doing so. We are at present exploring some of these possibilities using the 1974 version of the Bradshaw-Ferriss-Atwell boundary-layer program (Bradshaw & Unsworth 1974) which includes the lagged F-factor allowances for the effects of extra strain rates $\partial V/\partial x$, $\partial W/\partial z$ and div $\underset{\sim}{U}$.

7. CONCLUSIONS

This paper contains one piece of good news, that shear-layer interactions are comparatively easy to treat, and one relatively new concept, the FTSL approximation. The new measurements in highly-curved flows are bad news, however interesting, because they show that the simple formulae which successfully predict the effect of small extra strain rates fail for large or rapidly-changing strain rate. The difficulties with the measurement of intermittency and the formulation of length-scale transport equations are doubtless not insuperable but are typical of the minor inconveniences that complicate the already formidable problem of turbulence.

A more extensive review entitled "complex turbulent flows" (paying more attention to work done outside the author's department) is being prepared as one of a forthcoming series of review articles in ASME Journal of Basic Engineering: the aim is to publish in the June 1975 issue (Vol. 97, no. 2).

I am grateful to my former students Dr. I. P. Castro and Dr. R. B. Dean, and my present associates S/Ldr. C. Chandrsuda, Mr. J. Murlis and Mr. S. T. B. Young, for permission to quote their unpublished work.

REFERENCES

1. Bradshaw, P. 1973 AGARDograph 169.

2. Bradshaw, P. 1974 J. Fluid Mech. <u>63</u>, 449.

3. Bradshaw, P., Dean, R. B. & McEligot, D. M. 1973 J. Fluids Engg. <u>95</u>, 214.

4. Bradshaw, P. & Ferriss, D. H. 1971 J. Fluid Mech. <u>46</u>, 83.

5. Bradshaw, P., Ferriss, D. H. & Atwell, N. P. 1967 J. Fluid Mech. <u>28</u>, 593.

6. Bradshaw, P., Ferriss, D. H. & Johnson, R. F. 1964 J. Fluid Mech. <u>19</u>, 591.

7. Bradshaw, P. & Murlis, J. 1974 Imperial College Aero Rept. 74-04.

8. Bradshaw, P. & Unsworth, K. 1974 Imperial College Aero Rept. 74-02.

9. Brown, G. & Roshko, A. 1971 AGARD Conf. Proc. 93.

10. Castro, I. P.   1973 Ph.D. Thesis, Imperial College, London
     (available on microfiche from I.C.).

11. Castro, I. P. & Bradshaw, P.   1974 Submitted to J. Fluid
     Mech.

12. Dean, R. B.   1974a Ph.D. Thesis, Imperial College, London
     (available on microfiche from I.C.).

13. Dean, R. B.   1974b Paper presented at 5th Australasian
     Conference on Hydraulics and Fluid Mechanics.

14. Green, J. E., Weeks, D. J. & Brooman, J. W. F.   1973 RAE TR
     72231.

15. Klebanoff, P. S.   1955 NACA Rept. 1247.

16. Meroney, R. N.   1974 Imperial College Aero. Rept. 74-05.

17. So, R. M. C. & Mellor, G. L.   1973 J. Fluid Mech. 60, 43.

18. Weber, D. P.   1974 Ph.D. Thesis, Univ. of Illinois at Urbana-
     Champaign.

19. Winant, C. D. & Browand, F. K.   1974 J. Fluid Mech. 63, 237.

20. Winter, K. G., Rotta, J. C. & Smith, K. G.   1968 ARC R&M 3633.

DISCUSSION

KOVASZNAY: (Johns Hopkins University, Baltimore, Maryland)

I do not know whether you recall Blackwelder's measurements[1] on the rapidly accelerated boundary layer. One of the interesting results that came out is that the long lifetime of the largest structures, occurs equally in the rapidly accelerated boundary layer. What is also interesting is that the absolute value of the shear stress is pretty much persisting too, and a great deal of quantities associated with the large scale motion are all quite persistent. Now that lone persistence of rather large structures does not seem to appear in the type of transport equations that are typically written, say, for the length scale.

---

[1] Blackwelder, R. F., Kovasznay, L. S. G. Large scale motion of
     a turbulent boundary layer during relaminarization. J. Fluid
     Mech. 53, 61, 1972.

BRADSHAW:

The "time constant" that one gets out of these things (the equivalent of the stream-wise distance) is something like 10 shear layer thicknesses. If you have a strongly accelerating flow then the stream-wise distance is greater because you will be moving the flow faster. My guess is, if you look at it time-wise instead of space-wise--

KOVASZNAY:

That is, similar: it is similar in time-wise, not in physical space.

BRADSHAW:

Yes, but these equations do give you D/Dt (motion following the fluid along a streamline), so you can even apply these equations to unsteady flows, providing you use incantations about ensemble averaging. So, my guess is, you could probably sort it out this way. It might be a little messy to count on the fingers and get equivalent times and distances.

ALBER: (TRW Systems, Redondo Beach, California)

Have you been able to ascertain whether or not the dilatation factor can help to explain the reduced spreading rate of shear layers with Mach number as has been suggested by Oh[2] at NASA Langley.

BRADSHAW:

I have not thought very much more about this, and I do not think Dr. Oh has progressed beyond the work[2] that he put in the AIAA Journal. Certainly that paper explains the change in spreading rate with Mach number. I am not altogether sure that the average dilatation that he has used is really the appropriate one. This is done for numerical simplicity. The unfortunate thing is that the modeling also predicts, as far as I can see, a large change in spreading rates with density ratio because you still get a big div $\underline{u}$ even if you have a low Mach number, isobaric flow with a big density change across it. So this does not seem to be the

---

[2]   Oh, Y. H. Calculation of compressible turbulent free shear layers, AIAA J. 12, 401, 1974 (see also "Analysis of two-dimensional free turbulent mixing" paper presented at AIAA 7th Fluid and Plasma Dynamics Conf., June 1974).

complete explanation for the difference between spreading rate at given density ratio between a low speed flow and a high speed flow. I do not know whether Dr. Oh would like to comment.

OH: (NASA, Langley Research Center, Hampton, Virginia)

Yes, I tend to agree with you.

KLINE: (Stanford University, Palo Alto, California)

I want to ask a two-part question. You mentioned rotation in your presentation. We did not see any data for rotation, however. Three students of Jim Johnston, Halleen,[3] Lezius[4] and Rothe[5] have completed work on rotation. These flows look even more pathological, in some ways, than the cases you have reported. They observe things like doubling of the turbulence intensity followed by "saturation" in the form of large, Taylor-Görtler eddies on one wall, and on the other wall they see very thick, laminar-like layers. In between the measurements show large zones of negative eddy viscosity, infinite eddy viscosity--practically anything you want. Have you looked at any of those in terms of the kind of equations you proposed? The second part of the question is, if this kind of modeling does not work, have you any idea where one goes? What kind of modeling does one do?

BRADSHAW:

I will answer the second part first. I do not know; I am sure that this sort of modeling, which Hanjalic and Launder have pursued in a more detailed way[6] than anybody else, is the right way to go. Really, the only alternative to this sort of closure at second order, is closure at higher order or the numerical solution of the time-dependent Navier-Stokes equations. I doubt, on the whole, whether any sort of averaging other than Reynolds averaging is going to help us all that much. It may help us in some of the intermittency problems, but I do not think it is necessarily relevent here. So, I think we have just to fight away with length scale transport equations. George Batchelor says he does not think we are really entitled to call the D/Dt equations transport equations in any "control-volume" sense of the word, unless they refer to more-or-less conserved quantities, like energy or possibly Reynolds stress.

[3,4,5] See References in comment on paper of D. B. Spalding.

[6]   Hanjalic, K., Launder, B. E., A Reynolds-stress model of turbulence and its application to thin shear flows. J. Fluid Mech., 52, 609, 1972.

Coming to the first part of the question, you could possibly cope with the type of fully stabilized or highly destabilized flows that Jim Johnston has been working on, by the sort of length scale equation that I have mentioned. The real thing that seems to upset matters is fairly rapid changes in the streamwise direction. To cope with the "saturated" type of flow we need a good model for turbulent transport (diffusion), and we don't have one! However this is not as serious as the kind of fundamental obstacle that seems to be presented by the length scale equations I mentioned.

NAGIB: (Illinois Institute of Technology, Chicago, Illinois)

In regard to your remarks about the time sharing concept in shear layers interacting with each other in a confined flow, would you care to extend this also to free shear layers?

BRADSHAW:

Well, Tom Morel[7] has extended the calculation method that we use successfully in the case of ducts to the case of free shear layers. He needs to fiddle his empirical data from one shear layer to another, but the fiddling seems to be related more to the change in spreading rate, that is, if you like, the $\partial v/\partial y$ extra strain rate, than to the presence of the interaction. In other words, what I am saying is that even in the highly turbulent free shear layers, the effects of the interaction as such on the turbulent structure seems to be lost within the effects of the extra strain rate. But, it would be nice to have some more work on this, and I have a post-doctoral assistant just about to carry on an experiment extending what Bruce Dean[8] did in the duct to the case of two merging mixing layers at the end of the potential core of a two-dimensional jet. And we hope this will give some more data about the details of the interaction process in that case.

NEWMAN: (McGill University, Montreal, P. Q., Canada)

I wondered to what extent the difficulties you get into with length scale, as it is affected by mean distortion, are dependent upon your particular choice of length scale which is proportional to $(\overline{uv})^{3/2}$ rather than $(\overline{q^2})^{3/2}$.

---

[7]   Morel, T., Torda, T. P., Calculation of free turbulent mixing by the interaction approach. AIAA J. <u>12</u>, 533, 1974.

[8]   Dean, R. B. See reference in the paper.

BRADSHAW:

Not all that much, I think. I have probably not been making
it entirely plain, but the difficulty in the dissipation or length-
scale equation is found in thin shear layers where the ratio of
shear stress to turbulent intensity remains roughly constant. So,
I very much doubt if one could get away from this problem by choosing some other formulation of length scale. This contradicts something that Professor Spalding said this morning that he was seeking
a velocity scale-length scale combination that would fit the
gradient-diffusion transport model, but equally, I suppose, one
could try to seek a combination which would give you a reasonable
replica of what happens in the streamwise direction. Whether this
combination will be the same as the one which fits the gradient
transport model, I would not like to say!

MJOLSNESS: (Los Alamos Scientific Lab, Los Alamos, New Mexico)

Would the problems of the length scale equation be alleviated
by making the constant, little f in the equation, proportional to
the turbulence Reynolds number?

BRADSHAW:

I do not think one would want to make it proportional to the
turbulence Reynolds number because the length scale that one is
talking about is the length scale of the energy containing eddies
and the usual arguments about Reynolds-number-independence of the
energy-containing motion indicate that this is not going to depend
very strongly on the turbulence Reynolds number, at least at high
local Reynolds number, so I do not really see that this is a way
out of the problem.

NAGIB:

Regarding some of the large scale phenomena that have been
observed in the turbulent shear layers with curvature, like the
things that Professor Kline described, you feel that there is any
additional difficulty than what you have been able to handle?

BRADSHAW:

Yes! But enough is enough for the time being! I do not want
us to send people absolutely weeping from today's proceedings. You
are quite right; to cover flows with, say, steady streamwise vortices one will have to depart from the usual span-wise averaging,
one will have to consider lateral transport of momentum, and indeed
one will have to face problems equivalent in difficulty to the problems that you get in the corners of ducts dominated by secondary
flow.

ABBOTT: (Purdue University, West Lafayette, Indiana)

I think, while generally agreeing with your conclusions, that the extra ten terms referred to in your presentation may be associated with convected vorticity and vortex stretching which are second and third order terms respectively, compared with vortex diffusion. My comment applies, of course, only to thin shear layers very near a solid surface.

BRADSHAW:

Well, one still gets significant effects on the turbulence structure, even in thin shear layers. And this is the point of the factor of ten, everything is ten times worse than you expect it to be.

COHERENT STRUCTURES IN TURBULENCE

P. O. A. L. Davies

Institute of Sound and Vibration Research

University of Southampton, England

SUMMARY

This account summarizes contributions to a colloquium on coherent structures in turbulence held at Southampton in March 1974. It includes an account of work and of ideas held by the author and others over the past twenty years or more. Other relevant information can be found among the contributions to this meeting or in the current literature.

1. INTRODUCTION

The thought that large scale coherent structures containing concentrated regions of vorticity must play a dominant role in turbulent shear flow is certainly not new. This dominant role is all the more obvious if one recalls that such structures involve a substantial proportion of the turbulent energy. For example, Townsend (1) has reviewed the way in which turbulence can be modelled by an irregularly spaced assembly of ordered structures. He showed that point velocity measurements obtained from such a model, with only a single characteristic scale, would yield a broad wavenumber spectrum characteristic of turbulent flow.

On the other hand there has been a tendency to regard such structures, when they have been observed in the past, as representative of a transitional state in the development of turbulent flow (2). Recently, however, coherent structures have been clearly observed in flows that otherwise display all the characteristics of fully developed turbulence (3), (4), (5). Many examples of such observations were reported to the Southampton Colloquium (6).

In the light of current knowledge it seems reasonable, therefore, to regard turbulence as an assembly of repetitive ordered
structures which interact strongly with each other as they travel
downstream. Point velocity or other measurements, normally made
at fixed points, consist of a time history record obtained from a
succession of many such structures passing the measuring stations.
Since motions within those structures which are relatively remote
from each other become statistically independent, such time history
records display the statistical characteristics of a random process.

Experimental work employing time correlation analysis has provided much useful information on the average temporal and statistical
scales of turbulent flows (1) and on the average convection speed of
the array (7). However useful they are, such measurements yield
little information on the causal processes leading to the measured
state. Their value is thus limited to providing a description of
certain practical flow distributions but they offer little positive
information on how these observed flows may be controlled or modified, nor on how turbulent flow behavior can be predicted from first
principles. Furthermore, since the phase information is unavoidably
lost during data processing, statistical descriptions in the form of
power spectra are of limited value in predicting the magnitude of
the interactions between a turbulent flow and its boundaries.

As these limitations have been realized for some time, the
direction of some turbulence research has leaned towards more deterministic descriptions of turbulence. Experimental studies and techniques have benefited from the development of digital data processing
systems that can work in real time (8), (9). For some time there
have been numerous hints among statistical measurements which suggest
that bounded turbulent flow fields have an ordered structure at
length scales characteristic of the flow itself (1). This has recently been confirmed by several elegant examples of shear flow
visualization which illustrate not only these structures (5) but the
interactions between them (10), (11), (12).

This account traces the basic mechanisms which govern the
development, growth and interaction of the large coherent structures
found in turbulent shear flows (12). In particular it reviews the
significance of these recent discoveries in the deepening of our
understanding of turbulent mixing processes.

2. THE SOUTHAMPTON COLLOQUIUM

The Colloquium at Southampton was attended by over 100 active
workers in turbulence research of whom well over half came from
outside the United Kingdom. A majority of these had been concerned
with the deterministic rather than strictly statistical studies of

turbulent flow structure and many reported on the characteristics
and development of coherent structures they found there. Coherent
structures that were observed independently by several contributors
included the large scale, vortex-like eddies which dominate mixing
layer and jet flows, also the repetitive coherent vortex motions in
boundary layers and pipe flows.

The clearer picture of the large scale structure of turbulence
that has begun to emerge has also led to various attempts to model
the flow, some of which were reported to the colloquium. The results
showed that wave guide models can provide fair agreement with observations for some of the predicted modes of instability in the flow.
Other purely kinematic models involve studies of the behavior of
distributions of vorticity. It should be clearly understood that
the organized structure of such concentrated regions of vorticity
differs in many respects from that of the turbulent eddy or the wave
models of statistical theories of turbulence.

2.1   Coherent Structures in Free Shear Layers

The existence of coherent structures in shear flows can be most
readily observed in the mixing between parallel streams as described
by Brown and Roshko (5) and Winant and Browand (11). Observations
reveal that the interface which is initially a plane vortex sheet,
quickly rolls up by a Kelvin-Helmholz instability into coherent
line vortices. Subsequent growth of the mixing layer and line vortices occurs due to coalescence of these structures. Observations
also show, Lau (3), Davies (4) and Brown and Roshko (5), that such
vortices develop an irregular spacing whose probability density
approximates to a Rayleigh distribution. This and other characteristics of the flows where such structures are found display all the
characteristics of turbulence in terms of accepted statistical criteria. Earlier observations by Tani (2), had revealed similar
coherent interactions and growing structures during boundary layer
transition, but they had then been regarded as a transitional state
of the flow.

What is puzzling, perhaps, is that the flows in the two mixing
layer experiments apparently remain two-dimensional over the field
of observation, though the spacing of the individual vortices becomes increasingly irregular as they coalesce. The fact that such
clear details of this process have not been seen previously, may be
due to the care taken to provide steady uniform initial flow in
these experiments. One would, however, expect that ultimately a
three-dimensional structure would develop, since such line vortices
are unstable along their length to small perturbations, as is demonstrated by Hama (13). Indeed in figure 8 of Brown and Roshko's
paper (5) there are hints that such a development has just started.

What does emerge clearly from the two mixing layer experiments (5), (11) is that turbulent entrainment and mixing is largely performed by engulfment, particularly by coalescing vortices and also that viscosity plays a secondary role. This conclusion is supported by the agreement with observations provided by estimates of the entrainment rate based on an engulfment mechanism (5). It is also supported by the slow spreading of the interface between species observed in mixing gas streams (5), or in the slow rate of diffusion of the dye streaks seen in the water channel experiments (11).

Averaged measurements of the instantaneous velocity distribution in the water channel at different stages in the coalescing process (11) were obtained by phase averaged conditional sampling, using two reference probes either side of the mixing layer. It was found that large contributions to the turbulent shear stress occurred during the vortex pairing process. Taken together the observations suggest that the vortex pairing process plays a dominant role in both turbulent mass and momentum transfer.

This conclusion was also suggested by the observations of H. H. Fiedler and D. Korschelf (6) made with two dimensional and round jets. Similar coherent structures were found. Conditional sampling measurements in a mixing layer, with one stream heated, suggested that large scale line vortices dominated the bulk convection mechanism. V. W. Goldschmidt (6) described measurements of velocity and temperature conditioned with the turbulent interface of a plane jet. It was found that the interfaces of both the temperature and velocity fields were coincident. Further, that the higher mean turbulent transport rate for heat arose as a consequence of the relatively flatter temperature profile found in the turbulent region of the flow.

2.2   Observations in Turbulent Jets

The initial development of a circular jet bears a close resemblance to that of a free mixing layer. The development of vortex rings by the rolling up of the initial sheet has been described in much detail by Wille (14). Similar and additional details of the process were provided by several contributors to the Colloquium. A. J. Yule (12) traced the development of coherent motions in the first ten diameters of a circular jet. Initially the cylindrical vortex sheet rolled up into coherent ring vortices which grew by coalescence. These rings develop circumferential waves on their cores, an instability recently described quantitatively by Widnall (15) based on Hama's (13) work. This results in an ordered three-dimensional structure with smaller mixing jets being ejected from azimuthal positions between the wavy rings. The rapid development of such three dimensionality in the coherent structures makes them

more difficult to identify or study. Yule (6) stressed the importance of making simultaneous flow visualization and quantitative
velocity measurements, so that observed phenomena could be related
to measured velocity time histories. Such experiments showed that
observed coherent motions in the turbulent jets were directly
associated with distinct, repetitive features in the velocity signals. The importance of differentiating between turbulent and
unsteady laminar flow regions when investigating coherent structures
was also discussed and criteria for making this distinction were
suggested.

H. H. Bruun (12) described associated hot wire measurements
which used the characteristic velocity signatures discussed by Yule
(i.e. very large - ve or + ve peaks depending on radial position)
as triggering signals to educe the average motions of the large
scale coherent structures. It was found that the major part of the
turbulent energy, the Reynolds stress and statistical cross-
correlations obtained from continuous records for the mixing layer
region of the round jet were associated with these large eddy structures. More recently the observations have been extended to 20
diameters downstream with similar results.

The structure of jets subjected to some form of periodic forcing was described in contributions from Grenoble and Berlin. J. P.
Girard (6) presented observations on a round jet with a fluctuating
pressure field applied to the upstream chamber. While the structure
appeared more ordered, the initial jet spreading rate was also enhanced. More detail was provided by G. Binder (6) who used periodic
sampling and phase averaging of the hot wire velocity signals. The
results showed that the turbulence intensity grew more rapidly in
the ordered flow, the stronger signals being associated with the
strengthened vortex rings. The modifications induced by forcing
were no longer discernible after twenty diameters downstream.
E. Pfizenmaier (6) described experiments with a jet forced by a loud
speaker placed in the mixing chamber. His observation concentrated
on the initial development of the shear layer. At low forcing
Strouhal numbers the agreement with Michalke's (16) analysis was
good, but this was no longer so at higher Strouhal numbers.

2.3   Boundary Layer Structure

Clear evidence of the existence of similar ordered structures
in turbulent boundary layer flows was also reported to the Colloquium. In jets, mixing layers and wakes, development of coherent
structure usually commences at a well defined origin but the continued generation of vorticity at the wall increases the complexity
of boundary layer flows. The origin and growth of vortices near
the wall has been widely reported as wall bursts. One might surmise

that once they exist, regions of concentrated vorticity will always grow by migration and coalescence and produce the large-scale structures found in the outer part of the boundary layer reported recently by Offen and Kline (17). But, in boundary layer flow, identification of coherent structures is further complicated by the occurrence at the same streamwise station of a succession of smaller younger and larger older structures.

This problem was clearly recognized by a cooperative team of workers at the Max Planck Institute and the Ohio State University. Flow visualization experiments recorded by a camera moving with the flow (18) had revealed coherent motions in the form of transverse vortices which persisted for long downstream distances. The particle paths were analyzed to provide typical fixed point velocity time histories related to the vortices.

A digital pattern recognition procedure was developed which detected the characteristic slow deceleration in streamwise velocity followed by a more rapid acceleration. Due account was taken of the wide variation in time and amplitude scale in each pattern. When the procedure was applied in a boundary layer it was found that about half the recorded time history was 'recognized,' and for these portions of the signal the streamwise and transverse velocity were in antiphase, providing a finite contribution to the shear stress.

Associated measurements in the viscous region near the wall were described by Eckelmann (19). Coherent motions were deduced from observations of the velocity field and gradients near and at the wall. The results showed that the large streamwise perturbations travel towards the wall with a nearly constant velocity. This velocity agreed with the convection velocity of large coherent structures observed near the wall by R. L. Simpson (6) in a separating turbulent boundary layer. Eckelmann found that the Reynolds stress is highly intermittent, while close to the wall the streamwise velocity is highly correlated with the shear at the wall.

Other observations by W. W. Willmarth (20) were in general agreement with the results just described. More recent observations of the Reynolds stress showed that this was intermittent and accompanied by large low pressure areas at the wall (6). An investigation of the coherent structure of the boundary layer was described by A. Favre (6). Three point space time correlations of the temperature fluctuations with a heated wall were found to provide a more positive indication of the large scale structures. These structures are strongly three dimensional and compatible with the observations already described previously.

Studies of coherent structures in the outer region of a smoke filled turbulent boundary layer were described by Falco (21). Ciné

films of the flow revealed medium scale vortices which were typically of one-tenth the boundary layer thickness in scale though their
scale depended on the magnitude of the Reynolds number. Near transition they were effectively the large eddies of the flow. Simultaneous measurements of the velocity distributions revealed time
histories similar to those reported above (18) with a similar
pattern associated with the passage of a vortex.

The problems of identifying and studying coherent structures
in boundary layers are reduced if the flow can be made more orderly.
Johnston (22) achieved this by imposing Coriolis forces produced by
rotation of a channel flow. Ciné films of the resultant visualized
flow revealed large longitudinal vortices on the unstable side of
the channel. Here, however, the cores of the vortices were waxy so
that coalescence involved a complex entanglement in which the
details of the structure became somewhat obscure.

Some time ago Blackman (23) showed that the average of structure of the velocity field in an Emons (turbulent) spot was similar
in most respects to that of a fully developed boundary layer.
Assuming this is generally true D. Coles designed an experiment
where a laminar boundary layer was periodically tripped at a number
of spanwise positions. Preliminary results reported to the
colloquium by Roshko showed that the structures identified in the
flow were basically similar to those described above that were by
others in untripped boundary layers.

In another experiment involving periodic forcing, M. V.
Markovin (6) described the behavior of the shear layer behind a
rectangular obstacle buried in a laminar boundary layer. Acoustic
forcing at a frequency f in the precritical regime produced vortex
loops but no transition. Simultaneously applied frequencies f_1 and
f_2 produced vorticity waves with frequencies $f_1 + f_2$, $2f_1 + f_2$,
etc. and higher harmonics, but the transition to turbulence was
produced by the amplification of the very low level component,
$f_1 - f_2$. Direct forcing at this frequency did not produce transition and the possible generality of this non-linear phenomenon was
considered. Though the general validity of the results of such
experiments might be questioned, they are clearly of great value
in deepening our understanding of instability and transition.

2.4   Turbulent Pipe Flow

I. J. Wygnanski (6) described an experimental investigation of
artificially instigated pipe flow transition at lower Reynolds numbers in which phase averaged conditional sampling was used to educe
the average structures of 'turbulent puffs.' These events were
found to differ from the 'turbulent slugs' previously investigated

by Wygnanski (24) for higher Reynolds numbers. They consisted of
two co-rotating ring vortices with a smaller eddy between them
which represented the majority of the turbulent activity and shear
stress.

J. Sabot (6) discussed the investigation of fully developed
pipe flow by extensive space-time cross-correlation measurements.
The observations indicate that the core flow consists of narrow,
organized rotational structures derived from the pipe wall. The
core region was thus basically similar to the outer region of a
boundary layer and it could be considered to contain many uncorrelated 'bulges' or structures of different ages and originating
from different sectors of the pipe.

These observations suggest that the tendency for continued
growth of the vortices so clearly revealed in mixing layers also
occurs in all other shear flows, though the continued development
in scale may be constrained to some extent by rigid flow boundaries.
Whether unsheared turbulence exhibits the same behavior has not
been demonstrated, in fact observations of grid flow turbulence
indicate that growth in scale ceases after an initial adjustment
region. Shear flows are characterized by a distribution of vorticity which is predominantly of one sign while grid flows differ
in this respect; so one might expect a different behavior.

2.5   Models of Turbulent Flow

In the second invited lecture of the Colloquium, G. M. Lilley
discussed the modelling of large scale structure with emphasis on
stability analysis. It was proposed that the large scale organized
structures could be represented by wave packets with characteristics
derived from linear stability theory. There were a number of contributions to the colloquium on this topic which have been reported
in more detail elsewhere (6). One might summarize the progress
achieved so far by noting that there has been some marked success
in predicting the most likely modes and frequencies of the initial
flow instabilities.

Other models employed finite difference solutions of the complete or simplified Navier-Stokes equations. A. J. Grant (6),
working with the complete equations was able to predict a street
of vortex rings in the initial region of a circular jet. Associated
experiments concerned the role of vortices in entrainment mechanisms
and their effect on flame stability in jet diffusion flames. J. W.
Deardorff (6) presented an impressive numerical model of an atmospheric boundary layer. Temperature heat flux and humidity variations were included as was the rotation of the earth. Deardorff
described how conditional sampling and correlation methods could be
employed for comparisons between the model and observations.

Potential flow calculations of the development of an isolated vortex ring and of a turbulent jet were described by Davies (25). The results were illustrated by ciné films which showed alternating scenes of a visualized flow followed by its computed equivalent. In this kinematic model the cylindrical vortex sheet shed at the jet lip was represented by a row of periodically generated elementary vortex rings. Uniform flow through the jet orifice was maintained by an appropriate updated source distribution. Streaklines were simulated by arrays of periodically generated passive points.

The jet model showed that the vortex sheet developed a Kelvin Helmholz instability and rolled up into a train of vortex rings. Spreading of the mixing layer occurred by the coalescing of neighboring vortex rings into larger still coherent structures. The models were constrained to be axisymmetric and therefore the structure remained more orderly than that of a real flow. The effect of this constraint appears to be similar to that of forcing the jet by some initial disturbance which also encourages more order in the flow. The value of extending such models to provide a more realistic three-dimensional structure was discussed at the colloquium though it was agreed that the practical problems of maintaining a reasonable economy of computer time is a major constraint.

A. Leonard (26) described the initial results of numerical simulations of interacting three-dimensional vortex filaments which had the ultimate objective of modelling turbulent shear flows in a three-dimensional manner. Ciné films of the computed movements and distortions of the vortex filaments were shown for the interactions and instabilities of vortex rings, helical vortex filaments, trailing vortices and a round starting jet.

It appears that both the development of the Kelvin-Helmholz instability and the subsequent coalescence and migration of vorticity to provide the growth of the vortices can be illustrated by potential flow models (25). The way in which most transport properties may also be related to the potential flow fields of arrays of vortices has been demonstrated by Christiansen and Zabusky (27). That these purely kinematic models of the velocity field should predict quantitatively so many of the characteristic features of high Reynolds number flows is interesting. It also demonstrated the relatively minor role in their development played by viscosity, whose global effects can be replaced by appropriate distributions of vorticity.

The quantitative description of the detailed flow patterns in the coherent eddy structures and the investigation and quantification of their interactions presents a challenge to the experimentalist. Adequate conceptual tools are required to organize and interpret the new information in a meaningful and useful way, so that the understanding and modelling of turbulence structure can be improved.

3. DISCUSSION

There is some profit, perhaps, in making an initial attempt to correlate the new experimental information and ideas on coherent structure with existing statistically derived models of turbulent shear flow. The interaction and coalescence of large scale vortex-like eddies has been proposed as a characteristic element of the structures of jets, mixing layers, wakes and also of the outer boundary layer region and the core region of pipe flow.

The classical statistical turbulence is normally expressed in terms of a wavenumber spectrum $\phi(k)$. The theory defines individual component eddies within this structure as producing portions of the spectrum lying between wavenumbers k_r and $k_r + \delta k_r$. However, a single region of concentrated vorticity will always yield a broad wavenumber spectrum, the lowest wavenumber corresponding roughly to its physical extent while the highest corresponds roughly to its steepest local velocity gradients whose magnitude is limited by viscous diffusion.

Townsend (1) has demonstrated that an array of organized structures with a single characteristic scale also produced a broad wavenumber spectrum. Further, Base and Davies (28) have shown that a synthetic flow consisting of an array of randomly-spaced, identical vortices can reproduce both the probability and spectral characteristics of point turbulence measurements. Good matching involves only a judicious choice of their spacing and scale.

Although, there are fundamental differences between coherent structures and the classical statistical concept of an "eddy," one could always represent a homogeneous field of concentrated regions of vorticity by a group of statistical eddies. But it is known that the field for real shear flows is not truly homogeneous as the flow is bounded. Thus, perhaps a physically more meaningful and informative model of flow structure is provided by characteristic sequence of velocity or vorticity distributions. This can provide more insight than a spectral representation, although such a representation is convenient analytically. Further, since the energy of any component eddy is bounded, the wave system must be dispersive, so that the spatial extent of each wavenumber component k_r is restricted to lie within the flow boundaries. This restriction is avoided in the classical statistical theory by assuming that the turbulence is homogeneous.

3.1   The Turbulent Energy Spectrum

The concept of coherent structure may also clarify certain practical problems in the direct measurement of point wavenumber spectra in shear flow turbulence. Consider the velocity time history record at a fixed point in an unsteady flow consisting of a succession of irregularly spaced coherent structures. Since the velocity field must be continuous, one would expect to find a relatively high correlation between parts of the record lying close enough to be within a given structure. But one would expect a low, or negligible, correlation between parts arising from neighboring structures and zero correlation between those parts of the record from structures remote from each other. There seems little doubt that such a record will be ergodic and random, provided sufficiently long records are analyzed with an appropriately large time separation.

One can transform the frequency spectrum obtained from such a time history record to a wavenumber spectrum, either by assuming that the flow structure is frozen (Taylor's hypothesis), or by appropriate use of the phase velocity spectrum. It turns out, not surprisingly perhaps, that the phase velocity varies with frequency (29) though there are some uncertainties in its measurement. What evidence there is suggests that shear flow turbulence can never be taken as a frozen pattern, particularly for the high frequency components.

In shear flows the transformation of space correlation measurements of the velocity to a wavenumber spectrum reveals similar difficulties, since such correlation curves are antisymmetric and yield a complex spectrum. This suggests that the joint statistical characteristics of time histories taken from nearby points have properties which differ from those of the ergodic velocity time histories from single points. The strong interactions found between neighboring coherent structures suggests one reason for this lack of homogeneity.

However, if examined from the viewpoint of the statistical model, it is not difficult to see that information from such joint records might appear isotropic, particularly at wavenumbers (not scales) large compared with the characteristic wavenumbers of irregularly spaced coherent structures. Thus, although space-time correlations and spectral measurements permit the assignment of local wavenumbers and frequencies to the turbulence, strictly they must be regarded as local descriptions of the averaged properties of the information obtained from an array of vortices passing the point in question. Thus the transformation from a Lagrangian frame, in which such eddies exist, to the Eulerian one in which we are normally obliged to make quantitative observations, has local significance only.

The same difficulty applies to the interpretation of such averaged reduced quantities as eddy viscosity, eddy equilibrium and the turbulent energy balance, as any more than localized descriptions of observations which do not strictly represent actual properties of the moving turbulent flow. This is not to say that these concepts are not both useful and convenient methods of summarizing experimental data for practical application.

3.2   Entrainment and Mixing

Recent observations (5), (6), (11) suggest that the entrainment of ambient air by jets and mixing layers is accomplished by a potential flow engulfment process, which is controlled by the large concentrations of vorticity and their interaction. It appears that, in jet and mixing layers, entrainment is greatest during the later stages of coalescence. The fact that potential flow models furnish satisfactory flow kinematics shows that entrainment may be described by a potential flow mechanism. It is interesting to note that not only do interacting vortices promote mixing by entraining fluid, but they can also shed a significant proportion of the original potential flow associated with them. This is clear in the visualizations of Yule and of Fiedler (6) and it also occurs in the 'turbulent' vortex ring flow of Maxworthy (30).

4.   CONCLUSION

Such recent experimental evidence makes it quite clear that large scale coherent structures containing concentrated regions of vorticity do play a dominant role in shear flow development. Furthermore, great insight can be obtained into turbulent flow processes and turbulent mixing by studying the development of these structures and their interactions.

REFERENCES

1. Townsend, A. A., 1956. The structure of turbulent shear flow. Cambridge Univ. Press.

2. Tani, I. and Hama, F. R. 1953. Some experiments on the effect of a single roughness element on boundary layer transition. J. Aero. Sci. 20, 289-290.

3. Lau, J. D., 1971. Ph.D. Thesis, University of Southampton. The coherent structure of jets.

4. Davies, P. O. A. L., 1973. Structure of turbulence, J. Sound and Vib. 28, 513-526.

5. Brown, G. and Roshko, A., 1974. On density effects and large
    structure in turbulent mixing layers. J. Fluid Mech. 64,
    775-816.

6. Davies, P. O. A. L. and Yule, A. J., 1974. Coherent structures
    in turbulence, a Symposium, Southampton University, March,
    1974, to be published.

7. Fisher, M. J. and Davies, P. O. A. L., 1964. Correlation
    measurements in a non-frozen pattern of turbulence. J. Fluid
    Mech. 18, 97-116.

8. Kovasznay, L. S. G., 1971. The structure of turbulence in
    shear flow. AGARD CP-93, D1-D14.

9. Blackwelder, R. F. and Kaplan, R. E., 1971. Intermittent
    structures in turbulent boundary layers. AGARD CP-93, Paper 5.

10. Brown, G. and Roshko, A., 1971. The effect of density differ-
    ence on the turbulent mixing layer. AGARD CP-93, Paper 23.

11. Winant, C. D. and Browand, F. K., 1974. Vortex pairing: the
    mechanism of turbulent mixing layer growth at moderate Reynolds
    number. J. Fluid Mech. 63, 237-255.

12. Yule, A. J., Bruun, H. H., Baxter, D. R. J. and Davies,
    P. O. A. L., 1974. Structure of turbulent jets. ISVR Memo.
    506, University of Southampton.

13. Hama, F. R., 1963. Progressive deformation of a perturbed
    line vortex filament. Phys. Fluids 6, 526-534.

14. Wille, R., 1963. Growth of velocity fluctuations leading to
    turbulence in a free shear layer. AFOSR Tech. Rept., Hermann
    Föttinger Inst. Berlin.

15. Widnall, S. E. and Sullivan, J. P., 1973. On the stability of
    vortex rings. Proc. Roy. Soc. Lond. A332, 335-353.

16. Michalke, A., 1970. A note on the spatial jet instability of
    the compressible cylindrical vortex sheet. DLR-FB 70-51,
    Berlin.

17. Offen, G. R. and Kline, S. J., 1974. Combined dye streak and
    hydrogen bubble visual observations in a turbulent boundary
    layer. J. Fluid Mech. 62, 223-240.

18. Nychas, S. G., Hershey, H. C. and Brodkey, R. S., 1973. A
    visual study of turbulent shear flow. J. Fluid Mech. 61,
    513-540.

19. Eckelmann, H., 1974. The structure of the viscous sublayer
     and the adjacent well region of turbulent channel flow. J.
     Fluid Mech. 65, 439-569.

20. Wilmarth, W. W. and Lu, S. S., 1972. Structure of Reynolds
     stress near the wall. J. Fluid Mech. 55, 65-92.

21. Falco, R. E., 1973. Some comments on turbulent boundary layer
     structure. 12th Aerospace Sciences Meeting, AIAA Paper 74-99.

22. Johnston, J. P., Halleen, R. M. and Lezius, D. K., 1972.
     Effects of spanwise rotation on the structure of two-dimensional
     fully developed turbulent channel flow. J. Fluid Mech. 56,
     533-557.

23. Blackman, D. R., 1964. Ph.D. Thesis, University of Southampton.
     Wall pressure fluctuations in the laminar turbulent transition
     region of a boundary layer.

24. Wygnanski, I. J. and Champagne, F. H., 1973. On transition in
     a pipe, Part 1. The origin of puffs and slugs and the flow in
     a turbulent slug. J. Fluid Mech. 59, 281-335.

25. Davies, P. O. A. L. and Hardin, J. D., 1973. Potential flow
     modelling of unsteady flow. Int. Conf. on Numerical Methods
     in Fluid Dynamics, Southampton, Springer, Berlin.

26. Leonard, A., 1974. Numerical Simulation of Three Dimensional
     Vortices. Proc. 4th Int. Conf. on Numerical Methods in Fluid
     Dynamics, Boulder, Colorado, June 24-29, 1974, Springer, Berlin.

27. Christiansen, J. P. and Zabusky, N. J., 1973. Instability,
     coalescence and fission of finite area vortex structures.
     J. Fluid Mech. 61, 219-243.

28. Base, T. E. and Davies, P. O. A. L., 1967. Computer studies
     of vortex models to represent turbulent fluid flows. Aero.
     Res. Counc. (Lond.) Paper 29072, N519.

29. Davies, P. O. A. L. and Mercer, C. A., 1973. Phase velocity
     measurements using the cross power spectrum. Proc. Int.
     Symposium on Measurement and Process Identification by Corre-
     lation and Spectral Techniques in Measurement (Bradford) 27-36.
     Publ. Inst. of Meas. & Control.

30. Maxworthy, T., 1974. Turbulent vortex rings. J. Fluid Mech.
     64, 227-239.

LARGE SCALE MOTION IN TURBULENT BOUNDARY LAYERS

Leslie S. G. Kovasznay

The Johns Hopkins University

Baltimore, Maryland 21218

Experiments reported in the last five years confirmed more and more the existence of large scale coherent motion in turbulent boundary layers as well as in other turbulent shear flows. The principal experimental tool was the hot-wire anemometer followed by novel signal processing techniques. Conditional sampling and averaging, as well as the mapping out of double space-time correlations for different flow variables (velocity components, intermittency function, temperature, etc.) (Ref. 1), have indicated that the large scale flow structures (eddies) are relatively long lasting entities that may extend all across the layer's thickness in the direction of the highest gradient, namely perpendicular to the wall. They may be quite elongated in the flow direction and their inferred average life-times correspond to time intervals that allows them to travel downstream with the free stream velocity as much as 10-20 boundary thicknesses.

Unfortunately by using only the two-point statistics (two point double space-time correlation or any equivalent description) one still cannot distinguish between two possible models: on one hand generally large scale "blobs" that have grown by diffusion due to the smaller scale turbulence; or on the other hand, the flow may consist of smaller "blobs"; structures that during their existence may disperse by random walk and appear as large and more diffused structures.

Three point triple space-time correlations can distinguish between the above described two cases. In order to make a systematic study a cooperative program was established between present author and the group at IMST in Marseille (the authors of Ref. 2

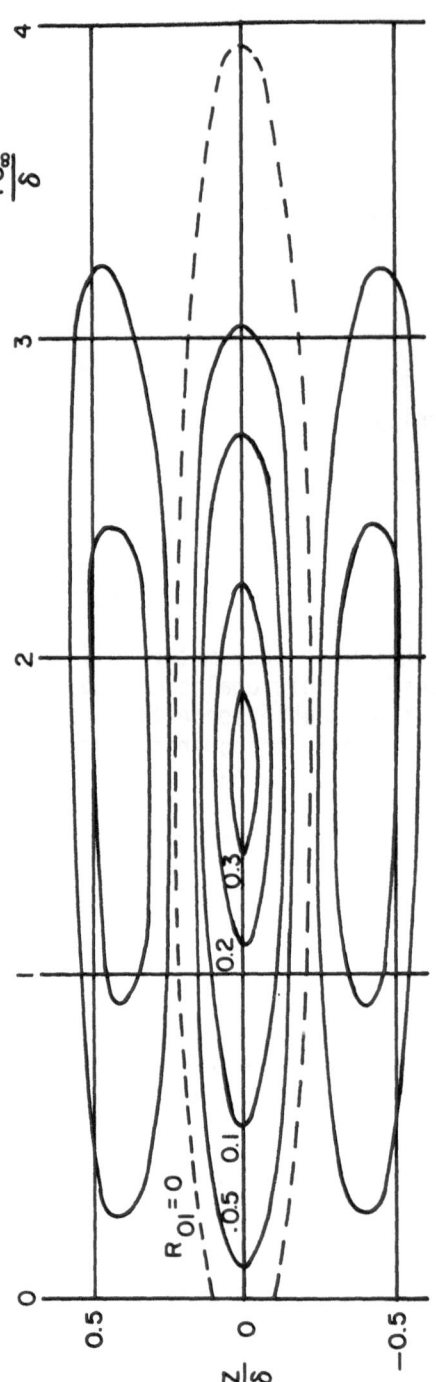

Figure 1.   Double Space-time Correlation R_{01};   X = 1.86;   y = y_0 = .034

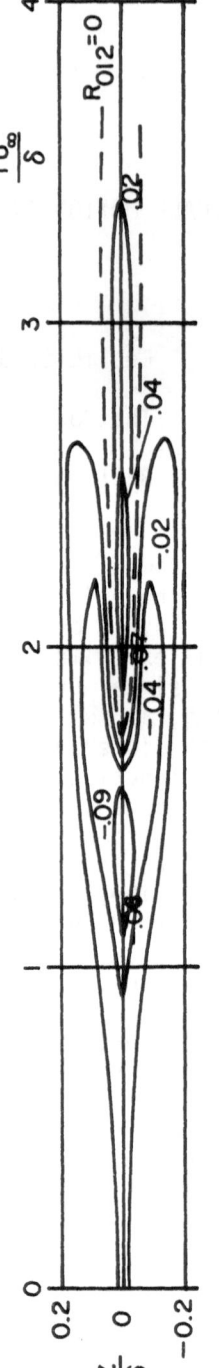

Figure 2.   Three Point Triple Space-time Correlation R_{012};
X = 1.86;   y = Y_0 = .034

and 3) under a NATO grant and the results are appearing in a
sequence of papers. The first two of those are already published
as Refs. 2 and 3.

        In order to simplify matters, it was decided to use a single
scalar variable namely, temperature produced by a moderate amount
of heating of the boundary layer and the double and triple space-
time correlations of the temperature fluctuations were chosen to
describe the large scale coherent structures. In order to minimize
the number of independent variables the following geometric configuration of the probes was used. All three probes were placed in
the same plane parallel with the wall. One probe was placed upstream at P_0 $(x = x_0$ $y = y_0$ $z = 0)$ and its signal was delayed by
a time T; the other two probes were downstream at P_1 $(x = x_0 + X$;
$y = y_0$ $z = Z)$ and P_2 $(x = x_0 + X, y = y_0;$ $z = -Z)$. The streamwise
distance of the probes was $X = 1.86$ δ where δ is the conventional
boundary layer thickness. In the measurements only two variables
were varied: the time delay T and the spanwise separation Z of the
two rear probes. One may plot "maps" of the double space-time
correlation

$$R_{01}(Z, T) \equiv \overline{v(P_{01}t) \; v(P_1, \; t + T_1)}$$

on the Z, T plane (Figure 1). Now turning to the three point
triple correlations, the triple correlation was defined as

$$R_{012}(Z, T) \equiv v(P_0,t) \; v(P_1, \; t + T) \; v(P_2, \; t + T)$$

and it can be plotted on the same Z, T plane (Figure 2). In addition the double correlation between the temperature at P_1 and P_2
without relative time delay was also measured (Figure 3).

$$R_{12}(Z) = \overline{v(P_1, \; t) \; v(P_2, \; t)}$$

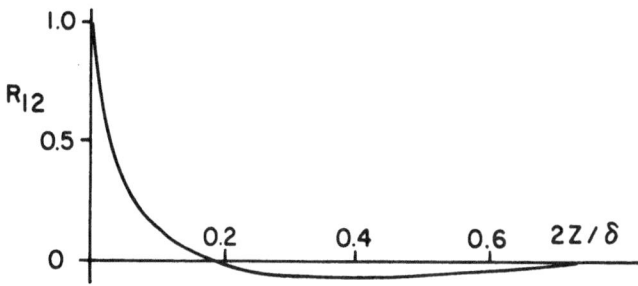

Figure 3.   Double Space Correlation X = 0; y = y = .034, T = 0

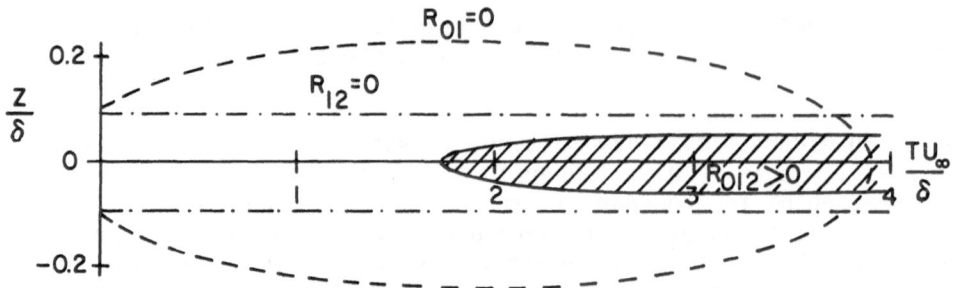

Figure 4. Zone of Positive and Negative Correlations. In Shaded
           Area all Three Correlations are Positive (Composite
           Drawing from Figures 1, 2 and 3)

One may easily outline a "portrait" of the turbulent patch by outlining the zero correlation contours for each one of the three
different types of correlation (Figure 4). By repeating the same
technique in different planes parallel to the wall, one may map
out the different zones of influence of the heated "blobs".

        The results were further refined by measuring the correlation
of truncated signals pinpointing the separate contributions to the
total double and triple correlation from all combinations of sign
in the fluctuations (see e.g., Refs. 4 and 5).

        The main conclusion is that the hot "blobs" are rather coher-
ent but they are much narrower in the span-wise (z) direction than
one would have inferred from double correlations alone. Furthermore, in the outer intermittently turbulent zone the double and
triple correlations are very similar to each other, thus strongly
supporting the pulse-like "chunky" structure of the flow. The
measurements may be considered as a direct evidence of the existence of large scale coherent structures in shear flow turbulence.

REFERENCES

1. Kovasznay, L. S. G., Kibens, V., Blackwelder, R. F., Jour.
    Fluid Mech. Vol. 41, p. 283 (1970).

2. Fulachier, L., Giovanangeli, J. P., Dumas, R., Kovasznay,
    L. S. G., Favre, A., Structure des perturbations dans une
    couche limite turbulente: zone interne Comptes Rendus de
    l'Academie des Sciences, Paris, t. 278 Serie B. p. 683 (1974).

3. Fulachier, L., Giovanangeli, J. P., Dumas, R., Kovasznay,
    L. S. G., Favre, A., Structure des perturbations dans une
    couche limite turbulente: zone d'intermittance, Comptes
    Rendus de l'Academie des Sciences, Paris, t. 278, Serie B,
    p. 999 (1974).

4. Willmarth, W. W., and Lu, S. S., Jour. Fluid Mech., Vol. 55,
    p. 65, (1972).

5. Brodkey, R. S., XIII Int. Congress of Theoretical and Applied
    Mechanics, Moscow, 1972 (article just about to be published
    in JFM).

DISCUSSION

GOLDSCHMIDT: (Purdue University, West Lafayette, Indiana)

There is some question in my mind whether we are looking at
the best co-ordinates for the system. We usually choose X and Y
because they are convenient and happen to be lined up with the wall.
Do you think there may be a situation where we should start to
search for things in other directions than normal and along the wall?

KOVASZNAY:

It is being done now in Marseille. It was not the usual
practice to make such measurements.

KLINE: (Stanford University, Palo Alto, California)

In continuation of what Vic Goldschmidt just asked, being
simple-minded, I have a prejudice for seeing what I am looking and
hence in allowing the visual studies to lead one to the next, hopefully more precise, measurements. In this instance there are not
only visual measurements by Runstadler et al. (JFM, 30, 4 Dec.
1967), but also isocorrelation plots by Favre et al. (Jour. Physics
of Fluids Supplementary Vol. for Kyoto Symposium, Sept. 1967) which
show that the lines of maximum correlation do not follow orthogonal
x, y, z lines. Consequently, I agree with Goldschmidt that it
would make more sense not to make x displacements primary. Moreover,
as you have suggested and as the data of Runstadler et al. (op. cit.)
and also the several papers by the group under Brodkey at Ohio State
show (e.g., JFM 37, 1969), the random walk portion of the motions
is one of the greatest difficulties in sorting out the coherent
structure. If this is so, then it follows that more progress should
be possible by following lines of maximum correlation, and hence I
fail to understand why you follow X-Y coordinates.

KOVASZNAY:

I reassure you, we are going in the other direction. We can
actually face the big probe a different way; that was the first
fact, which was pretty clear. You see, however, as soon as you
want the direction normal to the wall included, then, of course,
you will not have the symmetry you have here. But there are a lot
of other things that are actually rewarding.

KLINE:

Yes, I agree with you that three-point measurements of this
general sort should be much more rewarding in beginning to sort
out coherent structures. What I am trying to say beyond this,
and beyond what Vic Goldschmidt said, is that there are so many
different types of turbulent flows. Turbulence is not just a
single kind of base state, as we once thought--it is a whole array
of complicated phenomena. Hence I conclude that we need to optimize
the measurement efficiency by first finding the nature of the gross
structure and then designing experiments utilizing that information
of the sort you discuss.

KOVASZNAY:

I agree.

BRADSHAW: (Imperial College, London, England)

If the temperature signal is essentially a flat-topped on-off
signal, should there not be such statistical relations between the
"plus-plus," "plus-minus," "minus-minus" and other quadrant signals?

KOVASZNAY:

You need additional statistical assumptions to get that result.

BRADSIIAW:

So there is really no redundancy in these measurements that
you have made?

KOVASZNAY:

No. I think you can get probably some "anchoring" at a
distance from the wall where intermittency is about one half
($r = 0.5$). Look, if you have a very low intermittency factor r,
wherein the signals are mostly narrow pulses, then all order
correlations look alike, because only the coincidence gives you
positive correlations. And that is the tendency when you go

outward in the low intermittency region. I do not mean it is
exactly that, but that is my overall reaction. I get these very
high triple correlations, and remember, you never encounter such
high triple correlations anywhere else.

JONES: (University of Illinois, Urbana, Illinois)

Are you finding that you can effectively and spatially
separate a double correlation by making measurements in your other
directions? Since you completed the experiments in the plane
parallel to the wall, are you able to measure in the X_3- and X_2-
directions and infer correlation in the non-orthogonal direction?

KOVASZNAY:

Double correlations, we have done in that direction; it is in
our 1970 paper.* Double space-time correlations were done also in
the other direction.

JONES:

Are you finding that you can spatially separate them, such
that the double correlation on a diagonal may be expressed by
rather complete separability the double correlations in the
coordinate directions? This is illustrated in Figure A, in which
the two-dimensional case is presented.

KOVASZNAY:

If you mean by separation expressing these as a product, no,
you cannot do this. It has more "character," it is not just the
product of the three correlations in the three directions.

JONES:

I am not talking about the triple correlations.

KOVASZNAY:

No, the double. Look, just to give you an example: we
measured, for instance, things like vv', normal component at two
point stations. The distribution is quite skewed and it is not
just the product of the two unskewed correlations because there
is a great deal specific pattern, and there is no a priori reason
why one should be able to factorize.

---

*   Kovasznay, L. S. G., Kibens, V. and Blackwelder, R. F. (1970)
     J.F.M. 41 283.

JONES:

Well, we are finding this in pressure correlations.

KOVASZNAY:

Because the kinematic pressure can be regarded as a vector perpendicular to the wall.

JONES:

Pressure is scalar.

KOVASZNAY:

Can you measure it off the wall?

JONES:

No, not in this boundary layer.

KOVASZNAY:

That is what I meant. It is only a two dimensional space in which we can measure pressure, namely the X_1, X_3 plane at $X_2 = 0$. In that plane the fluctuations are reasonably homogeneous in both directions. But in X_2 (or Y) the coordinate perpendicular to the wall is still the direction in which the flow is not homogeneous. And, you get tilts in the correlation pattern because the maximum correlation is not along a line perpendicular to the wall.

SOME REMARKS ON A SYNTHETIC TURBULENT BOUNDARY LAYER

D. Coles and S. J. Barker

California Institute of Technology

Pasadena, California

Professor Murthy has kindly consented to insert me in the program at this time, like mayonnaise in a sandwich between two distinguished eddy chasers, Kovasznay and Roshko. My object is to establish my own credentials as an active eddy chaser, and to lay a foundation for the panel discussion this afternoon.

Eddy chasers are people who believe, to one degree or another, that most of the transport processes in a turbulent shear flow are associated with large, organized, almost deterministic structures. The problem is to find out what these structures look like and what their transport properties are. Ninety percent of the work on this problem has been done in the boundary layer, although there is an occasional case of serendipity like the photographs obtained by Brown and Roshko (1974) in the plane mixing layer. The problem with eddy chasing in the boundary layer is that if you look for a signature, either in terms of bursting, or longitudinal streaks in the sublayer, or local regions of intense energy at high frequencies, or bulges in the interface, or properties of the pressure signal, or whatever, it is very difficult to define the signature without knowing where the eddy is, and it is very difficult to find out where the eddy is without knowing what the signature looks like.

Now, I'm going to talk about a possible exit from this dilemma, involving something which I call a synthetic turbulent boundary layer. My own feeling is that the prototype large eddy is the turbulent spot. We know what these spots look like, what some of their properties are, how to produce them. The scheme in this particular experiment was to generate a regular array of turbulent

spots and send them down the boundary layer and see what happens.
The experiment was--I won't say it was done impulsively, but it
was almost ad-libbed, due to a particular, favorable combination of
circumstances. We had the model in the channel and bodies in a
laboratory course and the instrumentation working.

The whole experiment took five days, of which about 40 percent was productive. It was done in water using laser doppler
instrumentation. The model was a plate eight feet long with an
elliptical nose. In the absence of any disturbance the boundary
layer was still laminar at the trailing edge (the length Reynolds
number was about one million), except for the usual transverse
contamination. About a foot back was a row of small disturbance
holes where we could turn on intermittent jets normal to the
surface, using something like a big hypodermic needle to pump
water out. These holes could be manifolded together. For the
spot we used one hole, with enough time between spots so that they
were independent. For the synthetic boundary layer we had two
manifolds which were pumped alternately in a regular way.

The laser doppler system measured the streamwise velocity as
a function of time at one point. The nicest property of the laser
doppler velocimeter is that the instrument is linear, and that
really saves on computing costs. The frequency-modulated signal
from the photomultiplier was cleaned up by an amplifier and a
phase-locked loop and demodulated by a frequency-to-voltage con-
verter, and the reconstituted analog signal was run through a data
system. An integrating digital voltmeter integrated the signal
for one sixtieth of a second. This was done deliberately, to
remove the high frequencies and leave only the large-scale or
energy-containing part of the signal. Samples were taken at about
eleven per second. The data system wrote everything down on
magnetic tape and also generated the next disturbance at the right
time.

One of the byproducts of the research was a very useful idea
of what a turbulent spot really is. I claim, and I hope to con-
vince you, that a turbulent spot is a large U-shaped vortex which
moves down the plate with its ends slipping along the surface.
The first slide shows a spot using dye visualization. This photograph happens to be from Elder's paper (1960); we didn't get any
photographs of ours, but they looked exactly like this.

The next slide (Figure 1) shows a cross section through the
spot in a plan view. The main coordinate is time, not x. The
probe is outside the laminar boundary layer. At several lateral
positions we just accumulated data points for 64 spots and averaged.
What's plotted as velocity is simply normalized voltage and is
essentially raw data. The shape of the spot is shown projected on

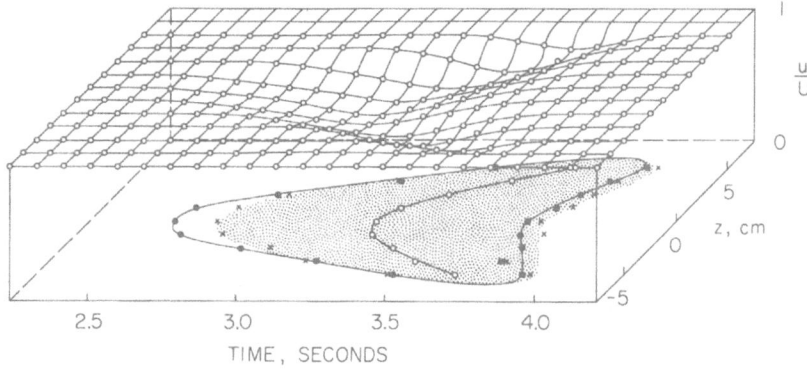

Figure 1. Section through a Turbulent Spot in a Plane Parallel
           to the Wall

the base of the figure. This shape was defined in two ways. The
solid circles show where the velocity is down by two percent from
the free-stream value. The x's and the shaded area show where the
rms low-frequency fluctuation level is two percent.

    There is one more line on the figure, drawn through the open
circles. These points are points of minimum velocity for each
traverse inside the spot. We believe that this line shows where
the vortex is, because it shows where the wall is doing best at
slowing down the fluid, and where the mixing is most effective.

    The next slide (Figure 2) is a similar cross section of the
spot on its plane of symmetry, with the probe at various distances
from the wall. The shape is defined in the same way, and so is
the line of minimum velocity, which turns out to be perpendicular
to the wall. The mean velocity profile along this line looks very
like the profile in a turbulent boundary layer.

    If we track the locus of minimum velocity or best mixing down
the flow, we get a celerity for the spot of 83 percent of the free-
stream velocity. Now in a coordinate system moving at this speed
we can imagine that the spot is steady and two-dimensional, and
compute a stream function. Of course, neither of these assumptions
is right, but I don't think there is any really serious error in
our conclusions.

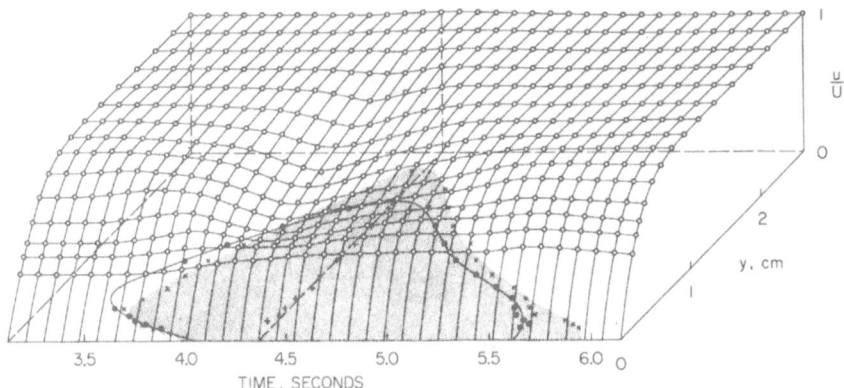

Figure 2.   Section through a Turbulent Spot in the Plane of
            Symmetry

These conclusions are shown in the next slide (Figure 3). The
spot turns out to be a large hairy vortex--in fact, that is all the
spot is. The turbulent part up in front is more or less like spray
being torn off a wave by wind. There are no scales on the figure,
because one coordinate is time and one is distance, but we estimate
that we have a distortion of about 5 to 1 for thickness to length.
The dashed line shows where the section was taken for the plan
view in Figure 1.

I call your attention to one streamline which comes in at the
front near the wall, goes around the vortex, and goes out the front
again. If we had really been doing this right, with x/t and y/t as
coordinates, this streamline would probably wrap up into the vortex.
This process must be what feeds the spot. The spot overruns the
laminar flow and sweeps up vorticity bearing fluid like a vacuum
cleaner. Furthermore, the transport of fluid away from the surface
inside the vortex loop, induced by the vortex, is probably what is
called "bursting" by the eddy chasers.

The next slide (Figure 4) shows the mean velocity profile in
the synthetic turbulent boundary layer, farther down the plate.
This velocity can be thought of as a sum of three parts. One is a
global mean, averaged over time and over lateral position. This
global mean profile is plotted in the figure, and it seems to look
like anybody else's profile for the low Reynolds number in question.
The second part of the profile is a periodic variation of about

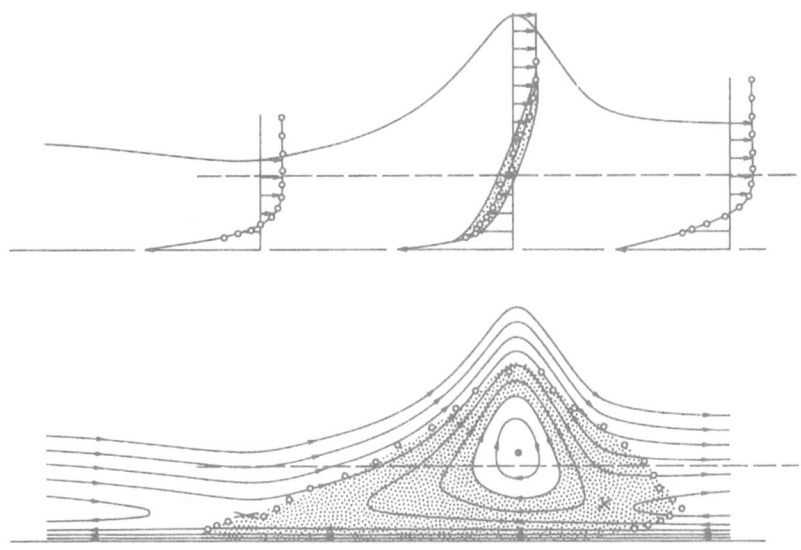

Figure 3.   Streamlines inside a Spot in Moving Coordinates Assuming
            Two-dimensional Steady Mean Flow

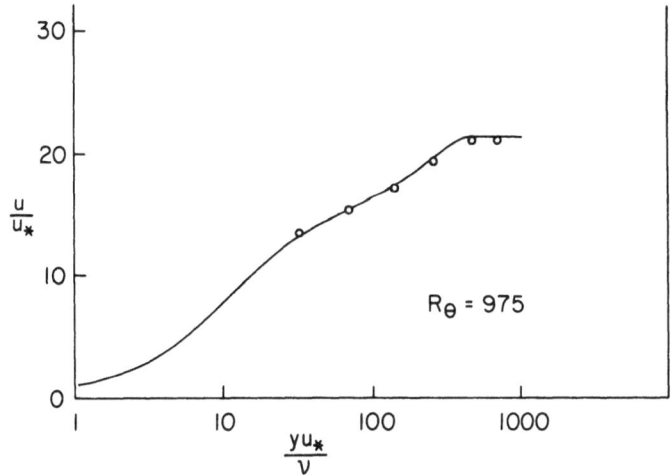

Figure 4.   Global Mean-velocity Profile in the Synthetic Turbulent
            Boundary Layer

plus or minus two percent--I mean periodic in space and time. This
is the part shown as a deviation with zero mean in the bottom half
of the next slide (Figure 5). One coordinate is time and the other
is distance across the flow.

    This is a plan view of what is happening about one third of
the way through the boundary layer. If I remember rightly, these
data are averages for 512 cycles of the spot pattern. There are
two complete identical cycles in the figure. The third part of
the profile is a low-frequency random fluctuation of about five
percent. This fluctuation is shown at the top in Figure 5. There
is also a fourth part, the high-frequency random fluctuations,
which we have thrown away. The signal we want is the periodic
part, and it is easier to dig this out of the whole signal by
eduction if the high frequencies are already gone.

    In the bottom half of Figure 5, I have shaded the region where
the periodic part of the mean velocity is negative. The idea is
still that this is where the mixing is most efficient, and therefore this is where the operators are. Notice that the shape is

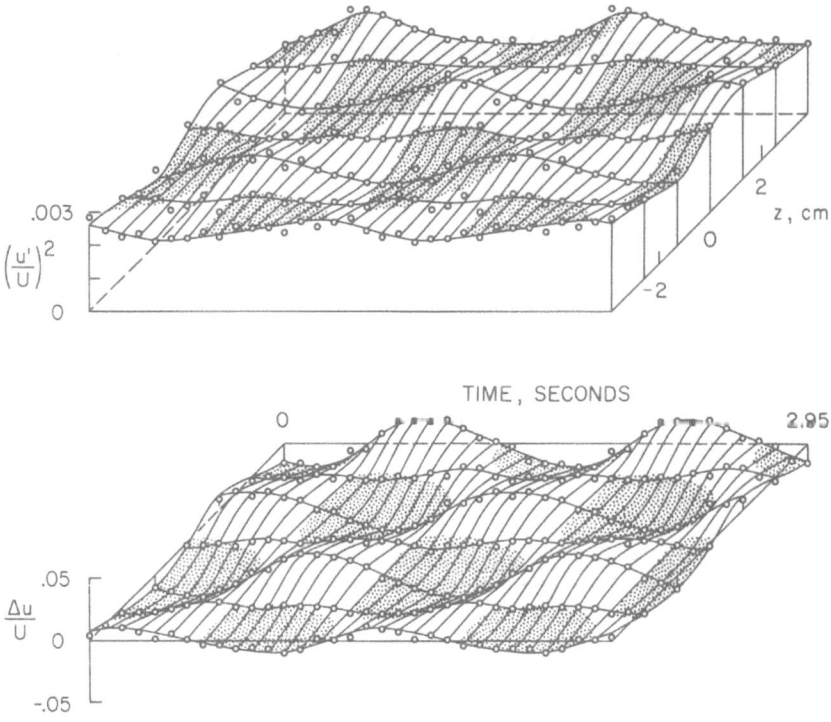

Figure 5.   Section through the Synthetic Boundary Layer in a Plane
             Parallel to the Wall

different; there is a blunt front end. We aren't sure this is
really so. The upper half of Figure 5 shows the mean square
fluctuation energy at low frequencies. The shaded areas are
transferred from the velocity plot down below. You can see an
interesting property of the operators--about the only property
we are sure of at the moment. When the probe enters a shaded
region or operator, with time increasing, the low-frequency energy
is a minimum. As the probe passes through the operator, the low-
frequency energy increases to a maximum at the rear, and then
decays until the next operator. Of course, it is really the
operators which are passing by the probe, and I regret that we
failed to measure their celerity due to simple stupidity.

The next slide (Figure 6) shows the same kind of data and
leads to the same kind of conclusions for a traverse normal to the
wall at a station near the center of one train of operators.
There is some indication that the operators lean downstream, since
they are encountered earlier when the probe is farther from the
wall. You can also see where the turbulent bulges in the interface must be.

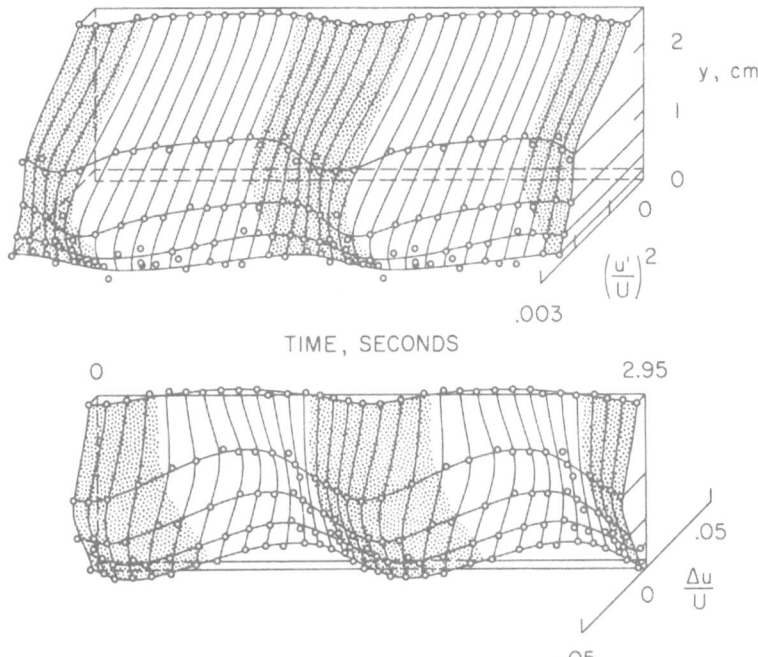

Figure 6. Section through the Synthetic Boundary Layer in a Plane
           Normal to the Wall

That is about all we know at the moment. You can be sure that
we intend to pursue this scheme of a synthetic boundary layer and
look at other properties of the turbulence. It is true we didn't
learn much about signatures in these preliminary measurements, but
we are satisfied that the approach is worth continuing, and we
intend to continue it.

REFERENCES

1. Brown, G. L., and Roshko, A.: 1974 On density effects and
    large structure in turbulent mixing layers. JFM 64, 775.

2. Elder, J. W.: 1960 An experimental investigation of turbu-
    lent spots and breakdown to turbulence. JFM 9, 235 (see
    Figure 2, Plate 1, facing p. 240).

DISCUSSION

KOVASZNAY: (Johns Hopkins University, Baltimore, Maryland)

At the end of the plate, how much of this periodic pattern
remains?

COLES:

I cannot tell you that because we only worked five days on
this whole experiment. It took one day to get that last figure.
One of the problems in constructing the synthetic boundary layer
is choosing the scales. The lateral scale you choose by how far
apart you put the disturbance holes, and the longitudinal scale
you choose by the period.

KOVASZNAY:

About how many boundary layer thicknesses?

COLES:

Our pattern was intended to be three boundary layer thicknesses across the flow and about ten along the flow. I was looking
at your "watermelon" plots when I chose these scales. It turned
out that three thicknesses was too much. The pattern actually
sideslipped a little, and where we made the measurements the period
across the flow was more like two thicknesses. Obviously, in water
you could string a hydrogen bubble wire out there and vary the
scales to get the most regular pattern in the intermittent region.
But that pattern could not remain regular, because the scales

change with distance. Eventually you would be in difficulty. The purpose of this experiment was to establish the organic connection between the turbulent spot, whose operational properties can easily be discovered, and the thing called a big eddy in the boundary layer. The connection should include properties associated with the big eddy, such as the longitudinal vortex structure in the sublayer, as well as bursting.

ROGERS: (Air Force Office of Scientific Research, Arlington, Va.)

This was a single component measurement? It was not u and v, was it?

COLES:

No, at the moment the measurements are single component. That is one of the things we obviously want to do. I would be very interested in the lateral component of velocity.

BRODKEY: (Ohio State University, Columbus, Ohio)

In an article on visual studies of a boundary layer in the Journal of Fluid Mechanics, by Nychas, Harshey and myself,[1] the authors came to the conclusion that the outer structure of a fully turbulent flow was not caused by what is happening inside but rather a vortex structure, observed in the outer flow, gave rise to the burst structure. It is amazing, the accord between those observations and your picture of the artificially structured boundary layer.

COLES:

I have always believed this. There is no stability problem in the sublayer. All these things you see are simply the signature of a large eddy.

---

[1] Nychas, S. G., H. C. Harshey and R. S. Brodkey (1973) J. Fluid Mech. 61, p. 513.

PROGRESS AND PROBLEMS IN UNDERSTANDING TURBULENT SHEAR FLOWS

A. Roshko

California Institute of Technology

Pasadena, California 91109

This discussion will be focussed on the plane turbulent mixing layer, featured in the illustration on the front of the program of this meeting. Another example of it is shown in Figure 1, while Figure 2 is a diagram which defines the basic parameters of this simplest of turbulent shear flows. Throughout this discussion the higher speed U_1 will always be in the upper part of the diagram. The speed ratio U_2/U_1 will therefore always be less than unity, but the density ratio ρ_2/ρ_1 may be less than or greater than unity. In high-speed flow the Mach numbers M_1 and M_2 would also be parameters. In fact, our interest in this problem was initially addressed to the question of how the characteristics of the flow depend on the density ratio, it having been supposed by many investigators that this was the parameter governing the observed, strong variations of spreading angle in the shear layer at the edge of a supersonic flow ($M_1 > 1$, $M_2 = 0$).

Our experiments, which were performed in a unique apparatus designed by Garry Brown, showed that, in fact, the effect of ρ_2/ρ_1 is not so strong as the supersonic experience had led one to believe, and that the observed strong effects for $M_1 > 1$ must be compressibility effects as distinct from density effects. These results along with some of those which I shall discuss today were initially reported at the AGARD Conference on Turbulent Shear Flows, held in 1971 (Brown and Roshko, 1971). A more recent and detailed account will appear shortly (Brown and Roshko, 1974).

A good idea of the influence of density difference in low-speed (incompressible) flow is obtained by comparing Figures 1 and 3. In both cases the velocity ratio $U_2/U_1 = 0.38$; in Figure 1 the

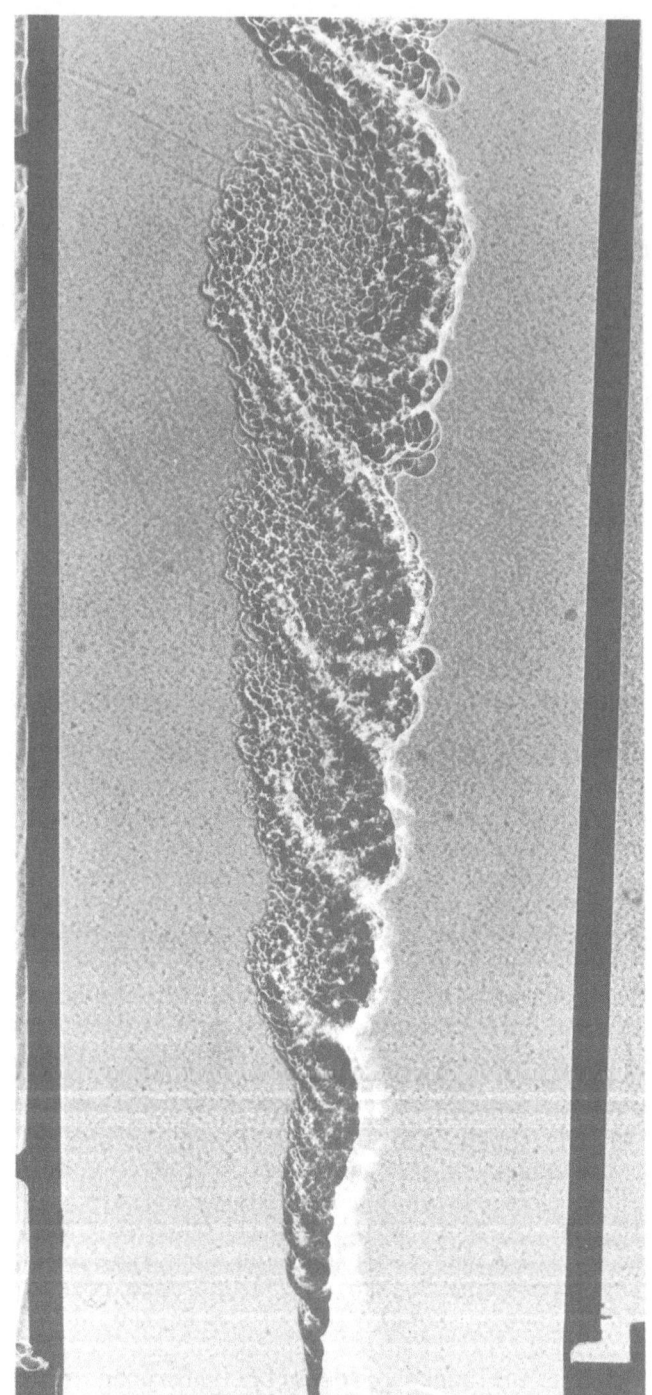

Figure 1. Spark shadowgraph of mixing layer between helium (upper) with U_1 = 1010 cm/sec and nitrogen (lower) with U_2 = 380 cm/sec. Pressure = 8 atm.

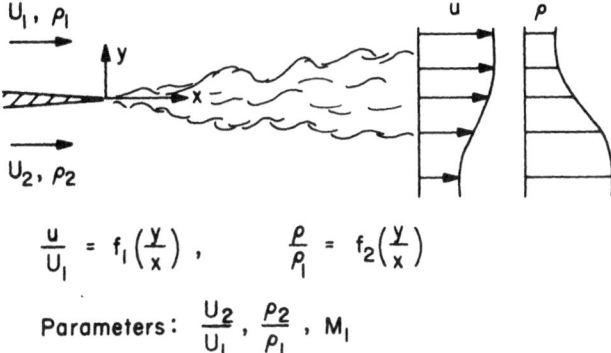

$$\frac{u}{U_1} = f_1\left(\frac{y}{x}\right), \qquad \frac{\rho}{\rho_1} = f_2\left(\frac{y}{x}\right)$$

Parameters: $\dfrac{U_2}{U_1}, \dfrac{\rho_2}{\rho_1}, M_1$

Figure 2. Plane, turbulent mixing layer. Parameters.

high-speed side is helium, the low-speed side nitrogen, so that $\rho_2/\rho_1 = 7$, while in Figure 3 the gases are reversed and $\rho_2/\rho_1 = 1/7$. Thus the density ratio in the two flows differs by a factor of 49, while the spreading rate changes by only a factor of two, approximately. The effects of compressibility, in a mixing layer at the edge of a supersonic flow, are very much larger.

But now I wish to turn to another feature which appears so prominently in the pictures to which we have been referring. That is the large wave-like or vortex-like structures, with which, I think, everyone is now fairly familiar. When we first saw them in our pictures they were a great surprise. At first we thought they were extraneous (for example, everyone suggests that the partition between the two streams may be vibrating) and we tried to get rid of them. Then we thought they might be associated in some way with the large density differences, but then found that they exist also in homogeneous flow ($\rho_2 = \rho_1$). Eventually we got used to them and realized that these coherent structures are part of the flow, that possibly they hold a key to the problem of turbulent mixing, and not only in the plane mixing layer but in all turbulent shear flows (Don Coles introduced another example of this possibility in the preceding talk).

Now I propose to discuss several features of these coherent structures in the turbulent mixing layer and to speculate about what role they may be playing in the turbulent processes, such as entrainment.

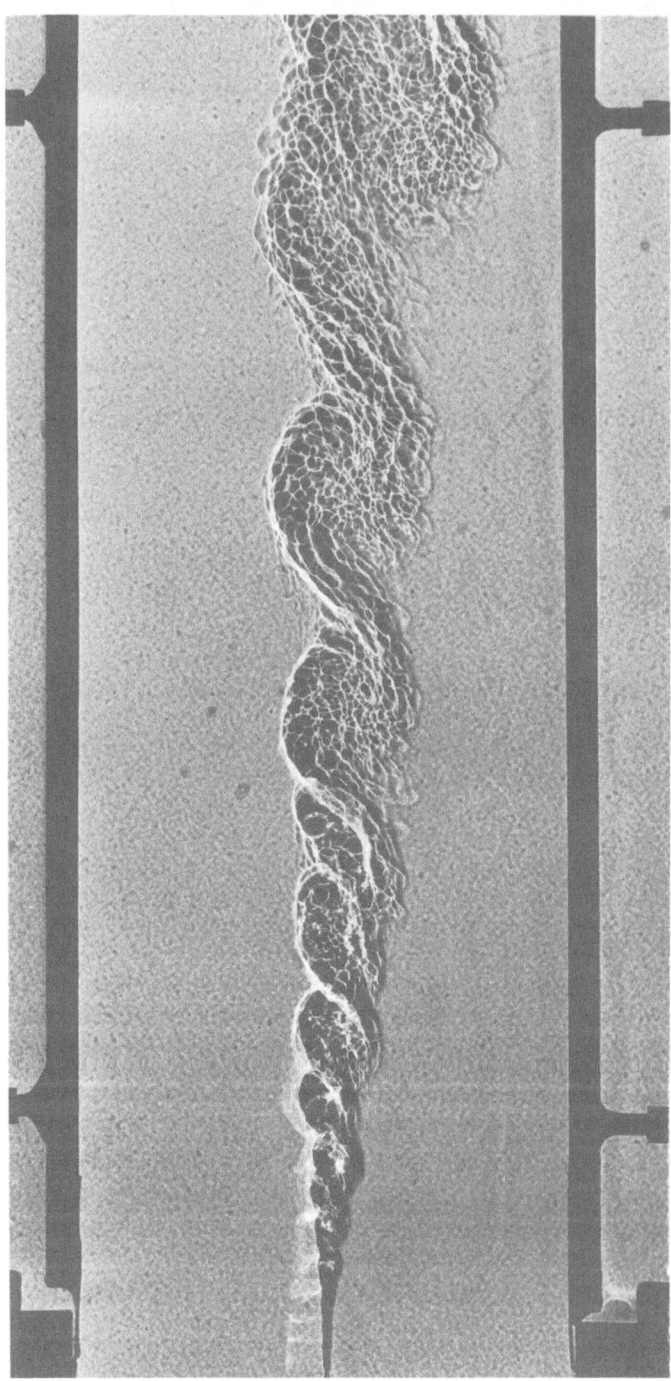

Figure 3. Spark shadowgraph of mixing layer between nitrogen
(upper) with U_1 = 880 cm/sec and helium (lower) with
U_2 = 330 cm/sec. Pressure = 4 atm.

One obvious feature on the pictures is that the sizes of the structures (let us call them vortices) and their spacings increase with distance downstream of the origin. On high-framing-rate shadow movies the coherence of the vortices is exhibited in that it is possible to identify and follow individual ones for many frames and to plot them on an x-t diagram such as that shown in Figure 4 (frame number corresponds to time). A striking feature is the fairly constant speed with which the vortices travel (approximately midway between U_1 and U_2), but even more striking is the finite lifespan of each vortex. Each lifespan begins and terminates rather abruptly; coincident with two (or more) nearly simultaneous terminations is the beginning of a new lifespan. This process has been studied in detail at USC by Winant and Browand (1974) in a mixing layer in water. They observed in detail how two succeeding vortices suddenly coalesce and form a new, larger one. They called the process "pairing" (in our movies we occasionally observe three vortices coalescing nearly simultaneously).

On reflection, it is clear that some such process must be occurring, so long as coherent structures exist. They are moving at practically constant speed and therefore, during the lifetime of a vortex and its neighbor, the spacing has to be pretty well constant as they move downstream. On the other hand, the similarity properties of the mean flow require that the spacing should increase downstream, linearly in the mean. Since the spacing of each particular pair is pretty well fixed, the way out of the dilemma is for the changes to occur discontinuously. A corollary is that these events must occur irregularly at various points up and down the layer so that, on average, the scale grows smoothly (and linearly). At any one station (x) the scales of the vortices coming by are dispersed about the mean value appropriate to that station. The dispersion is shown in the histogram in Figure 5. Actually, instead of plotting the distribution of spacings at one station, we have taken advantage of the expected similarity and have normalized the spacings with x, so that the data from all locations are included on the one plot. A similar distribution was obtained by Winant and Browand, who measured time intervals between successive vortices, instead of spacings. From their and our measurements, as well as those of Jones, Planchon and Hammersley (1973), it appears that the average spacing $\bar{\ell} = 3\delta$, for a variety of cases, where δ is the vorticity thickness.

While the instabilities leading to vortex amalgamation have yet to be understood fully, it seems likely that they provide the mechanism that produces "turbulence" from discrete, coherent structures. Not only dispersion but also three-dimensionality needs to be explained. On this latter point, Peter Bradshaw yesterday showed some pictures which suggest that the pairing may actually be spiral--at least this is one possibility for getting three-dimensionality into the picture.

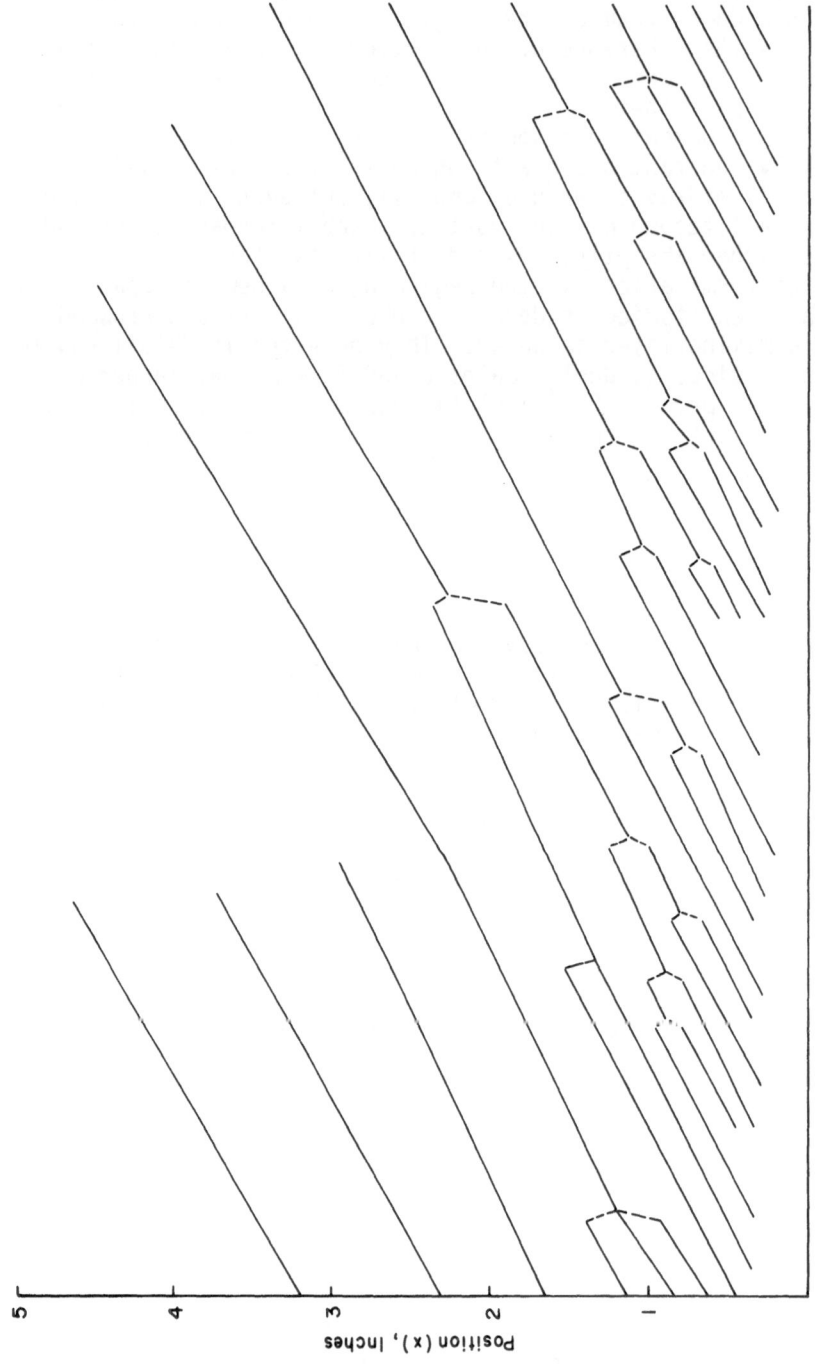

Figure 4.   Trajectories of vortices in x-t plane.   Flow similar to that
in Fig. 1, with U_1 = 1060 cm/sec, U_2 = 400 cm/sec, p = 7 atm.

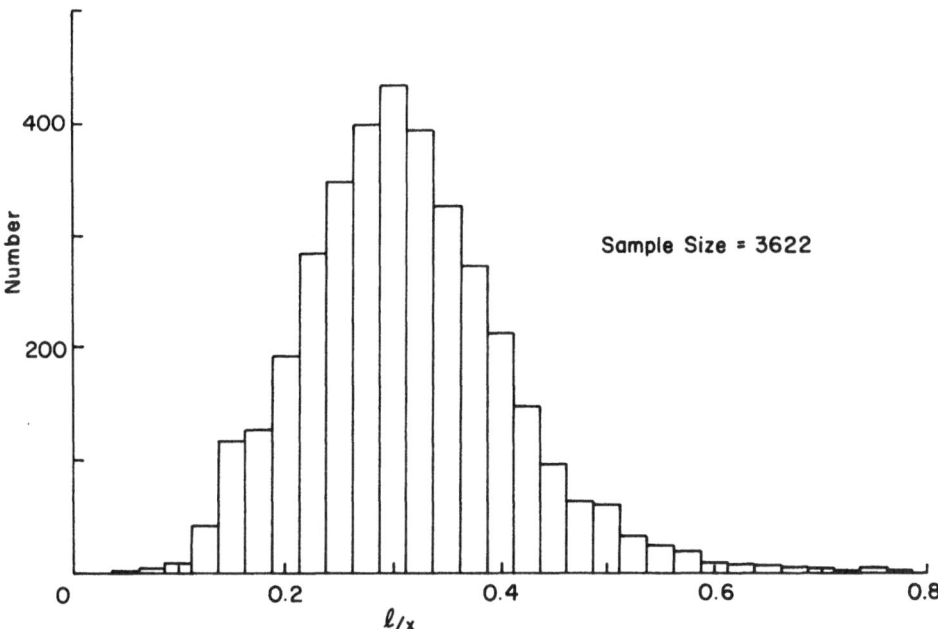

Figure 5.   Distribution of vortex spacings ℓ normalized by x.

Connected with the peak in the distribution of spacings (or frequencies of vortex pairs on the histogram, one might expect to find a peak in the energy spectrum of, say, $\overline{u'^2}$. We were troubled at first that there had been no reports of peaks in energy spectra measured previously in mixing layers. However the results of Spencer and Jones (1971) appeared shortly thereafter, and they displayed a distinct spectral peak which corresponds well to the peaks observed by Winant and Browand and by us. Apparently this peak stands out better (from the lower wave numbers) in a spectrum when $U_2/U_1 \neq 0$, as in Spencer and Jones' experiments. But even for $U_2 = 0$ one can now discern, with hindsight, something of a peak, at the right place, in the spectra obtained by Wygnanski and Fiedler (1970).

Jones, Planchon and Hammersley estimated that the energy in $\overline{u'^2}$ associated with the spectral peak is about 30% of the total. What accounts for the large amount of energy at wave numbers below the peak, that is, scales much larger than the coherent vortex scales? I believe that these large scales are associated with the amalgamation events, which appear to contribute importantly to all aspects of the turbulent development.

In spite of these suggestions as to how coherent structures might interact to produce a flow with "turbulent" properties, I think that many will remain unconvinced that the pictures we have been exhibiting are of "truly turbulent" flows. Let me say a few further words on this matter. In Figure 6 are shown pictures of the flow at several Reynolds numbers, obtained by changing the pressure level and/or the velocity level in the experiment. It may be seen that, although the small-scale structure is enhanced at the higher Reynolds numbers (partly from greater optical effect), the large, coherent structures are similar in all cases. The highest Reynolds number (corresponding to the same picture as in Figure 1) is 10^5, based on downstream distance to about the middle of the picture. Pictures at Reynolds number an order of magnitude higher, obtained by replacing the helium on the high-speed side by air, still showed the structures quite clearly (Brown and Roshko, 1971, 1974).

An interesting question that might be asked next is "how do these structures differ from the very similar ones that occur in the transition region of a shear layer?" Some beautiful pictures of that region obtained by Freymuth (1966) clearly show the similarity between the structures we have been discussing and those in the transition region. The difference is that, in the transition region, there is only one value of the spacing, namely the most stable wave length that was selected by the laminar layer. In this region, the scales have not yet become dispersed, as they are in the turbulent region farther downstream, where there is no memory of the initial region. Nor has there been time yet for three dimensional instabilities to come into play. If we assume that the shear layer does not forget about its initial conditions until the average vortex spacing is, say, ten times the spacing in the transition region then, relating the latter to the initial momentum thickness, we find that the required downstream distance is about 1000 of those momentum thicknesses, a value suggested by Bradshaw (1966) some time ago.

Because of the similarity of the coherent structures to those in the transitional range, the latter may be a better place to look for their basic forms, uncluttered by the irregularities which develop further downstream. This, you will recall, was the theme of Don Coles' experiment to synthesize a turbulent boundary layer from Emmons' spots.

Now let us consider some other aspects of the role of the coherent structures.

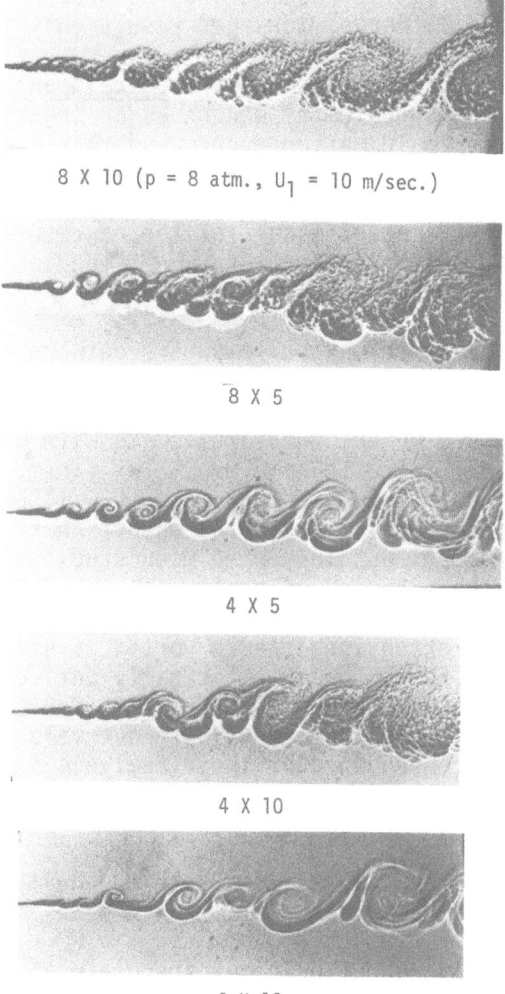

Figure 6. Effect of Reynolds number on mixing layer between
helium (upper) and nitrogen, $U_1/U_2 = \sqrt{7}$.

First, consider the growth of the mixing layer. Winant and Browand associated this with the pairing process, that is, with the increase in vortex size which results from the amalgamation of two (or more) vortices. Indeed, on our shadowgraphs the extent of the mixing zone is fairly well defined by the envelope of the vortices. In this view, it is assumed that no growth of the vortices occurs during their lifespans, changes result only from pairing.

A similar view can be taken of the entrainment process; that is, that most of the entrainment occurs as an ingestion or "gulping" process, during the amalgamation event, and that each coherent vortex convects more or less passively between such events. This is rather different from an older view of the entrainment process which holds that entrainment occurs at the edges of the mixing layer by a kind of "nibbling" process (i.e., the interface at the edge "propagates" into non-turbulent fluid).

Actually, the answer may be somewhere between these two views; possibly both processes are at work, and it is not clear at the moment how much growth of individual vortices and how much entrainment into them occurs before they amalgamate with others. From our movies it seems that there is some growth at first, just after a new vortex is formed. But more measurements will be needed to define this. Moore and Saffman (1974) assert that growth must occur before amalgamation. They have been studying, analytically and numerically, the behavior of vortices interacting in a straining flow and find that finite vortices in a row can exist only if the maximum rate of extension of each vortex is small enough compared to the vorticity. From an empirically derived quantitative criterion for this, they deduce a lower limit for the spacing of vortices in a row (in good agreement with our value for mean spacing). A necessary consequence of their analysis is that continual growth by turbulent entrainment is necessary, that pairing is not even essential.

Thus, some important questions as to the behavior and role of the coherent structures remain.

Nevertheless, we cannot help but speculate about the possible existence and form of basic structures in other turbulent shear flows. Don Coles has already made a suggestion for the turbulent boundary layer. What about wakes and jets? These are already more complicated than the mixing layer. The latter is, basically, a layer of vorticity of one sign, while a plane wake or plane jet is composed of two layers of opposite vorticities. The vortices that form in each layer can interact not only with similar ones in the same layer but also with vortices of opposite sign in the other row. (von Kármán made this problem famous!) The vortices which are shed from bluff bodies and the interaction phenomena at the end

of the potential core of a jet may be clues to the basic structures
in wakes and jets. In those early regions they are still tied to
the initial geometry (for example, the vortex shedding frequency is
still quite definite, not dispersed). What can be said about them
further downstream when all memory of the initial region is lost?
In this respect it may be useful to recall a little known work of
Taneda (1959), who described the reforming of the vortex street far
downstream but with spacings larger than those in the initial
region at the cylinder. Papailiou and Lykoudis (1974), here at
Purdue, have also described coherent vortices in the far-downstream
region of a wake and have associated them with entrainment.

Proceeding in this way, we might consider even more complex
flows. Yesterday Peter Bradshaw was talking about some very complicated situations. When you approach one of these from the
point of view of Reynolds-equation modeling you try to account for
anomalies by fixing up some terms in the equations, look for a way
to inject this new experience into the equations. We "eddy chasers"
would try to see, if we can, what the big structures are doing. An
example of such a more complicated (but not too complicated) flow
is given in the work of Rebollo (1973) in our laboratory, who
studied the behavior of the mixing layer in an adverse pressure
gradient. We had realized that for the case $\rho_2 U_2^2 = \rho_1 U_1^2$, both U_1
and U_2 outside the layer could be allowed to vary and the equality
of dynamic pressures would be preserved. Thus $U_2(x)/U_1(x)$ would
remain constant and a similarity flow in a pressure gradient could
possibly be established. It turns out that it is like a Falkner-
Skan flow in that $\frac{x}{U_1}\frac{dU_1}{dx} = \frac{x}{U_2}\frac{dU_2}{dx} \equiv \alpha$ must be a constant. After
some heroic tinkering, Rebollo was able to set up such a flow with
$\alpha = -0.18$ and to measure some of its properties. The layer still
spreads linearly but at a rate different from that at constant
pressure. It is greater, but even greater than what would be predicted by extrapolating from the constant-pressure case with simple
eddy-viscosity arguments. (The spreading rate is increased by
about 50% and the maximum shear stress is about doubled.) Now I
don't know whether the existing turbulent models would predict that
or not, but it would be an interesting exercise. From the point of
view of an eddy-chaser, the shadowgraphs of the flow (Figure 7),
from Rebollo's thesis, are interesting. We see that there is a
stretching of the vortices outward from the center of the layer.
These deformations must be due to the divergence away from the
dividing streamline in such a decelerating flow, and they may be
involved in the enhanced growth rate, shear stress, etc.

Finally, I would like to turn to the problem that is perhaps
the main concern of the workshop. That is the problem of chemical
mixing and reaction in turbulent flow. It is a problem that we are
trying to study in our laboratory. Actually, if a reaction occurs

it considerably complicates the situation, unless perhaps the reactants are very dilute. For insight into the reaction problem, we are trying to determine how much inert <u>molecular</u> mixing occurs inside turbulent mixing regions such as those we have been discussing. Turbulent entrainment brings together "lumps" of fluid from the opposite sides, but how much molecular inter-diffusion occurs? This is what determines how much reaction can occur (between very dilute reactants). It could be different in gases and in liquids because, although the turbulent mixing rates are the same in the two cases, the molecular diffusion rates (Schmidt numbers) are very different, as pointed out by Professor Brodkey yesterday.

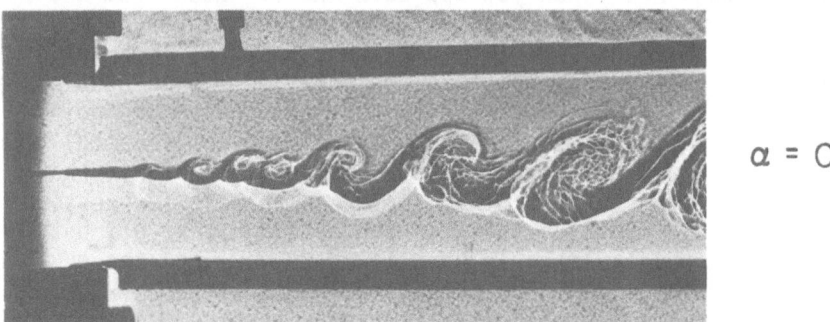

$\alpha = 0$

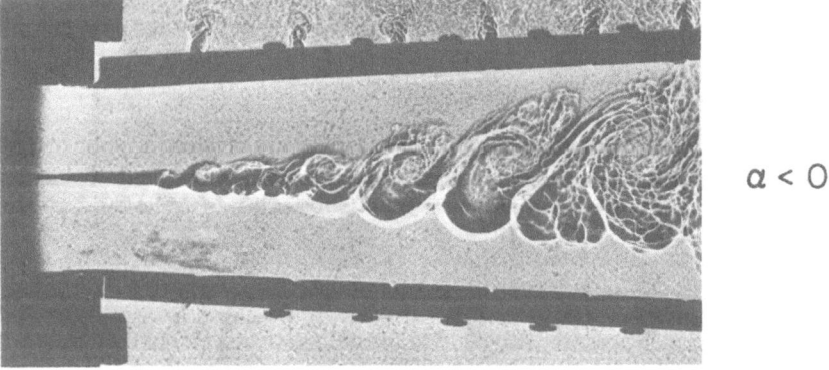

$\alpha < 0$

Figure 7.   Effect of adverse pressure gradient ($\alpha = -0.18$) on mixing layer between helium (upper) and nitrogen.

But even to determine how much is molecularly mixed is not very simple. In principle, it requires a concentration sampling probe with perfect resolution. A probe for this purpose was built by Brown and Rebollo (1972). Its response is fast but its spatial resolution is somewhat uncertain (perhaps a few thousandths of an inch). It reveals very large fluctuations of concentration even in the middle of the mixing layer, suggesting that the flow there is not very well mixed, in the molecular sense.

We are still trying to sort this out, but in the meantime we have tried to set upper limits on possible <u>reactant flux rates</u> in our mixing layers simply from the mean concentration and mean velocity profiles. The approach is illustrated in Figure 8, on which the profiles of mean velocity and of mean concentration (C_2) are shown together with the (normalized) fluxes of C_2 and C_1 (recall that $C_1 = 1 - C_2$). We reason that in the upper part of the profile where the flux UC_1 is much higher than UC_2 it is the latter that controls the "reaction" rate (in a simple binary reaction, $M_1 + M_2 = M_{12}$), while in the lower part UC_1 is controlling.

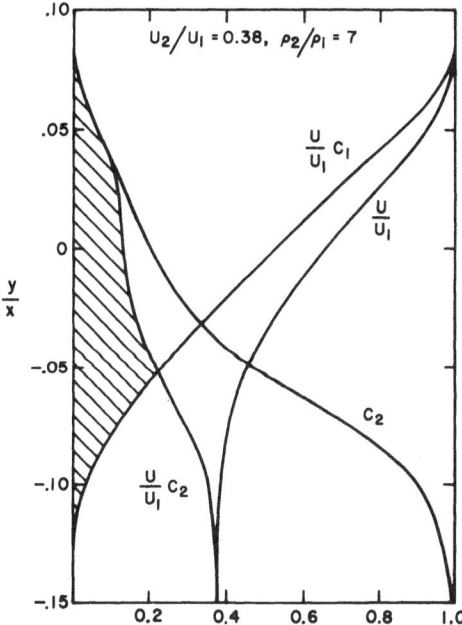

Figure 8. Concentration and concentration-flux profiles. $U_2/U_1 = 0.38$, $\rho_2/\rho_1 = 7$.

Thus, the maximum reaction flux profile is that outlined by the shaded region whose integral defines the mean flux of reactant product.

Figure 9 illustrates the case where the two densities are reversed, while Figure 10 is for homogeneous flow. The latter figure was obtained by interpreting the measurements of Sunyach (1972), who measured temperature profiles in a homogeneous mixing layer with $U_2 = 0$. (We converted his temperature profile into a "concentration" profile). These estimates of flux rate do not include a possible contribution from turbulent diffusion terms $\overline{u'c'}$, but these would have opposite signs in upper and lower parts of the layer and would tend to balance in the integral. More serious is the fact that the mean profiles give no information about the possible "unmixedness" at the molecular level. Nevertheless, they may serve some useful purpose as upper limits. The results are summarized in the following table, where the "reactant" flux has been normalized with $U_1 x$. Also shown for reference is the spreading angle α_{viz} defined by the visual boundaries of the mixing

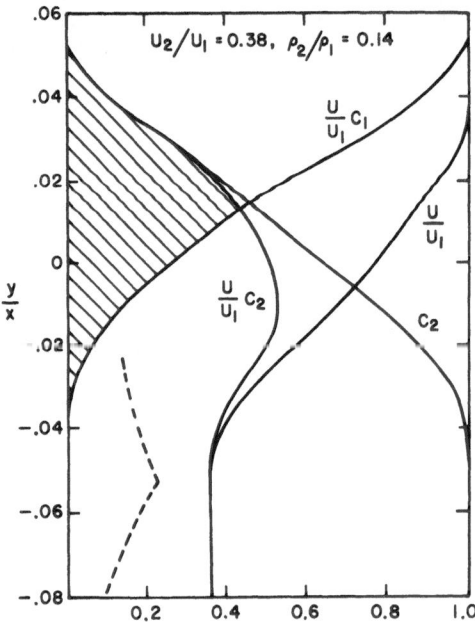

Figure 9.   Concentration and concentration-flux profiles. $U_2/U_1 = 0.38$, $\rho_2/\rho_1 = 1/7$.

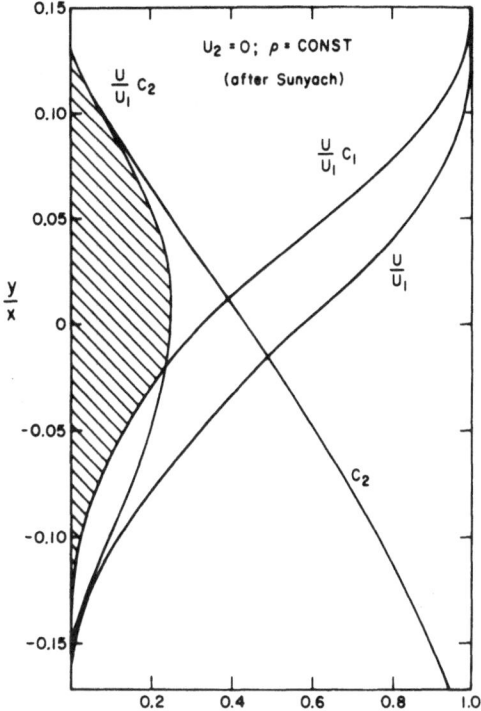

Figure 10. Concentration and concentration-flux profiles. $U_2 = 0$, ρ = constant. (based on data in Ref. 11)

layers in the shadowgraphs, as well as an estimate of the normalized flux of fluids from the two sides that have become "entangled" by the entrainment process.

U_2/U_1	ρ_2/ρ_1	α_{viz}	"Entangled" Flux*	"Reactant" Flux*
0	1	0.38	.104	.036
.14	7	0.40	.053	.029
.38	7	0.21	.031	.020
.38	1/7	0.12	.017	.017

* Normalized with $U_1 x$

REFERENCES

1. Bradshaw, P. 1966 The effects of initial conditions on the development of a free shear layer. J. Fluid Mech. 26, 225.

2. Brown, Garry & Roshko, Anatol 1971 The effect of density difference on the turbulent mixing layer. Turbulent Shear Flows, AGARD-CP-93, 23-1.

3. Brown, Garry & Roshko, Anatol 1971 On density effects and large structure in turbulent mixing layers. J. Fluid Mech. 64, 775-816.

4. Brown, G. L. & Rebollo, M. R. 1972 A small, fast-response probe to measure composition of a binary gas mixture. AIAA J. 10, 649.

5. Freymuth, Peter 1966 On transition in a separated laminar boundary layer. J. Fluid Mech. 25, 683.

6. Jones, B. G., Planchon, H. P. & Hammersley, R. J. 1973 Turbulent space-time correlation measurements in a plane two-stream mixing layer at velocity ratio 0.3. AIAA Paper No. 73-225.

7. Moore, D. W. & Saffman, P. G. 1974 The density of organized vortices in a turbulent mixing layer. To be published.

8. Papailiou, D. D. & Lykoudis, P. S. 1974 Turbulent vortex streets and the entrainment mechanism of the turbulent wake. J. Fluid Mech. 62, part 1, 11-31.

9. Rebollo, Manuel 1973 Analytical and experimental investigation of a turbulent mixing layer of different gases in a pressure gradient. Thesis for the Ph.D. degree, California Institute of Technology.

10. Spencer, B. W. & Jones, B. G. 1971 Statistical investigation of pressure and velocity fields in the turbulent two-stream mixing layer. AIAA Paper No. 71-613.

11. Sunyach, Michel 1971 Contribution a l'étude des frontiéres d'écoulements turbulents libres. Thesis for D. Sc. degree, L'Université Claude Bernard de Lyon.

12. Taneda, S. 1959 Downstream development of the wakes behind cylinders. J. of the Physical Society of Japan, 14, no. 6, 843-848.

13.   Winant, C. D. & Browand, F. K. 1974 Vortex pairing: the
      mechanism of turbulent mixing layer growth at moderate
      Reynolds numbers. J. Fluid Mech. 63, 237.

14.   Wygnanski, I. & Fiedler, H. E. 1970 The two-dimensional
      mixing region. J. Fluid Mech. 41, 327.

ACKNOWLEDGMENT

      This paper was based on research work supported by the
Department of the Navy, Office of Naval Research, under Contracts
N00014-67-A-0094-0001 and N00014-67-A-0226-0005.

DISCUSSION

BRADSHAW: (Imperial College, London, England)

     First, I would like to refer to the spectrum data on the mix-
ing layer which we produced in 1964, Journal of Fluid Mechanics,
Vol. 20. It shows a very full set of U, V, and W spectra, and some
correlation measurements for an axisymmetric mixing layer. I mention this not to get it on the record, but because it might be
useful now for sorting out the data on high Reynolds number.

     Two, the shadowgraph technique, that Roshko uses, necessarily
gives spanwise averages. Therefore, it will show up the two dimensionality of the structure very much more strongly than any other
method.

     The third point is about the helical pairing of vortices in
mixing layer transition which I mentioned yesterday. Helical pairing involves stretching of the vortices and probably the formation
of intense shear layers which will rapidly break down into turbulence.

     It strikes one that there may be various other forms of pair-
ing apart from the ordinary type of uneventful helical pairing and
maybe other ways in which more than two vortices can come together,
a whole spectrum of perversions, in fact.

     The fourth point is that we must be careful about applying the
same sort of concept of instability to all the shear layers: mixing
layers, jets and wakes are inviscid-unstable, and it is unlikely
that the same sort of phenomena would occur in boundary layers which
are inviscid-stable. On the question of wakes, another reference
which might be handy on the large eddy structures is Grant, Journal
of Fluid Mechanics, 1958. He had a pretty good look at large eddies

in a wake and although a part of his work could be interpreted as
the persistence of spanwise vorticity, I do not think he concerns
himself with that in itself. And the last point, I certainly agree
with Prof. Roshko that the retarded shear layer measurements would
be difficult to reproduce with a prediction method. This is in
fact, an example of the difficulty which I talked about yesterday,
in which one would like to have a (shear stress)[1.0] term in the
length-scale transport equation, but for dimensional reasons one
is only permitted a (shear stress)[0.5] term. It seems to me fairly
clear now that the effect of the vertical divergence, dv/dy, on the
turbulence structure becomes important. It is responsible apparently for the difference in structure of the jet and the wake, and in
the retarded mixing layer $\partial v/\partial y$ is very large.

DONALDSON: (ARAP, Princeton, New Jersey)

        I would like to point out the very interesting behavior of two
vortices of the same sign which rotate about each other. The problem is related to the coalition of eddies discussed by Professor
Roshko, since two vortices rotating about themselves break down
(or coalesce) to form a larger eddy. The phenomena can be noted
by studying the behavior of the two vortices of the same sign shed
from each side of a high aspect ratio flapped wing. Such a wing
with an idealized lift distribution is shown in Figure 1.

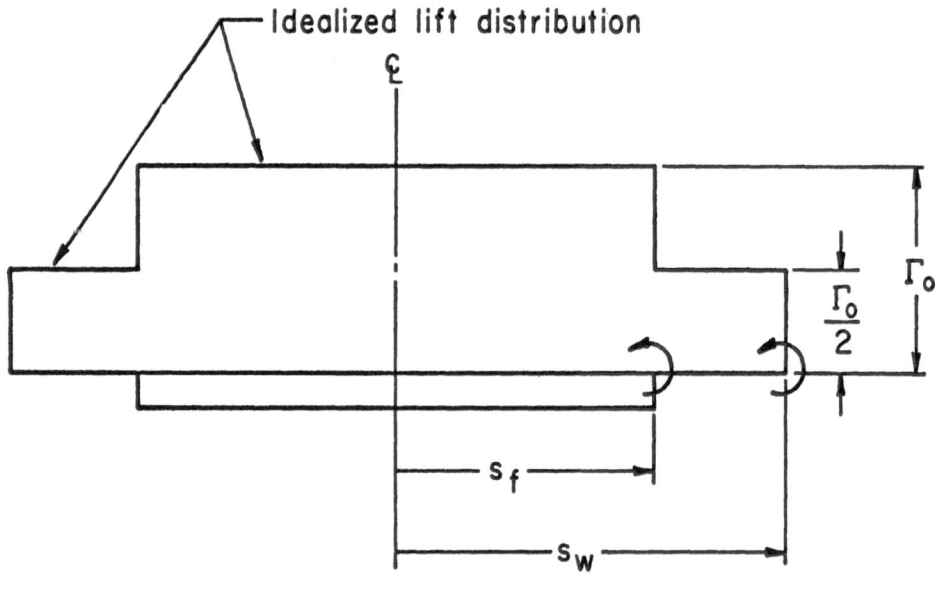

Figure 1

For the case shown, two vortices are shed from the right-hand side of the wing. Both are of strength $\Gamma_0/2$. One is shed at the semi-span of the flap s_f and the other at the wing tip or at the semi-span s_w. The two vortices rotate about each other, with the center of rotation, in this case, at $(s_f + s_w)/2$. This center of rotation descends at an oscillatory rate due to the influence of the two vortices shed from the other side of the wing.

As the vortex pairs rotate about one another in a real fluid, they strain each other and eventually break down to form one larger vortex with the vorticity which was initially tightly centered in two locations now widely spread over the larger vortex. If the pairs of vortices themselves are initially far apart, they will rotate about each other for a long time before breaking down into a larger vortex. If they are initially close together, they will break down into one vortex rather rapidly.

We are attempting to study such phenomena via second-order closure schemes. Figure 2 shows a computer-generated calculation of vortex rotation and breakdown under such circumstances. The lines shown are isopleths of constant smoke intensity (if a Gaussian distribution of smoke had been injected into each vortex at the wing, i.e., x/b = 0). In these figures, x is the distance behind the wing, b is the wing span, A is the wing aspect ratio, and C_L is the wing lift coefficient.

A final observation is related to Anatol's question as to whether present modeling techniques might properly capture the effects of strong adverse pressure gradients. I believe this is related to the resolution of the problem of whether or not an invariant equation for the scale Λ_1 used in second-order closure techniques can be found. Λ_1 is essentially a measure of the integral scales. We are working on this problem, and I am hopeful that we will be able to construct an invariant scale equation. Constructing an invariant equation is not difficult mathematically; constructing one which is invariant in the sense that the same equation with the same parameters allows one to compute all known measured flows with equal and adequate accuracy is quite another matter.

STREHLOW: (University of Illinois, Urbana, Illinois)

The last two slides which Roshko showed indicate to me some asymmetry in the mixing layer, but your pictures, for example, your helium and nitrogen pictures showed very strong asymmetry in that there was a continuous gradient region at the bottom in the nitrogen layer, and you were pulling nitrogen off with the helium into the rest of the flow. I would like to know if you thought any more about that and could make a little comment about the asymmetry.

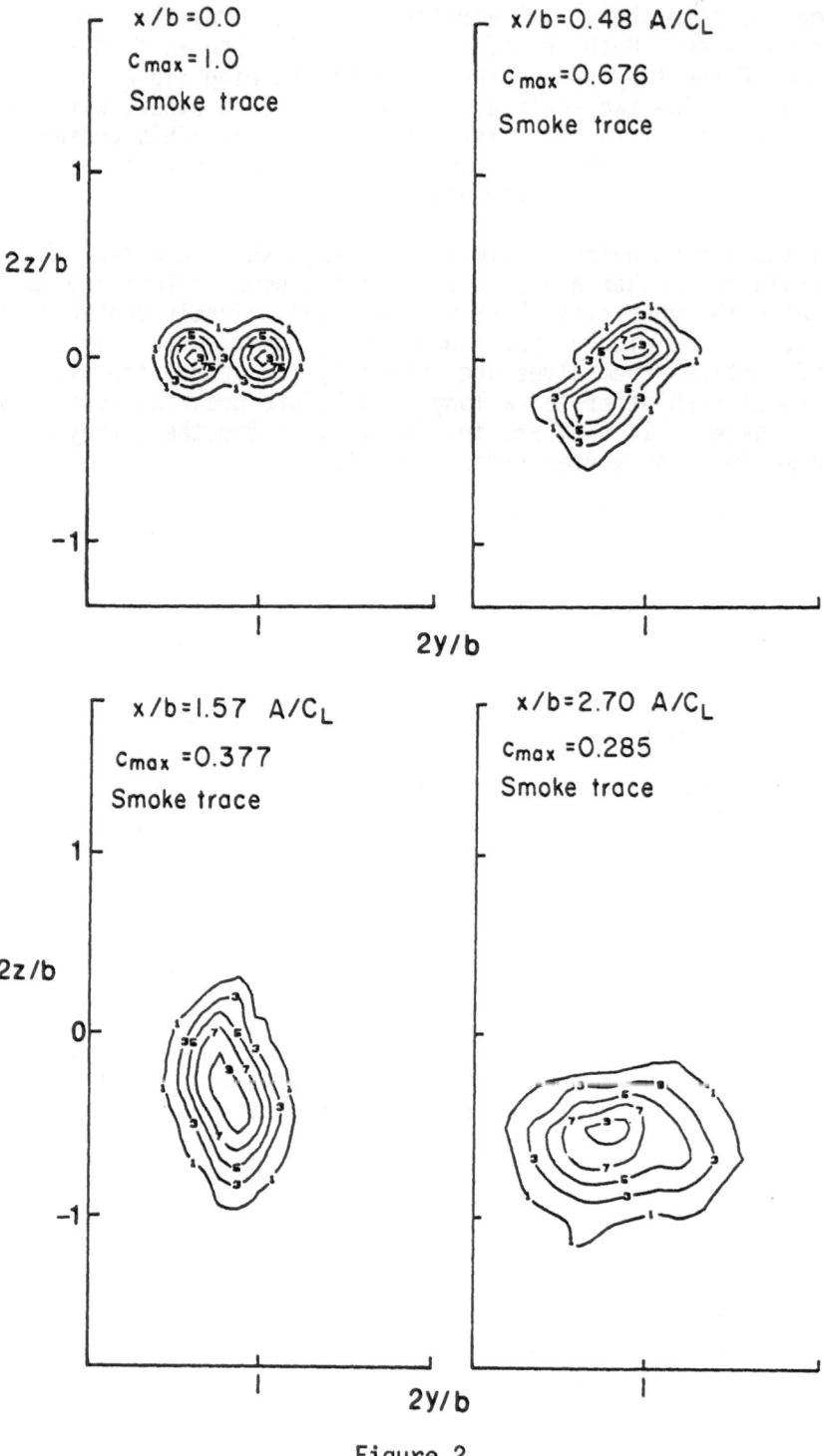

Figure 2

ROSHKO:

Are you thinking about the dipping of those layers?

STREHLOW:

No, I am thinking of the fact that the nitrogen-helium interface is asymmetric. The bottom interface is essentially a continuous line of high density fluid and at a point in the turbulent wake, the flow was not continuous. It essentially consisted of composite of vortices.

ROSHKO:

That is an optical effect. I believe that the basic structures look rather like those in the picture of Winant and Browand (1974) or Freymuth (1966). In our pictures an asymmetry is introduced by the shadow optics: the bright lines (on the nitrogen side) seem closer together than they really are, tending to close them up in some cases.

STREHLOW:

I do not agree. This looks almost like the familiar phenomenon of the heavy fluid being lifted off by a high velocity light fluid.

NEWMAN: (McGill University, Montreal, P. Q., Canada)

I would like to comment on Dr. Roshko's statement that the rate of growth increases with adverse pressure gradient in his self preserving flows. We have measured both axisymmetric and two-dimensional wakes and jets of uniform density in streaming flow which are made self preserving by suitably adjusting the adverse pressure gradient. They maintain self preservation of both the mean velocity and the turbulent stress tensor. For jets it is indeed true that the rate of growth increases as the jet becomes stronger and the associated adverse pressure gradient to maintain self preservation increases. However for wakes the reverse is true. As the wake becomes deeper the required adverse pressure gradient becomes milder while the measured rate of growth increases from the smallest values which exist for small deficit wakes.

BEVILAQUA: (Aerospace Research Labs, Dayton, Ohio)

It has been recognized for some time now, since the discovery of intermittency, that there is a correlation between the turbulent mixing rate and the occurrence of "large eddies." I would like to suggest a reason why the vortex-like character of these eddies has remained hidden for this time, especially in the turbulent wake.

The effect of a vortex street is to cause a periodic variation in the direction of the mean velocity vector, without changing its magnitude. Such fluctuations cannot be detected by a cylindrical hot wire sensor parallel to the axis of the vortices, since the circular symmetry of the filament limits its response to changes in the magnitude of the velocity vector. Here at Purdue we have investigated the fluctuating components of the velocity in the turbulent wake of a cylinder, and observed a marked peak corresponding to the Strouhal frequency in the spectra of the v'-component. At the same point in the wake, the u'-component appears to be fully turbulent.

O'BRIEN: (State University of New York, Stony Brook, New York)

     With respect to the mixing and reaction in a mixing layer, it is easy to say what information you need to solve that problem. One requires the joint probability density function of velocity and temperature at each point and the joint probability density functions of temperatures at noncoincident points. What I would like to urge is that in trying to build a framework in which to place your efforts to understand the simple mixing problem it might be wise to do it through those kinds of quantities rather than the traditional moments because that is exactly the same information you will need to solve chemically reactive mixing layers.

ROSHKO:

     We are trying to improve our techniques for measuring our density-air concentration histories in order to get such information.

ENSEMBLE-AVERAGED LARGE SCALE STRUCTURE IN THE TURBULENT MIXING LAYER

F. K. Browand

Department of Aerospace Engineering

University of Southern California

Earlier work by Winant and Browand (1974), at Re $\approx 10^3$, has suggested that the large scale structure in the mixing layer consists of lumps of vorticity, or simply vortices, aligned across the flow. The vortices arise initially from instability of the laminar flow, and thereafter interact by rolling around one another and coalescing in pairs to form a disturbance with twice the spacing and half the frequency. After four such sequential pairings in a distance of 1.5 meters, the mixing layer has grown in thickness by a factor of 16 and is no longer thin compared to the scale of the apparatus.

Vortex pairing, as it was called, seems to be the most important interaction, although other interactions take place. Occasionally a vortex is simply elongated or strained between two neighbors. When this happens, the vorticity associated with the central vortex is shared --- sometimes equally, but not always --- by the two neighbors.

As a continuation, I will describe measurements made using a conditional sampling technique intended to display the spatial structure of these large vortices (Browand and Weidman, in preparation). The first requirement for the application of conditional averaging is the existence and recognition of a desired event. In this case, the desired event is either pairing or no pairing, and recognition was handled in the following way. Figures 1a, 1b show the instantaneous output from two hot-film probes placed slightly above and below the mixing layer formed downstream from a splitter plate in a water channel. Superimposed is a photograph of the flow using dye to mark the vorticity carrying fluid. The instant the

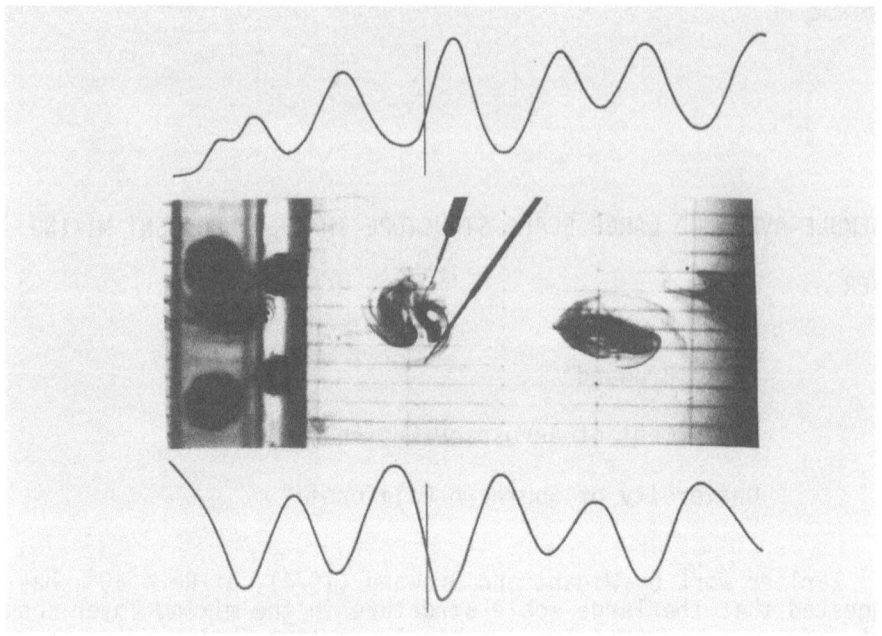

(a) Flow Field and Velocity Fluctuation, Pairing has
     Occurred Upstream

Figure 1. Voltage - time traces from two hot-film probes placed
           above and below the mixing layer, superimposed on a
           photograph of the flow (time of photo is marked by a
           bar). Dye marks the fluid with vorticity. Flow is
           left-to-right, lower layer faster.

photograph was taken is marked on the hot-film traces. Flow is
from left to right, lower layer faster, and the Reynolds number at
the point of measurement is about 300 based on the velocity difference and the maximum slope thickness. The pairing process is
intermittent in space and time and, at the measurement position,
one observes the whole sequence of interaction. Figure 1a represents a portion of time for which pairing has already occurred
upstream. Signals from the two hot-films are equal in amplitude
and almost exactly out of phase --- as would be expected if, at
this moment, the vortices were simply convected past the probes
with little interaction between neighbors. Figure 1b, at another
time, shows a pairing in progress. The hot-film signals show a
distinctive phase change during the interaction. By placing
thresholds upon the amplitude and the phase of the two reference

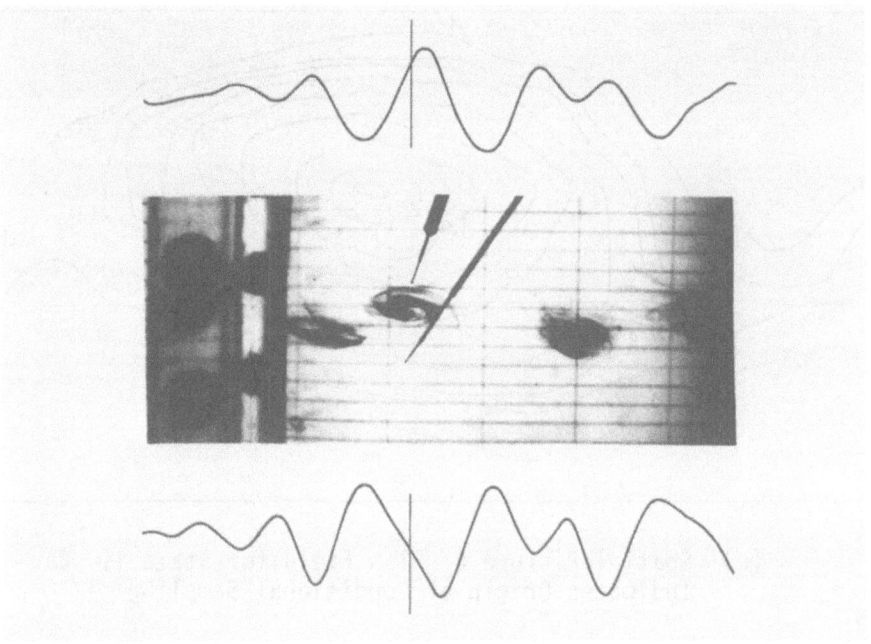

Figure 1. (b) Flow Field and Velocity Fluctuation, during Pairing

probes, conditions corresponding to Figures 1a and 1b can be detected. These are referred to respectively as state I and state II.

Measurements of horizontal and vertical velocity were obtained in digital form at 13 vertical locations across the mixing layer using a split-hot-film probe as the sensor. The time record was conditionally sampled to obtain either state I (2% of total time record), or state II (4% of total time record). In either case, data at thirty increments in time were separately sampled and averaged. The total time intervals --- thirty time increments --- corresponds to the characteristic period of vortex passage. Approximate spatial vorticity distributions were obtained for state I and state II. For this determination, time was replaced by x/U_C, where U_C is the vortex convection speed. The derivatives of the velocity data, $\partial u/\partial y$ and $\partial v/\partial x$ were smoothed by a least squares polynomial fit. Contours of constant vorticity, normalized by the maximum time averaged vorticity, ζ_0, are shown in Figures 2a, 2b. The uncertainty is estimated to be $\pm$ .12 ζ_0. In Figure 2a, state I, the vorticity is clearly lumped with a single maximum. The maximum value at the core is about 50% greater than the time

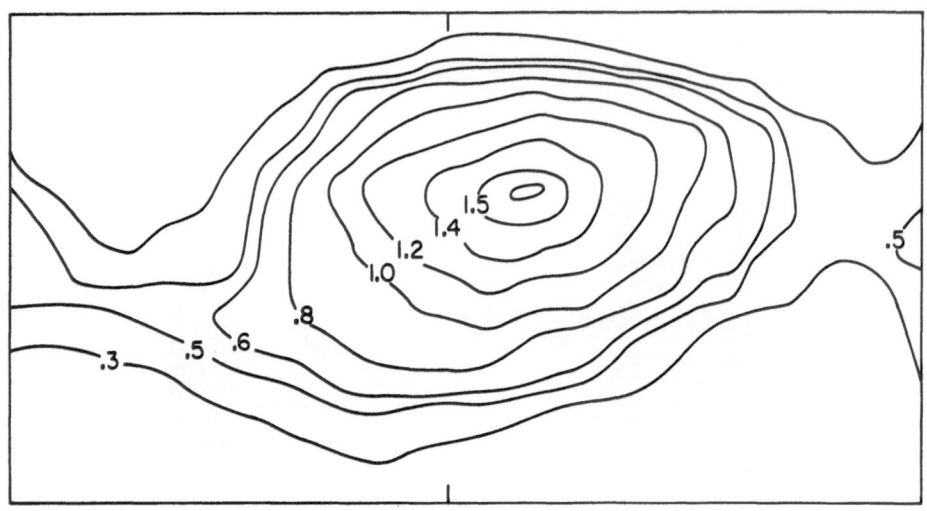

(a) Spatial Picture of Flow Field for State I. Bar
     Indicates Origin of Conditional Sampling

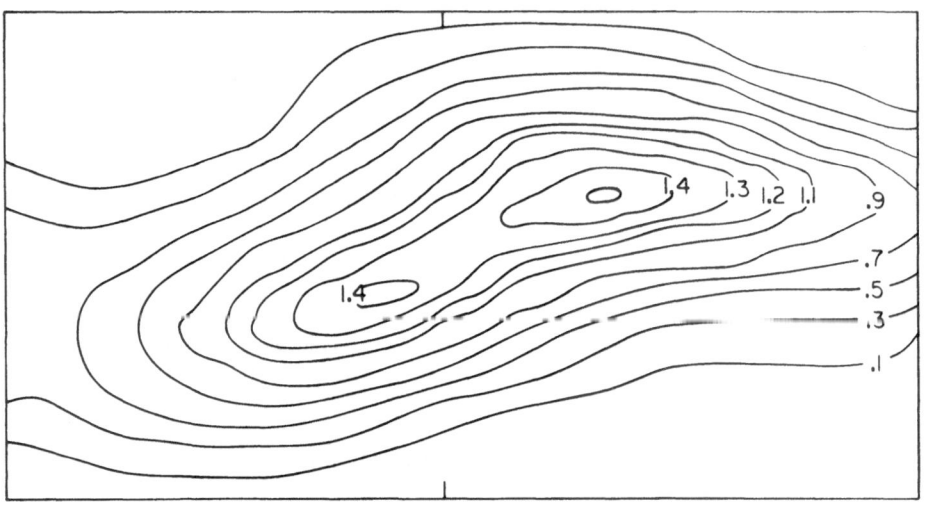

(b) Spatial Picture of Flow Field for State II

Figure 2. Contours of constant vorticity, normalized by the maxi-
           mum time averaged vorticity, ζ_0

mean velocity. The vorticity gradually increases from the outer regions towards the core, and no significant region of constant vorticity appears to be present. The "tails" which connect the vortex with its upstream and downstream neighbors are seen at the left and right boundaries of the picture. Figure 2b shows the vorticity distribution for a pairing in progress (state II). As expected, two vorticity maxima are identifiable, and again peak values are roughly 50% greater than the maximum time mean vorticity.

The urgent question is the behavior of the mixing layer at large Reynolds numbers, $\theta(10^5 - 10^6)$ say. Evidence exists, Brown and Roshko (1974), Winant and Browand (1974), to indicate that the large scale features have many similarities with their lower Reynolds number counterparts. The technique developed here, or refinements of this technique, will provide a useful means of studying large scale structure at higher Reynolds numbers.

The present work was supported under Contract N0014-67-A-0269-0031.

REFERENCES

1. Brown, G. L. and Roshko, A. 1974   On Density Effects and Large Structure in Turbulent Mixing Layers. J. Fluid Mech. (to appear).

2. Winant, C. D. and Browand, F. K. 1974   Vortex Pairing: the Mechanism of Turbulent Mixing-layer Growth at Moderate Reynolds Number. J. Fluid Mech. 63, 2.

SOME OBSERVATIONS ON THE MECHANISM OF ENTRAINMENT

P. M. Bevilaqua

Air Force Aerospace Research Laboratory

Dayton, Ohio

The turbulent entrainment process has been variously described as due to the effect of the small scale turbulence at the interface (Corrsin and Kistler, 1955; Townsend, 1966), the action of mixing jets which engulf volumes of fluid in bulk (Grant, 1958) and, most recently, a process of vortex pairing (Winant and Browand, 1974). A simple visualization study undertaken to examine the fundamental mechanism directly will be described in the present article. This experiment was conducted in the wake of a sphere towed through a water-filled trough, so that the "free stream" was actually motion-less fluid. It was thus possible to mark a volume element of the free stream with dye, and follow its motion as it was injested by the wake. The common technique of injecting dye behind the model was used to make the wake itself visible. Motion pictures of the mixing process were made for subsequent study. Figure 1 is a tracing from this filmstrip.

The surprising result of this experiment is that none of the suggested mechanisms adequately describes the process observed. As the turbulent interface approaches a marked element, the element deforms as though it will be pushed aside by the advancing surface wave. It seems that the interface is material. However, when the element is actually touched by the wave surface, it is swirled into the wake by a large scale rotational motion associated with the wave. The newly injested fluid is not immediately diffused through-out the wake by the turbulence, but continues to rotate with the large eddy. Homogenization occurs gradually, as this eddy diffuses.

The rotational motion of the fluid inside the wave is not always present to the same degree, but the marked volume was never

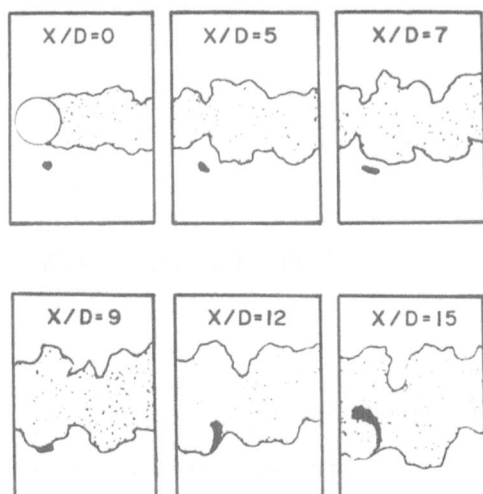

Figure 1.   Entrainment of an Element of the Freestream by an
            Axisymmetric Turbulent Wake

seen to be entrained on the upstream (nearest the wake generating
sphere) side of the wave. Occasionally a droplet would be
entrained by a first wave, whose rotating motion could be seen to
decay, and then the entire volume of now "old" wake fluid would be
re-entrained by a new, more energetic eddy coming from the interior
of the wake. The preliminary results of this experiment have been
reported by Bevilaqua and Lykoudis (1971).

Further indication for the nature of the surface wave may be
found in the work of Papailiou and Lykoudis (1974). They observed
that the vortices shed from a cylinder remain coherent for greater
distance downstream, and at higher Reynolds numbers than generally
suspected. In fact, vortices were seen beyond 300 diameters downstream at Reynolds numbers greater than 20,000. Initially at
least, these vortices occur in a kind of vortex superlayer which
separates the turbulent core from the laminar free stream, see
Figure 2. Motion in the external flow can be seen to be part of
the vortex rotation.

This finding, considered together with the character of the
entrainment process observed in the wake of the sphere, indicates
that the vortices shed by the body have the primary role in determining the structure of the flow and the rate of entrainment. The
nature of the entrainment mechanism suggested by these observations
is as follows. The motion of a boundary wave, which is the exposed
side of a large rotating eddy, displaces the surrounding fluid,

Figure 2.   Large Eddies in the Wake of a Cylinder at x/d ~ 150,
             Re = 6,000. The Motion is from Right to Left

much as air is displaced by a swelling ocean wave. Away from the
turbulent interface this motion resembles the flow around a solid
obstacle; at the interface the action of viscosity retards the
motion relative to the wave, so that the fluid near the surface is
then entrained into the turbulent core by the rotation of the eddy.
The interface for the turbulent vorticity is kept relatively sharp
by the action of the eddy in bringing high intensity turbulence to
the surface from the interior of the wake. Since the effects of
viscosity are not confined to the interface, there is no superlayer
in the usual sense.

     The occurrence of these vortices is compatible with previously
reported wake phenomena and indeed, accounts for some of them.
Grant (1958) reported an approximate periodicity in the appearance
of the mixing jets and associated a circular motion with each of
them. It now seems probable that the mixing jets are part of the
vortex system and actually consist of fluid that has been entrained
by the diffusing vortices. Gartshore (1965) found that the rate of
entrainment in turbulent shear flows was correlated with the size
of the largest eddy. A pattern of rotation was seen within the
billows of a turbulent boundary layer by Blackwelder and Kovasznay
(1970). A similar motion was associated with an "average bulge" in
the wake of a disk by Oswald and Kibens (1971).

     The motion associated with the large eddies at the free
surface of other turbulent shear flows resembles that in the

turbulent wake. While these eddies may originate with an instability of the mean velocity profile, their role in the entrainment
process is probably similar. The traditional notion that the large
eddies disintegrate into small scale turbulence does not seem to be
correct. The key to understanding the mechanism of entrainment is
the dynamics of the relatively persistent large eddies.

REFERENCES

1. Bevilaqua, P. M., and Lykoudis, P. S.: 1971 Mechanism of
   entrainment in turbulent wakes, AIAA J. $\underline{9}$, 1657-1659

2. Blackwelder, R. F., and Kovasznay, L. S. G.: 1970 Large
   scale motion of a turbulent boundary layer, Johns Hopkins
   Univ., Dept. of Mech., Interim TR 2

3. Corrsin, S., and Kistler, A. L.: 1955 Free stream boundaries of turbulent flows, NACA TR 1244

4. Gartshore, I. S.: 1965 Experimental examination of the
   large eddy equilibrium hypothesis, J. Fluid Mech. $\underline{24}$, 89-98

5. Grant, H. L.: 1958 Large eddies of turbulent motion,
   J. Fluid Mech. $\underline{4}$, 149-190.

6. Oswald, L. J., and Kibins, V. R.: 1971 Turbulent flow in
   the wake of a disk, Univ. of Mech., TR00280

7. Papailiou, D. D., and Lykoudis, P. S.: 1974 Turbulent
   vortex streets and the mechanism of entrainment, J. Fluid
   Mech. $\underline{62}$, 11-31

8. Townsend, A. A.: 1966 Mechanism of entrainment in free
   turbulent flows, J. Fluid Mech. $\underline{26}$, 689-715

9. Winant, C. D., and Browand, F. K.: 1974 Vortex pairing:
   the mechanism of turbulent mixing layer growth at moderate
   Reynolds number, J. Fluid Mech. $\underline{63}$, 237-255

SUPERSONIC FREE TURBULENT MIXING LAYERS

Youn H. Oh

NASA Langley Research Center

Hampton, Virginia

Experimental investigations indicate that the spreading rate of supersonic free turbulent mixing layers is substantially lower than that of incompressible flows. In the 1972 Working Conference on Free Turbulent Shear Flows at NASA Langley,[1] no level of turbulent closure schemes succeeded in predicting adequately the observed reduced spreading rate in high Mach number, high Reynolds number free shear layers without introducing significant empiricism or "data fitting." This is contrary to the common optimism for the extension of incompressible methods to high Mach number flows which saw considerable success with wall shear layers in the past. A predictive method which will remedy this situation has been developed.

The basic approach taken herein, as shown in Figure 1 is a mean turbulent energy closure method with a length scale function. The new feature of the method is the inclusion of the effects of the pressure-dilatation term which appears in the compressible form of the turbulence energy equation.

The pressure-dilatation term does vanish in incompressible flow, but there is no reason for it to be negligible in compressible flow. In fact, it is often suspected to be responsible for the observed reduced turbulence level in adiabatic high Mach number mixing layers. The large density variation accompanying a high Mach number flow also was suspected to have something to do with the reduced mixing. However, crucial tests conducted by Brown and Roshko[2] with low speed binary mixing indicate that the density variation along may not explain the observed phenomenon. Mixing nitrogen and helium gases at a static pressure of 7 atmospheres,

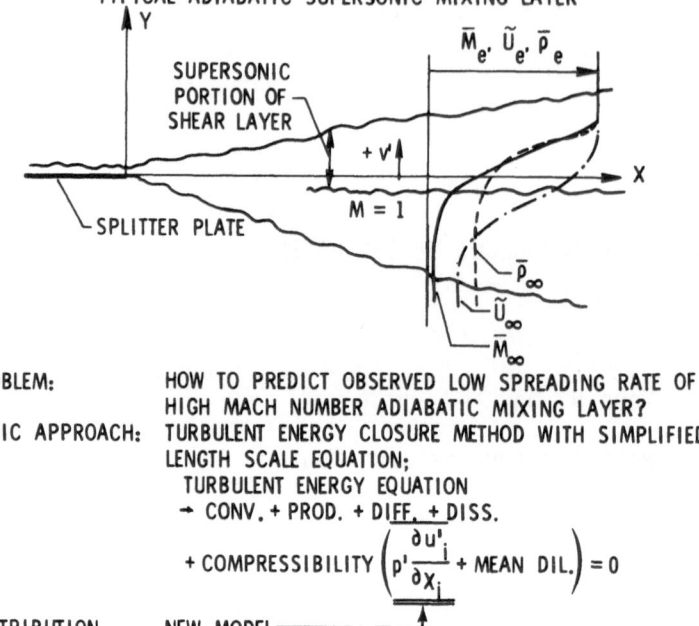

Figure 1.   Free Turbulent Shear Layer

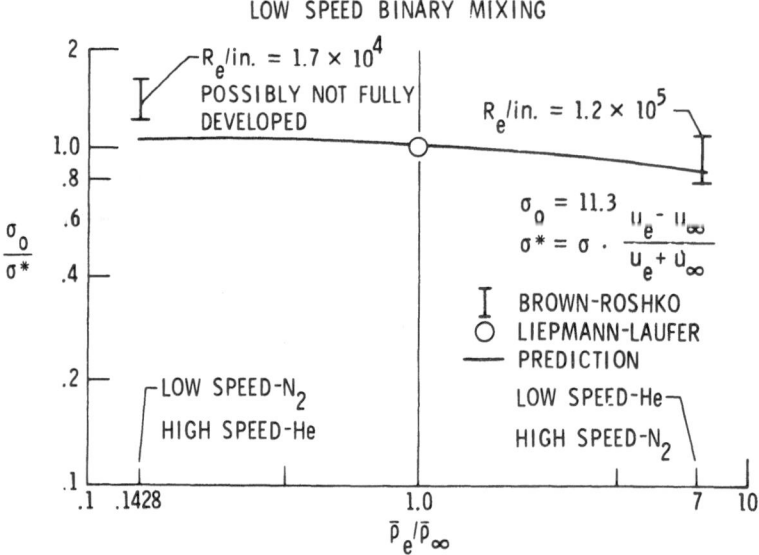

Figure 2.   Effect of Density Ratio on Spreading

they varied the density ratio by a factor of 49, but found very little change in σ as shown in Figure 2.

The present modeling of the pressure-dilatation term is briefly described in Figure 3. The expression that McDonald et al.[3] derived from the continuity equation is used as the starting point for the modeling. The conditions set out for the model are: (1) The pressure-dilatation correlation has to vanish for constant density flow. (2) The effect of this term has to be insignificant for the heterogeneous low-speed mixing flow as Rhosko et al.'s experiment shows. (3) The term should significantly decrease the turbulence shear in adiabatic high Mach number free shear layers. To this end the following hypothesis has been introduced: "When and only when the local mean Mach number is greater than unity, some fraction of the turbulence energy is generated by migrating eddy shocks.[4] In other words, some eddies would be decelerated by passing through weak compression shocks which are almost Mach waves but have finite strength. Also, only pressure fluctuations arising from these kinds of disturbances significantly correlate with velocity and density." The resulting model is as shown in Figure 3, where c is a modeling constant, $\bar{e}$ is the turbulent kinetic energy and $\bar{u}_s$ is the local average sonic speed.

This model indicates that the pressure-dilatation correlation may act like a turbulence energy sink or source depending on

FROM CONTINUITY EQ. $\quad \overline{p' \dfrac{\partial u'_i}{\partial x_i}} \approx - \dfrac{\overline{p'\rho'}}{\bar{\rho}} \dfrac{\partial \bar{u}_i}{\partial x_i} - \dfrac{\overline{p'v'}}{\bar{\rho}} \dfrac{\partial \bar{\rho}}{\partial y}$

MAIN ASSUMPTIONS

   1. EDDY SHOCKS EXIST
   2. THESE SHOCKS ARE MAJOR CONTRIBUTORS FOR $\overline{p'v'}$
   3. QUANTIFICATION OF 1, 2

FINAL MODEL $\quad \overline{p' \dfrac{\partial u'_i}{\partial x_i}} = -c \dfrac{\sqrt{\bar{M}^2-1}}{\bar{M}} \bar{e}\left(\bar{\rho}\dfrac{\partial \bar{u}_i}{\partial x_i} + \text{SIGN}\left\{\dfrac{\partial \bar{M}}{\partial y}\right\}\bar{u}_s \dfrac{\sqrt{\bar{M}^2-1}}{\bar{M}} \dfrac{\partial \bar{\rho}}{\partial y}\right),$

$$\bar{M} \geqslant 1$$

$$= 0, \bar{M} < 1$$

CHARACTERISTICS OF MODELED TERM

   • PREDICTS TURBULENCE ENERGY DECREASE WHEN $\dfrac{\partial \bar{\rho}}{\partial y}$ AND $\dfrac{\partial \bar{M}}{\partial y}$ ARE SAME SIGN
   (IN AGREEMENT WITH DATA)

   • PREDICTS TURBULENCE ENERGY INCREASE WHEN $\dfrac{\partial \bar{\rho}}{\partial y}$ AND $\dfrac{\partial \bar{M}}{\partial y}$ ARE OF OPPOSITE
   SIGN (NOT YET PROVEN-NO DATA YET AVAILABLE)

Figure 3. Modeling of Pressure-Dilatation Term

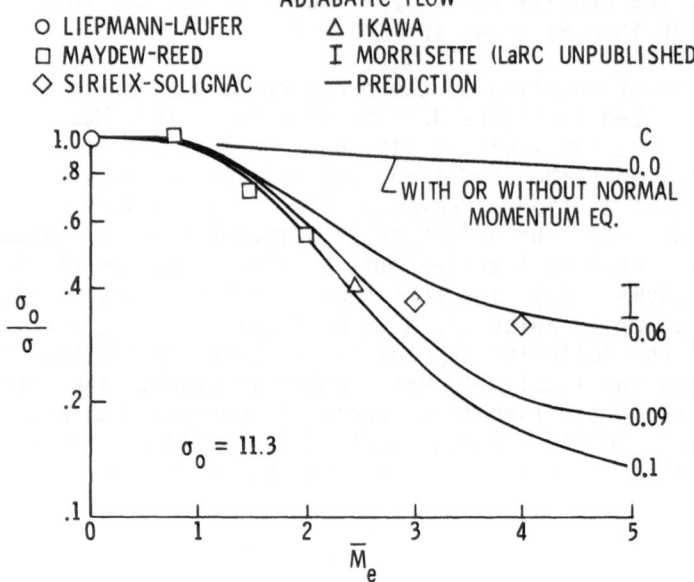

Figure 4.   Effect of Mach Number on Spreading

whether the transverse gradients of mean Mach number and mean density have the same or opposite signs. It should be noted that this modeling assumes that the important pressure fluctuations in supersonic flow are non-isentropic.

Experimental data are compared in Figure 4 with predictions using different values of the modeling constant c. A detailed report on this work has been presented at the 7th Fluid and Plasma Dynamics Conference[5] of the AIAA.

REFERENCES

1.   Anon, "Free Turbulent Shear Flows," Vol. I - Conf. Proc.,
     NASA SP-321, 1973.

2.   Brown, G., and Roshko, A., "The Effect of Density Difference
     on the Turbulent Mixing Layer," AGARD CP-93, 1971, pp 23-1 to
     23-12.

3. Shamroth, S. J., and McDonald, H., "Assessment of a Transi-
    tional Boundary Layer Theory at Low Hypersonic Mach Numbers,"
    NASA CR-2131, 1972, also, United Aircraft Res. Lab. Rept.
    L-911207-11, July 1972.

4. Phillips, O. M., "On the Generation of Sound by Supersonic
    Turbulent Shear Layers," Journal of Fluid Mechanics, Vol. 9,
    Part 1, pp 1-28, 1960.

5. Oh, Y. H., "Analysis of Two-Dimensional Free Turbulent
    Mixing," AIAA Paper No. 74-594, June 1974.

STUDIES RELATED TO TURBULENT FLOWS INVOLVING FAST CHEMICAL REACTIONS

Paul A. Libby

University of California, San Diego

La Jolla, California 92037

SUMMARY

We report on a series of investigations concerned with turbulent flows involving chemical reactions in the simplest chemical
system, fuel-oxidizer resulting in a single product. The conditions of the flow are assumed to be such that at a molecular level
the reactions are infinitely fast. In this limiting case the properties of the turbulence determine the extent of chemical reaction.
The physical picture of the chemical aspects of the flow which
results from the assumption of "fast chemistry" and the experimental
evidence to support this picture are emphasized. The mathematical
consequences appropriate for the case of highly dilute reactions is
then developed; it is shown that the crux of the problem of describing analytically the mean composition field resides in knowledge of
rather detailed properties of a synthetic scalar quantity ζ whose
behavior can be related to that of a passive scalar in turbulent
flows, for example, temperature or the concentration of helium in
helium-air mixtures. In particular, we show that if at each point
in the flow in question we know the probability density function of
ζ, then the mean composition and the mean rate of creation of each
species is completely and readily determined. With this relation
between ζ and the behavior of passive scalars in mind, we review
recent data on the pdf's of temperature and of helium concentration
in order to indicate the various possibilities which must be taken
into account in an adequate analysis. The role of intermittency is
emphasized in this regard. The theoretical difficulties of an
a priori calculation incorporating these possibilities are pointed
out. Nevertheless, we show the results of some calculations of a
two-dimensional mixing layer with fuel in one stream and with

oxidizer in the second stream. The results show the expected
finite reaction zone. We conclude by emphasizing the problems
needing further attention before the analysis of even this simplest
chemical can be considered to be in good order.

1.   INTRODUCTION

One of the largely unsolved problems connected with turbulent
flows of interest to engineers relates to the effect on turbulence
on chemical reactions. This problem arises in many practical
applications from the wakes of reentering vehicles to chemical
lasers to combustion chambers. The corresponding fluid dynamic
aspects of these applications, although not completely treated by
current phenomenology, can be considered to be better understood
than the chemical aspects.

From the point of view of the engineer concerned with making
a prediction of turbulent flow properties when chemical reaction
takes place, the crux of the difficulty concerns representation of
the mean chemical creation terms, usually denoted $\bar{w}_i$, the time mean
rate of production of species i per unit volume per unit time. We
can describe the instantaneous terms $\dot{w}_i$ and in the usual fashion
can take the time average of them. This process introduces difficulties when the molecular rate laws involve Arrenhius type expressions with high activation energies. In addition even in simpler
cases there results a variety of correlations between the fluctuations of concentration of the various participating species. Given
this situation and the need for some answer, it has been common
practice for the engineer to ignore turbulence effects and to
replace $\bar{w}_i$ with mean concentrations and to insert mean fluid properties into the instantaneous relation. This practice appears to
have been done, not out of conviction as to its correctness, but
out of ignorance of a viable alternative.

In recent years, there has been a quickening of activity
devoted to the description of chemical reactions in turbulent flows
with explicit effects of turbulence taken into account. However,
the considerable modelling required and the several questionable
assumptions which must be employed to obtain a closed set of equations suggest that it continues to be worthwhile to study a simple
system in depth in order to establish a firm basis for its analysis.
In this spirit we have been concerned with a particularly simple
chemical system under a special set of circumstances. It consists
of two reactants, possibly simulating fuel and oxidizer, leading
to one product in a one-step, irreversible reaction. The conditions
of the flow are assumed to be such that at a molecular level, the
reactions are infinitely fast. We term this the case of "fast
chemistry." It has been studied in the past presumably because it

should be the first problem properly handled in the full sequence leading to resolution of the problem of the effect of turbulence on chemical reactions in complex systems. Our efforts and point of view have been closely parallel to those of O'Brien and his co-workers [1, 2].

Here we shall first discuss the physical picture of the flow which derives from the assumption of infinitely fast chemical reactions and the experimental evidence to support it. This discussion seems appropriate at the present time because much of the existing literature relating to turbulent flows with chemical reaction emphasizes the mathematical aspects at the expense of physical content. Next the mathematical consequences of fast chemistry are described. The interesting result which derives from these consequences is that the crux of the problem resides in the need to define at each point in the flow the probability density function for a passive scalar quantity which must be considered synthetic but whose behavior can be inferred from appropriate experimental data, e.g., knowledge of the pdf of concentration of a passive contaminant or of temperature can be applied directly. In this regard, several points are especially important to note relative to these introductory remarks. First, past studies of this case of fast chemistry have not benefitted from experimental data on the pdf's of scalars and have generally made assumptions concerning that function of an unrealistic nature. Recently the methods of digital analysis applied to experimental turbulence research have led to means for obtaining pdf's and have permitted more realistic considerations. Second, it is unusual in engineering problems involving turbulence to require as much detail as is implicit in a pdf, which in principle gives the mean value and all the moments of the fluctuations. In most problems the engineer is satisfied to know the mean and the intensity of the fluctuations. Third, although the point of view leading to the treatment of chemical effects by means of the probability density function is perhaps central to understanding the phenomenology involved, there are perhaps other approaches to the analysis of the problem. (Recently, in an as yet unpublished contribution, Professor Frank Marble has proposed a strain model for the case of "fast chemistry" discussed here in order to avoid the pdf approach.) Finally, while experimental data on passive scalars can be employed by analogy to make predictions of chemical behavior in the flows studied experimentally, the main utility of such data is to assist the development of theories permitting a priori calculations of other flows.

Finally, we describe some calculations which are clearly provisional but which indicate the main features of the flows with chemical reactions of the sort treated here.

2.   THE PHYSICAL PICTURE

To simplify the flow and thereby to expose essential aspects, we consider a turbulent shear flow with highly diluted reactants in a background gas of uniform properties. Two reactants lead by a one-step reaction to a single product according to

$$M_1 + M_2 \rightarrow M_3 \tag{1}$$

and involve two atomic species. Thus M_1 and M_2 may be thought of as hydrogen and oxygen and M_3 as water reacting in air diluted with nitrogen.

Under circumstances of pressure and temperature such that the equilibrium constant related to the reactions of Equation (1) is "large," we are naturally led in laminar flows to the flame sheet model, due originally to Burke and Schuman [3] in 1928. According to this model the product $Y_1Y_2 \cong 0$, where Y_i is the mass fraction of species i, and a flame sheet separates the flow into a portion with M_1, the product and diluent present and into another portion with M_2, product and diluent present. In brief the two reactants do not coexist. In early calculations of turbulent shear layers [4, 5] this model is applied essentially with turbulent transport replacing laminar transport; as a result the flame sheet is sharp as in the laminar case and presumably located at some mean position within the reaction zone of the real flow. Although such calculations have apparently been useful for engineering calculations, the resulting picture of the flow and chemical processes is not credible; for example, we expect the reaction and heat release to be diffused in a reaction zone as the turbulence causes the flame sheet to oscillate. One of the aims of the present research is to provide more realistic phenomenology.

The notion of a large equilibrium constant with the consequences that $Y_1Y_2 \cong 0$ can be applied to turbulent flow. The nature of the "flame sheet" depends on the flow. In some circumstances we can think of contiguous eddies with one reactant, product and diluent in one eddy and with the other reactant, product and diluent present in the adjacent eddy. In this picture the product is formed at the sharp interfaces between such eddies as reactants diffuse to the boundaries of the eddies.

This view of successive eddies with one reactant present may be the appropriate one for the transient experiment reported by Gibson and Libby [6]. A weak solution of acetic acid in a beaker was rotated by a magnetic stirring bar. Bursts of weak base solution were added. A single-electrode, conductivity probe whose output depends on the product of the acid-base reaction, a salt, ammonium acetate, was immersed in the beaker. If our ideas of the formation of product at the interface between contiguous eddies with one reactant present are correct, the signal from such a probe should be initially spikey and should indicate a gradual increase of the background level of salt as turbulent strain smears out the interfaces and as the reaction goes to completion. This is indeed found to be the case as shown in Figure 1 taken from Reference 6, which contains a detailed discussion of these results including the effects of spatial resolution of the probe. This simple experiment reinforces our notions of reacting surfaces and indicates that theoretical estimates of the thickness of the reaction zone given in Reference 7 are not unreasonable.

In flow situations of more practical interest, the physical picture to associate with the idea of non-coexistence of reactants may be that suggested in Figure 2. We consider a two-dimensional mixing layer with one reactant, perhaps M_1, in the faster moving stream with a x_1-velocity component of U, and with the second reactant in the slower moving stream with $u_1 = rU$, $0 \le r < 1$. As shown, there are two interface surfaces between the turbulent fluid and the external, irrotational flow; these are the well-known interfaces reported by Corrsin and Kistler [7]. The intermittent nature of the turbulent flow plays an important role in our considerations so we discuss some of the features of intermittency, which in the present context implies the percent of time at a given space point the flow is turbulent. We denote the intermittency by $\gamma : \gamma = 1$ in a fully turbulent flow, $\gamma = 0$ in the external potential flows.

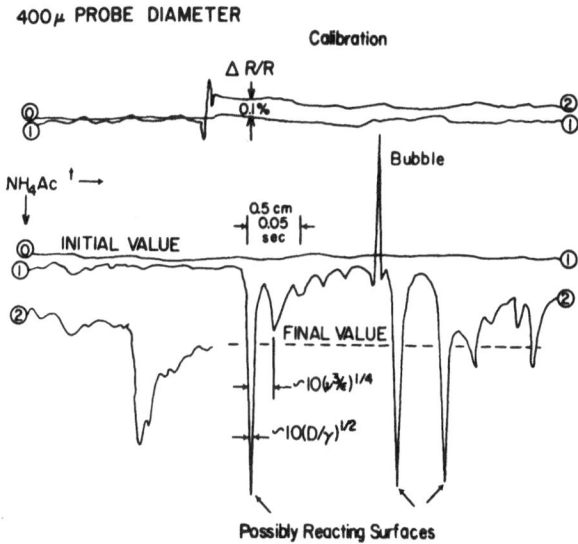

Figure 1. The Production of Salt from a Weak-acid, Weak-base Reaction (From Reference 6)

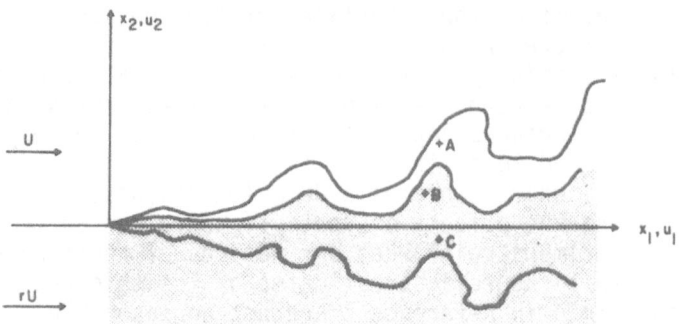

Figure 2.   The Two-dimensional Mixing Layer between Reacting
           Streams

It is probably worth noting here that we have shown in Figure
2 the two interfaces to have "overhangs," i.e., cases wherein the
interface position at a given x_1, x_3 is double valued in x_2. We do
so because LaRue and Libby [8] found that in the downstream edges
of the interface such overhangs occurred in roughly 40% of such
edges whereas they occurred only 8% of the time on the upstream
edges. In terms of our two-dimensional mixing layer we expect this
result to imply that on the side of the faster moving stream overhangs are frequent on downstream edges and that on the side of the
slower moving streams frequent on upstream edges. There appears to
have been no direct experimental confirmation of this expectation.

We also show in Figure 2 another interface; this is between
two turbulent fluids in contrast to the other interfaces but between
fluids with only one reactant present. This is the flame sheet now
viewed as a convoluted oscillating surface. Our previous consideration concerning the production of product apply; in addition the
theoretical estimation of Gibson and Libby [6] gives some indication of the structure of the interface in terms of the turbulent
strain. The picture in terms of overhangs, i.e., the extent of
multiple values of the flame sheet location in x_2 at a given x_1, x_3,
is unclear but overhangs are expected. It is the oscillation of
this flame sheet interface across a finite zone of the mixing layer
that results in a reaction zone. As a way of thinking of the phenomena we can consider the mean rate of destruction of species M_1
at a given space point x_1, x_2, x_3 within the reaction zone to be
represented by a pulse train in time corresponding to crossings by
the flame sheet of the point in question and by an incremental
destruction of M_1 by each crossing. Thus we could write

$$\dot{w}_1 = -\dot{w}_{1n}\, \delta(t - t_n), \quad t_n = t_1, t_2, \ldots, t_N$$

where t_n is the sequence of crossing times and $\dot{w}_{1n}$ is the incremental destruction of M_1 due to the crossing at $t = t_n$. Taking the usual time average, we would get (note that $\dot{w}_{1n}$ is non-dimensional)

$$\overline{w}_1 = \lim_{T \to \infty} \frac{1}{T} \int_0^T \dot{w}_1 dt = \lim_{\substack{T \to \infty \\ N \to \infty}} \left(\frac{N}{T}\right) \left(\frac{1}{N} \sum_{n=1}^{\infty} \dot{w}_{1n}\right) = - f_{I_f} \overline{w}_{1n}$$

where f_{I_f} is the crossing frequency of the flame sheet and $\overline{w}_{1n}$ is the ensemble average of the incremental destruction of M_1. If $\overline{w}_{1n}$ is roughly independent of location within the reaction zone, Equation (1) suggests that the quantity important for engineering calculations, $\overline{w}_1$, will have a bell-shaped distribution across the reaction zone, will have a peak at the most probable location of the flame sheet, and will effectively vanish at the ends of the reaction zone seldom reached by the flame sheet. Although this way of thinking of the two-dimensional mixing layer applies to other flow situations and may be conceptually useful, it does not appear to lead to a strategy for calculations of the flow.

The extent of the reaction zone and its position within the flow depends on the concentrations of the two reactants in the two streams. The increase of one reactant relative to the other tends to drive the reaction zone away from the increased reactant. It is conceivable that the flame sheet can be made essentially contiguous with the interfaces between the irrotational flow and the turbulent fluid. This fact may be useful in permitting information concerning the statistical behavior of interfaces to be interpreted in terms of flame sheet behavior.

It will be important for further developments of our discussion to consider the probability density functions of the mass fractions of the two reactants at several points in the mixing layer. The probability density function (pdf) for species M_1 should be interpreted in the present context as giving the percentage of time that the concentration Y_1 is between Y_1 and $Y_1 + dY_1 \simeq Y_1 + \delta Y_1$ at a given point in the flow; mathematically we think of $P_1(Y_1; x_1, x_2, x_3)$. It may also be helpful to relate the pdf to the time history of the concentration which builds up the pdf after sufficient time. Accordingly, we show in Figure 3 the relevant pdf's and time histories in three distinct regions of the flow; external to the reaction zone near the faster-moving stream, Point A; within the reaction zone, Point B; and external to the reaction zone near the slower-moving stream, Point C.

Several preliminaries are indicated; the concentration of reactant M_1 in the faster-moving stream is taken to be Y_{11}, that of reactant M_2 in the slower-moving stream to be Y_{22}. The

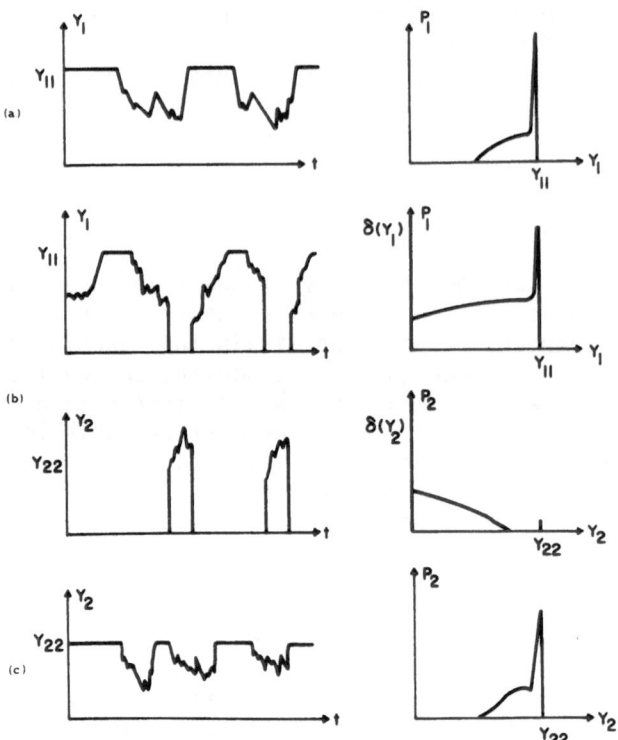

Figure 3.   The Schematic Representation of the Probability Density
            Functions and Related Time Histories of Concentration at
            Selected Points in the Two-dimensional Mixing Layer.

            a) Near the Edge of Faster-moving Stream.
            b) In the Reaction Zone.
            c) Near the Edge of the Slower-moving Stream.

concentrations Y_{11} and Y_{22} are arbitrarily taken to be such that
the reaction zone is principally in the upper portion of the mixing
layer but distinct from the upper interface. Qualitatively the
same results apply when the reaction zone is in the lower portion
of the layer. The flow is assumed to be intermittent at all three
points under consideration. Thus at Points A and B a percent of
time equal to $1-Y_A$ and $1-Y_B$ respectively, the concentration of M_1
equals Y_{11}; at Point C, a percent of time equal to $1-Y_C$ the concentration of M_2 equals Y_{22}. Reactant M_2 does not exist at Point A
and M_1 does not exist at Point C. Because M_1 and M_2 are consumed
in the reaction zone, their concentrations within the turbulent
fluid will be less than Y_{11} and Y_{22}. Although we are not particularly interested in the detailed behavior of the product, similar

considerations apply to M_3; if we assume M_3 is absent in the two external streams, it will be present only in the turbulent portions of the flow.

In Figure 3a we sketch the time history of reactant M_1 at Point A and the related pdf; in Figure 3c we do likewise for reactant M_2 at Point C. Attention is particularly drawn to the spikes in the pdf's at $Y_1 = Y_{11}$ and $Y_2 = Y_{22}$; although we shall later for practical reasons idealize these spikes into delta functions, they have a finite width due to concentrations close to Y_{11} and Y_{22}. These contributions are from the structure of the interface and possibly from recently engulfed external fluid not completely obliterated by the turbulent strain but within the turbulent fluid, i.e., not necessarily in the superlayer. In our subsequent discussion, we shall for simplicity, neglect the later contribution and associate the structure of the spikes near the bounds in concentration with the superlayer.

Note that the pdf corresponding to concentrations within the turbulent fluid are not shown to be Gaussian or near-Gaussian. As we shall see in detail below when we discuss related experimental data, the pdf's are known to be highly skewed. In this regard it is probably appropriate to emphasize the importance of the physical bounds on variables such as concentration and of the implications of those bounds on the pdf's of concentration; here $0 \leq Y_1 \leq Y_{11}$, and $0 \leq Y_2 \leq Y_{22}$ and thus $P_1 \equiv 0$ for $Y_1 < 0$, $Y_1 > Y_{11}$, $P_2 \equiv 0$ for $Y_2 < 0$, $Y_2 > Y_{22}$. The significance of these simple, physically obvious notions can be appreciated by the following consideration: at a point in the flow where the mean value of Y_1 is $\overline{Y}_1$, one can always find an intensity of fluctuation of M_1, $\overline{Y_1'^2}$, such that the bounds 0, Y_{11} require a non-Gaussian distribution of Y_1. It is the neglect of this point which makes suspect the accuracy of some of the early calculations of the effect of turbulence on chemical reactions, based as they are on a Gaussian or two-sided Gaussian distribution of the concentration of reactants. O'Brien [2] has emphasized the necessity in an adequate theory of providing for highly skewed pdf's of the several concentrations.

In the reaction zone, for example at Point B, the situation is more complex. Since we have assumed that the reaction zone is in the upper portion of the mixing layer, two types of values of the reactant M_1 occur just as at Point A; i.e., values equal to and close to Y_{11} corresponding to the external flow and to the interface structure and a range of values clearly within the turbulent fluid. However, part of the time the flame sheet will be outside of Point B and the M_2 will then exist there. The values of reactant M_2 will be in turbulent fluid and will thus be distributed across some range of values. Finally, if the pdf's of M_1 and M_2 correspond to the percent of total time that a value of

Y_1 and Y_2 respectively prevails at a given point and if we recall the non-coexistence of M_1 and M_2, then whenever M_1 exists at Point B we must make entries at $Y_2 = 0$ and vice versa. In addition, just as there is a structure to the pdf associated with the other two interfaces due to the superlayer, there will be a structure to the entries near Y_1, $Y_2 = 0$ due to contributions from the structure of the reaction zone. These characteristics are shown schematically in Figure 3b.

3.   THE PROBLEM OF PREDICTION

With the physical aspects of turbulent shear flows involving fast chemistry set forth we now consider the quantitative treatment of such flows. (Our presentation is based on Reference 9; similar treatments are given elsewhere, e.g., see Lin and O'Brien [2] but the details and exposition here are distinctive.) The assumption of highly dilute reactants and uniform properties of the background fluid implies that attention may be focused on the conservation equations of the species and elements. In the usual notation of Cartesian coordinates the former equations for the mass fraction of species i are

$$\frac{\partial Y_i}{\partial t} + \frac{\partial}{\partial x_k}(u_k Y_i) = \nu \frac{\partial^2 Y_i}{\partial x_k \partial x_k} + \dot{w}_i, \quad i = 1, 2, 3 \tag{2}$$

where we make the benign assumption for present purposes that a single diffusion coefficient applies to the molecular transport of all quantities. If, as we have assumed, the reactants involve two elements, we introduce the element mass fractions

$$Z_1 = Y_1 + (W_1/W_3)Y_3$$
$$Z_2 = Y_2 + (W_2/W_3)Y_3 \tag{3}$$

where W_i is the molecular weight of species i. Related to the conservation of elements and in fact suggesting the definitions given by Equations (3) are relations among the creation terms

$$\frac{\dot{w}_1}{W_1} = \frac{\dot{w}_2}{W_2} = -\frac{\dot{w}_3}{W_3} \tag{4}$$

and

$$\frac{\partial Z_i}{\partial t} + \frac{\partial}{\partial x_k}(u_k Z_i) = \frac{\nu \partial^2 Z_i}{\partial x_k \partial x_k}, \quad i = 1, 2. \tag{5}$$

The mean of Equation (5) in the boundary layer approximation with the usual orientation of coordinates (cf., Figure 2) is

$$\frac{\partial}{\partial x_k} (\overline{u}_k \overline{Z}_i) \cong - \frac{\partial}{\partial x_2} (\overline{u_k' Z_i'}), \qquad i = 1, 2 \tag{5a}$$

where to close the equations some representation of, or additional equations for, the mean flux of the element mass fraction Z_i, i.e., $\overline{u_2' Z_i'}$, is needed.

These relations do not explicitly account for the assumption of fast chemistry, i.e., of the non-coexistence of reactions M_1 and M_2. To do so we define a new variable*

$$\zeta = Z_2 - (W_2/W_1)Z_1. \tag{6}$$

From Equation (5) we have

$$\frac{\partial \zeta}{\partial t} + \frac{\partial}{\partial x_k} (u_k \zeta) = \nu \frac{\partial^2 \zeta}{\partial x_k \partial x_k} \tag{7}$$

which is the conservation equation for a passive scalar. The mean of Equation (7) subject to the same comments as those given above for Equation (5a) is

$$\frac{\partial}{\partial x_k} (\overline{u}_k \overline{\zeta}) \cong - \frac{\partial}{\partial k_2} (\overline{u_2' \zeta'}). \tag{7a}$$

We shall return to Equations (7) and (7a) below but we now focus on the utility of $\zeta(x_1, x_2, x_3, t)$. From Equations (3) note that

$$\zeta = Y_2 - (W_2/W_1)Y_1 \tag{6a}$$

and we can now conveniently impose the assumption of fast chemistry; if at a particular point in space and time, $\zeta > 0$, then since Y_1 and Y_2 do not coexist at that point, $\zeta = Y_2$. Correspondingly, if at another point in space and time $\zeta < 0$, for the same reason, $\zeta = -(W_2/W_1)Y_1$.

---

* Mr. Peter Bradshaw has pointed out that there would be an appealing symmetry without alteration of the essentials if we had defined $\zeta \equiv (Z_2/W_2) - (Z_1/W_1)$; in order to preserve the development in [9] we retain the earlier asymmetric definition. In addition we note that if one works in terms of $\hat{\zeta}$ as defined in Equation (10), no difference arises.

These features of the variable ζ can be employed in the determination of the mean composition in a turbulent shear flow as follows: suppose from an appropriate solution of the mean of Equations (5) for $i = 1, 2$ we know at a given point in space the mean element mass fractions, $\bar{Z}_1$, $\bar{Z}_2$. Then the mean of Equations (3) permit the mean values of two of the three species to be determined in terms of the third; thus, e.g.,

$$\bar{Y}_2 = \bar{\zeta} + w\bar{Y}_1$$

$$\bar{Y}_3 = w(\bar{Z}_1 - \bar{Y}_1)$$

(8)

where $w \equiv W_2/W_1$ is the ratio of molecular weight of the two reactants. Thus if we know $\bar{Y}_1$ at the given point in space as well, we can determine the entire mean composition at that point. In addition if we consider the mean of Equation (2) for $i = 1$ and if we can estimate in that equation the x_2-derivative of the one correlation with velocity significant in turbulent shear flows of the boundary layer type, i.e., $\partial/\partial x_k(\overline{u_k'Y_1'})$, then we can determine $\bar{w}_1$. This emphasizes that mean rate of destruction of reactant M_1 is determined after the mean composition is obtained and not as part of the determination of the mean composition. The limiting behavior to be associated with "fast chemistry" as defined here should be noted; we can write for dilute reactants in general $\bar{w}_1 = -k\overline{Y_1Y_2}$ where k is a rate constant. For fast chemistry $\overline{Y_1Y_2} = 0$ but $k \to \infty$ so that $\bar{w}_1$ can take on any value, in particular it takes on the value consistent with the conservation requirements on $\bar{Y}_1$.

To find Y_1 at the given point in question we employ the properties of ζ by taking the mean of the values of ζ when $\zeta < 0$; more precisely we have

$$-w\bar{Y}_1(x_1,x_2,x_3) = E(\zeta|\zeta<0) = \int_{\zeta_{min}}^{0} P_\zeta(\zeta;x_1,x_2,x_3)\zeta \, d\zeta \qquad (9)$$

where $\zeta_{min} < 0$ is the smallest possible value ζ may take on, and where P_ζ is the pdf of ζ at the space point in question. Equation (9) replaces the earlier statements $\overline{Y_1Y_2} = 0$ which lead to a fixed flame sheet in turbulent shear flows and which disregard the effect of turbulence on the chemical behavior of the flow.

Having established the utility of the variable ζ and the need to know sufficiently about it to permit estimation of the conditioned expectation $E(\zeta|\zeta < 0)$, we can consider other aspects of ζ. The variable ζ is not a directly measurable physical quantity but because it is a passive scalar, measurements of such scalars as temperature, helium concentration in helium-air mixtures, etc. can be directly related to ζ. This may be seen more clearly if we

refer to two turbulent shear flows with identical velocity fields
(cf. Figure 2); in one we have reactant M_1 in concentration Y_{11} in
the faster moving stream and the reactant M_2 in concentration Y_{22}
in the slower moving stream. In the other we have an absolute
temperature T_1 in the faster moving stream, T_2 in the slower. Two
variables

$$\hat{\zeta} = \frac{\zeta - Y_{22}}{-wY_{11} - Y_{22}}$$

$$\theta = \frac{T - T_2}{T_1 - T_2}$$

(10)

will have identical initial and boundary values

$$\hat{\zeta}(0,x_2>0,x_3,t) = \theta(0,x_2>0,x_3,t) = \hat{\zeta}(x_1,x_2\to\infty,x_3,t)$$

$$= \theta(x_1,x_2\to\infty,x_3,t) = 1$$

$$\hat{\zeta}(0,x_2<0,x_3,t) = \theta(0,x_2<0,x_3,t) = \hat{\zeta}(x_1,x_2\to\infty,x_3,t)$$

$$= \theta(x_1,x_2\to\infty,x_3,t) = 0.$$

With identical velocity fields, implying identical behavior of the
two interfaces, the behavior of θ including its pdf will be identical with that of $\hat{\zeta}$.*

Before passing on to other matters it is perhaps worth noting
that Equations (10) suggest

$$\hat{\zeta} = \frac{(\zeta/-wY_{11}) + (Y_{22}/wY_{11})}{1 + (Y_{22}/wY_{11})}$$

(10a)

Actually the parameter $C \equiv (Y_{22}/wY_{11})$ completely determines the
mean composition in these flows involving highly dilute reactants;
it is unnecessary to specify either the individual concentration
in the two streams or the individual molecular weights. It is
this parameter which determines the location of the reaction zone
within the mixing layer. In this form the relevant conditioned

---

* This point of view of the direct relevance of measurable scalar
  quantities to the determination of ζ is due to Lin and O'Brien
  [2]. Earlier the author had taken a more conservative but less
  imaginative, less constructive position (cf. [9]). In [2] the
  authors actually use data on the pdf of temperature in a round
  heated jet to calculate the behavior of such a jet if it contains
  one of the two reactants, the other being in the quiescent
  external region.

expectation is $E(\hat{\zeta}|\hat{\zeta} < C/(1+C))$ so that if $C \ll 1$ (Y_{11} "large"), small fluctuations of $\hat{\zeta}$ from its value in the slower-moving stream, namely zero, involve the presence of reactant M_2 and chemical reaction. In this case the reaction zone occurs on the side of the slower moving stream. If $C \gg 1$ (Y_{11} "small"), only on the faster moving side of the mixing layer will M_1 and M_2 be present at the same point at different times and the reaction zone is on the faster-moving side. In accord with our earlier discussion, we see that if C is sufficiently large, the flame sheet and the lower interfacial surface may be coincident.

4. SOME RELEVANT EXPERIMENTAL RESULTS

Having established the connection between the behavior of passive scalars and chemical reactions of the type considered here and the desirability of having the pdf's of scalars in turbulent shear flows, we give some recent pertinent results of temperature measurements in the wake of a heated cylinder from LaRue and Libby [8] in order to indicate the phenomena which must be handled by an adequate theory. We reinforce our earlier remark by stating that only in recent years as tape recording and the techniques of digital analysis have been employed in experimental turbulence research has it been readily possible to provide data on the pdf's of flow variables within turbulent flows. The absence of such data has clearly hindered the proper analysis of reacting flows even with simple chemistry.

We show in Figure 4 experimentally determined pdf's of temperature relative to the ambient temperature in the wake of a heated cylinder. Because we are interested only in the qualitative nature of these functions, we present the pdf's directly as given in Reference 8 where a standard representation is used.* Figures 4a-c correspond to different transverse positions in the wake; the first is close to the wake axis where over 99% of the time the fluid is turbulent. There we see that the pdf is well represented for most purposes by a Gaussian distribution. Clearly the boundary $\theta = 0$ on the left is essentially ineffective. The next two figures correspond to decreasing values of intermittency and show increasing dominance of the spike corresponding to $\theta = 0$, i.e., to the external flow and to small temperatures presumably associated with the

---

* The notation is as follows: θ is the instantaneous temperature relative to the ambient temperature; $\overline{\Theta}$ is the mean temperature; θ' is the root mean square of the temperature fluctuations. The dotted line on the left of these figures is the location of the gate in θ used to discriminate potential from turbulent fluid.

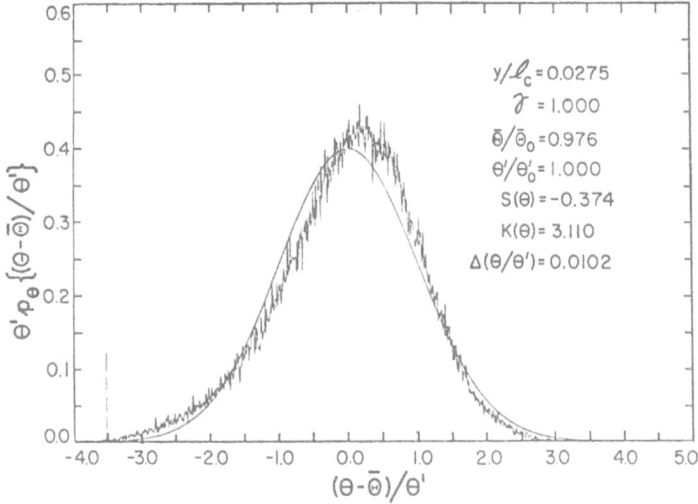

Figure 4a

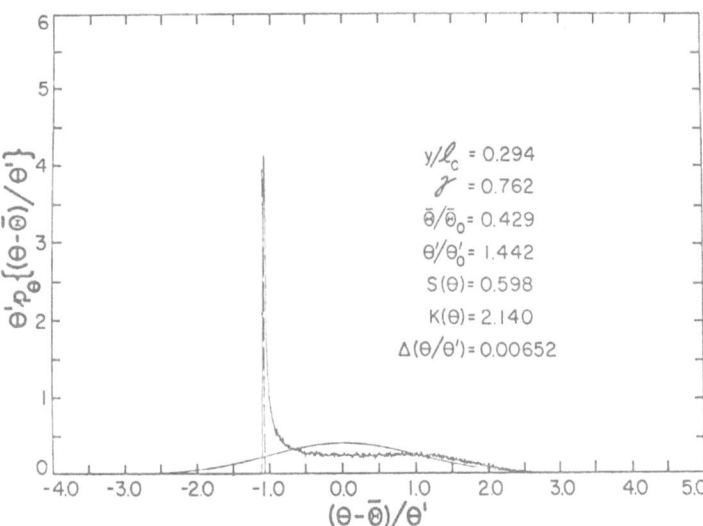

Figure 4b

Figure 4. The Probability Density Function for the Non-dimensionalized and Scaled Temperature in the Wake of a Heated
Cylinder (From Reference 8)

a) On the Wake Axis.
b) Off-axis where the Intermittency is 0.76.
c) Off-axis where the Intermittency is 0.51.

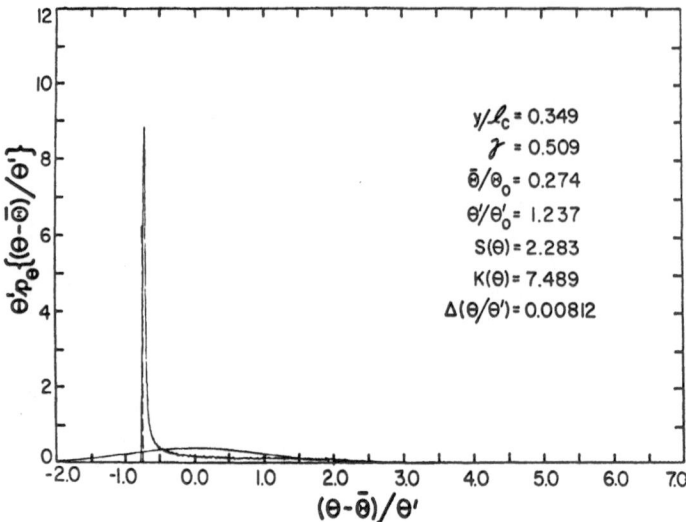

Figure 4c

interface. It is interesting to note in connection with Figure 4c
that such a pdf can be reasonably represented by a delta function
at $v = 0$ and an "equally-probable" plateau within the turbulent
fluid. Such a simple representation is shown in [9] to lead readily
to predictions of the mean concentration distributions. This model
may be considered appropriate when the reactant concentrations force
the reaction zone to be near the limits of the turbulent fluid. In
these cases the boundary value of $\theta = 0$ and the importance of intermittency and non-Gaussian behavior are clearly of great physical
significance. It should be emphasized that there may be circumstances in which the reaction is fully embedded in the middle of
the turbulent fluid. In these cases intermittency and non-Gaussian
behavior may be unimportant.

We now turn to a second set of data which involve a passive
scalar and which are useful in clarifying our ideas. Mr. Richard
Stanford and the present author are using hot-wire anemometry and
digital techniques with a sensor consisting of an extension of the
Way-Libby helium probe [10]. Simultaneous measurements of two
velocity components, u_1, and u_2, and of helium mass fraction c with
good space and time resolution are obtained. The techniques of
calibration, data collection, data reduction and accuracy assessment are rather elaborate and cannot be described here; however,
they are presented in detail in References 10 and 11. It suffices
for present purposes to state that there are obtained at each
measuring station in the flow from these techniques three time

series, one each for u_1, u_2 and c, from which the pdf's as well as mean values of each variable, moments of the fluctuations of each variable, and the various correlations of the fluctuations of the several variables can be obtained. A considerable effort is devoted to the assessment of accuracy; on the basis of the results thereof, we believe the instantaneous mass fraction of helium to be accurate to within $\pm$ 0.001. We shall thus interpret as "air" any concentration of helium less than 0.001.

Our present experiments are being carried out in a porous tube 75 cm long and 7 cm in diameter; turbulent air is blown through one end and pure helium is injected through the porous cylindrical surface. The flow is shown schematically in Figure 5. We show there the interface between pure air and dilute helium. For the flow conditions of our experiment this interface is obliterated upstream of the end of the tube, i.e., all the air has been contaminated with some helium by the end of the tube. However, at x/L = 0.5, 0.75, two out of three measuring stations, the interface can be clearly identified as we shall describe below. It is across this interface that pure air is entrained in fluid with some helium present, diluting the helium being injected in a pure state through the porous wall.

We also show in Figure 5 the instantaneous position of another contour of constant concentration, say $c = c_0$. This contour can be interpreted to be analogous to a flame sheet since it directly corresponds to $\zeta = 0$ (cf. Equation (9)). The dynamics of, and the

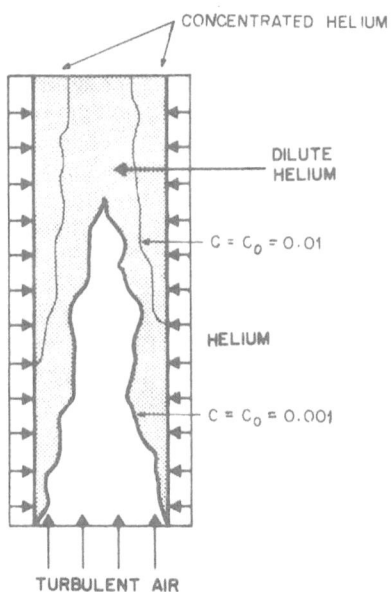

Figure 5. Helium-air Mixing in a Porous Tube

velocity and strain field associated with this contour, would thus
appear to be of interest in connection with the study of turbulent
reacting flows. In what follows we focus on the air-dulute-helium
contours; i.e., we let c_0 = 0.001. However, our data can be
interpreted for larger values of c_0.

Consider the experimental data at x/L = 0.75; on the axis we
find air 96% of the time. (An indication of the accuracy associated with these data is obtained by the results at this station on
the axis, r/R = 0, at x/L = 0.5; roughly a quarter of a million
terms in the time series for c, obtained after eight hours of cali-
bration and data collection give $\bar{c}$ = -0.0011, $\overline{c'^2}$ = $0.355(10^{-3})$.)
At radial positions approaching the wall the percentage of time we
find fluid with helium present increases. At r/R = 0.25 air is
present 81% of the time while at r/R = 0.5, 0.75, air is present
only 26% and 1% of the time respectively.

Consider the time histories and pdf's of helium at the radial
stations discussed above. In Figure 6 we show these data for the
most interesting location r/R = 0.5. Appendix I provides additional experimental results. Figure 6a shows the concentration
history for roughly 0.2 second of data, the unconditional mean concentration of 0.0069, and the "gate," c = 0.001, which can be used
to establish a zero-one discriminating function to identify fluid
on each side of the interface. The spikey nature of the concentration is indicated; e.g., a maximum helium concentration six times
the mean is noted. Figure 6b shows the corresponding pdf based on
a quarter of a million data points, i.e., 500 traces such as that
shown in Figure 6a.

It is clearly of interest in understanding the nature of the
flame sheet interface to investigate the characteristics of the
contours of constant concentration; as an example, consider the
air-dilute-helium interface and some preliminary results from the
data shown in Figure 6. First, we develop a zero-one function as
indicated above and then use this to obtain zone averages, crossing frequencies and point statistics.* By these means it is found
that the mean axial velocity in the air is 12% higher than in the
dilute helium, 8.87 m/sec versus 8.71 m/sec; and the radial velocity in the air is 33% less than in the dilute helium, -0.241 m/sec
versus -0.296 m/sec. These results are consistent with a priori
ideas of the behavior of slower moving helium when injected into a

---

* We use the standard verbiage of conditioned analysis. The reader
  may consult Kovasznay, et al. [12] for a detailed discussion of
  these techniques which are now in widespread use in experimental
  turbulence research.

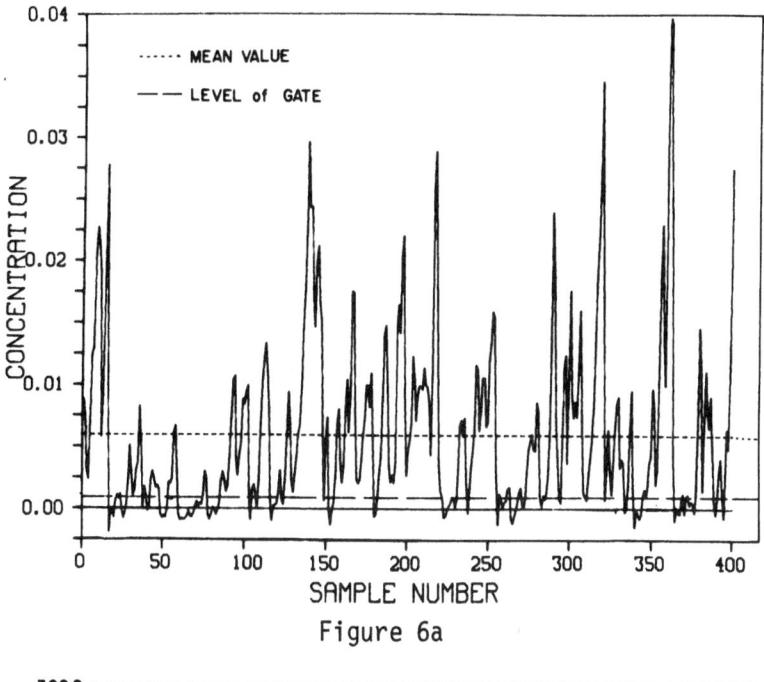

Figure 6a

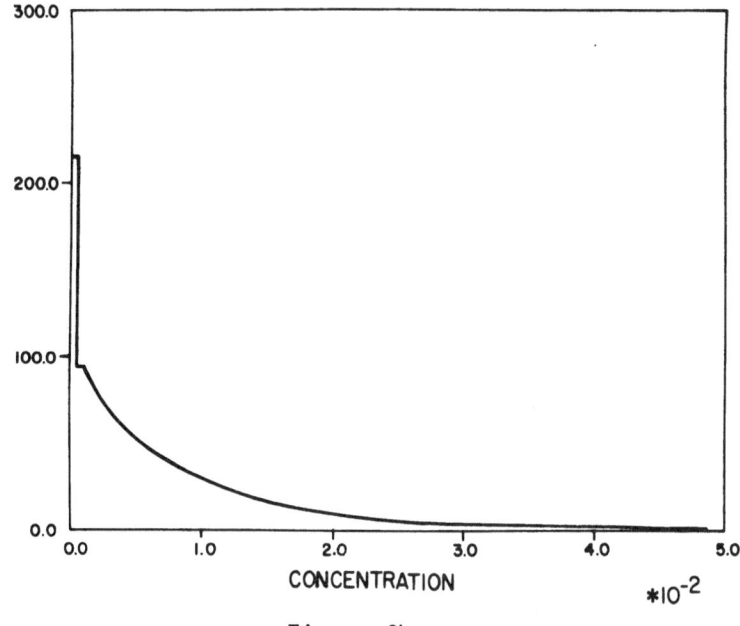

Figure 6b

Figure 6. Experimental Results for the Time History of Helium Concentration and Probability Density Function: x/L = 0.75, r/R = 0.5

   a) Time History of Concentration
   b) The Probability Density Function

faster-moving airstream near the axis of the pipe. It is this sort of differential velocity field on the two sides of the flame sheet interface which determines the turbulent strain entering the estimates of the structure of the interface (cf. Reference 6).

By counting the number of times the discriminating function changes from zero to one in a given period, we can find the crossing frequency f_I and can determine a Strouhal number $f_I D/\bar{u}$, where D is the tube diameter and $\bar{u}$ the unconditioned mean axial velocity; we find $f_I D/\bar{u}$ = 1.68.

Finally, we have determined the ensemble mean values of the axial and radial velocity components at the crossings; when the probe crosses a front from air to dilute helium, the axial and radial velocities are 8.86 and -0.179 m/sec respectively. When the probe crosses from dilute helium to air, the corresponding velocities are 8.88 and -0.343 m/sec respectively. A little reflection indicates that these results are consistent with the expected rolling of large turbulent structures of dilute helium under the influence of the wall shear and the faster moving airstream near the axis.

These same considerations and techniques can be applied to contours of constant concentration for a range of c > 0.001. It would be especially useful to obtain on such contours, as well as on the interface, a measure of the velocity field from which the instantaneous turbulent strain can be estimated. Mr. Stanford and the present author have underway an experiment involving a multiple probe (two u-c probes, closely spaced) which hopefully will provide such data.

5.   FURTHER REMARKS ON PREDICTION

We now turn attention to the problem of estimating at each space point the pdf of ζ so that we may compute the conditional expectation shown by Equation (9). Several preliminary remarks are indicated; if in the flow of interest, at a particular space point, data on the pdf of a passive scalar are actually available, e.g., as in Figures 4 or 6b, then following Lin and O'Brien [2], and Equations (10) or equivalent, we may use these data to compute the mean values of $\bar{Y}_1$ for a variety of values of the concentration parameter, C; if, in addition, means for estimating $\bar{Z}_1$ are employed, e.g., by use of Equation (5a), then at that space point all mean concentrations and the mean rate of creation may be computed.

The nature of an a priori calculation of the mean concentrations in terms of the pdf of ζ is clarified if we consider a simple model for such a pdf. There have been several such models. We

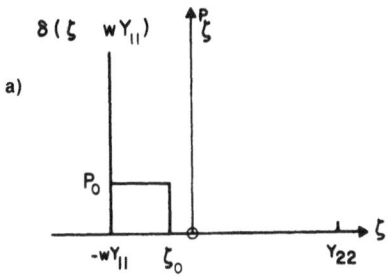

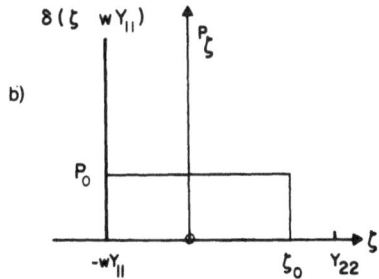

Figure 7. A Highly Idealized Model for the Probability Density
Function of ζ

a) External to the Reaction Zone
b) Within the Reaction Zone

suggested earlier in connection with Figure 4c that if the reaction
zone is near the outer edges of the shear layer, the pdf of ζ can
reasonably be represented by a delta function to describe the
intermittency and a constant value P_0 to represent the values of ζ
within the turbulence. In [9] this is termed the "equally probable
model" and is shown schematically in Figure 7 for the case when the
reaction zone is near the high speed edge of a mixing layer where
in the irrotational flow $\zeta = -wY_{11}$. A similar analysis applies
when the reaction zone is near the other edge.

To exploit this model we assume that at the space point under
consideration the intermittency $\bar{I}$ is known. In addition we assume
that suitable phenomenology for passive scalars has been brought
to bear so that at this point $\bar{\zeta}$, $\overline{\zeta'^2}$ and $\bar{Z}_1$ are known. These
assumptions perhaps supplemented by others as well would generally
have to be made for any pdf model. Now from the nature of
$P_\zeta(\zeta;x_1,x_2,x_3)$ we readily compute:

$$P_0 = \overline{I}/(\zeta_0 + wY_{11})$$

$$\overline{\zeta} = -(1-\overline{I})wY_{11} + \frac{\overline{I}}{\zeta_0 + wY_{11}} \frac{1}{2}\left(\zeta_0 - (wY_{11})^2\right) \tag{11}$$

$$\overline{\zeta'^2} = (1-\overline{I})(\overline{\zeta}+wY_{11})^2 + \frac{\overline{I}}{\zeta_0 + wY_{11}} \left(\frac{1}{3}\left(\zeta_0^3 + (wY_{11})^3\right) - \frac{1}{2}\overline{\zeta}\left(\zeta_0 - (wY_{11})^2\right)\right.$$
$$\left. + \overline{\zeta}^2(\zeta_0 + wY_{11})\right)$$

The first of these defines ζ_0 in terms of $\overline{\zeta}$ and $\overline{I}$, namely

$$\zeta_0 = \frac{2}{\overline{I}}\left(\overline{\zeta} + (1-\tfrac{1}{2}\overline{I})\,wY_{11}\right). \tag{12}$$

If $\zeta_0 < 0$, then at the space point in question there never exists any reactant M_2, no reaction takes place, and $\overline{Y}_1 = -\overline{\zeta}/w$. If $\zeta_0 > 0$, reactant M_2 is present some of the time, in fact a fraction of the time equal to $\int_0^{\zeta_0} P_0 d\zeta$. In this latter case $\overline{Y}_1$

$$\overline{Y}_1 = (1-\overline{I})Y_{11} + \frac{\overline{I}}{\zeta_0 + wY_{11}} \frac{(wY_{11})^2}{2w}. \tag{13}$$

Clearly the condition $\zeta_0 = 0$ determines the edges of the reaction zone in this case.

This simple "equally probable model" may be reasonably accurate for those cases in which the reaction zone is near the edge of shear layer. It is clearly inaccurate when the reaction zone is fully embedded within the turbulent fluid. More important for present purposes is the demonstration of how information on fluid mechanical quantities $\overline{I}$, $\overline{\zeta}$, $\overline{\zeta'^2}$ can be employed along with an assumed form for the pdf of ζ to calculate chemical behavior.

Having indicated the situation concerning the quantitative treatment of the turbulent shear flows involving the simple reactions under consideration, we turn to approaches which may in due course provide appropriate procedures for a priori calculations and be realistic. An approach involving direct calculation of $P_\zeta(\zeta;x_1,x_2,x_3)$, perhaps along the lines pursued by Lundgren [13, 14] for the pdf of velocity in turbulent flows, is appealing and may in due course be developed. Whether the complexity in the pdf of ζ suggested above as being necessary for complete realism can ever be obtained by a direct calculation is, of course, unknown. For the present we follow a more modest path and after [9] take the following view: If we are able to calculate, with intermittency taken into account, the mean and first few moments of the fluctuations of ζ, say $\overline{\zeta'^2}$, $\overline{\zeta'^3}$, $\overline{\zeta'^4}$, etc., then we can estimate the pdf of ζ yielding these quantities and can carry out the conditioned calculation associated with Equation (9).

To exploit this view we neglect the structure associated with the spikes corresponding to limiting values of the concentrations and replace the spikes with delta functions. The assumed functional form for the pdf of ζ must, on the one hand, be capable of describing Gaussian and near-Gaussian distributions and on the other, highly skewed distributions.

With these requirements in mind consider a point in the upper portion, $x_2 > 0$, of the mixing layer (either points A or B in Figure 2) so that $\zeta = -wY_{11}$ in the external flow; we take that

$$P_\zeta(\zeta;x_1,x_2,x_3) = (1 - \overline{I})\delta(\zeta + wY_{11})$$

$$+ Ae^{-a^2(\zeta-\zeta_0)^2}\left[1 + a_1(\zeta-\zeta_0) + a_2(\zeta-\zeta_0)^2 + \ldots\right] \tag{14}$$

where A, ζ_0, and as many a_i coefficients as desired and needed are functions of x_1,x_2,x_3 such that the computed values of $\overline{\zeta}$, $\overline{\zeta'^2}$, $\overline{\zeta'^3}$ etc. at each space point are in accord with Equation (14). A corresponding form applies to points in the lower portion, $x_2 < 0$, of the mixing layer; in particular the first term on the right side is replaced by $(1 - \overline{I})\delta(\zeta-Y_{22})$.

Equation (14) provides the requisite flexibility in the pdf of ζ. We see this as follows: if $\zeta_0 = \overline{\zeta}$, i.e., if the peak and mean coincide, then depending on the intensity $\overline{\zeta'^2}$ and the location of $\overline{\zeta}$ relative to the bounds, $-wY_{11}$, Y_{22}, a full Gaussian, or a truncated Gaussian is obtained from Equation (14). If $\zeta_0 \neq \overline{\zeta}$, and if, e.g., $\zeta_0 > Y_{22}$, then the pdf of ζ within the turbulent fluid and within the bounds on ζ will vary monotonically. In addition, Equation (14) provides for incorporation of additional information on the behavior of ζ as it is available by the inclusion of additional a_i coefficients.

To indicate how Equation (14) may be used, we review some of the results given in [9]. Assume that only $\overline{I}$, $\overline{\zeta}$ and $\overline{\zeta'^2}$ are known, as we did earlier in the "equally probable model," and thus that a_i, i = 1, 2, ... must be taken to be zero; determine A, a, and ζ_0 such that at each point in the flow P_ζ is consistent with the known values of $\overline{I}$, $\overline{\zeta}$, $\overline{\zeta'^2}$. We have from the properties of the pdf

$$I = \int_{-wY_{11}}^{Y_{22}} P_\zeta d\zeta$$

$$\bar{\zeta} = \int_{-wY_{11}}^{Y_{22}} P_\zeta \zeta d\zeta$$

$$\overline{\zeta'^2} = \int_{-wY_{11}}^{Y_{22}} P_\zeta \zeta^2 d\zeta - \bar{\zeta}^2$$

so that some calculation leads to

$$\bar{I} = \frac{1}{2} \tilde{A} \pi^{\frac{1}{2}} \left[\text{erf}\left(\tilde{a}(C-\tilde{\zeta}_0)\right) + \text{erf}\left(\tilde{a}(1+\tilde{\zeta}_0)\right) \right]$$

$$\frac{\bar{\zeta}}{wY_{11}} = -(1-\bar{I}) + \tilde{\zeta}_0 \bar{I} - \frac{\tilde{A}}{2\tilde{a}} \left[e^{-\tilde{a}^2(C-\tilde{\zeta}_0)} - e^{-\tilde{a}^2(1+\tilde{\zeta}_0)^2} \right] \tag{15}$$

$$\frac{\overline{\zeta'^2}}{(wY_{11})^2} = -\left(\frac{\bar{\zeta}}{wY_{11}} - \tilde{\zeta}_0\right)^2 + (1-\bar{I})(1+\tilde{\zeta}_0)^2 + \frac{\bar{I}}{2\tilde{a}^2} - \frac{\tilde{A}}{\tilde{a}}(C-\tilde{\zeta}_0)e^{-\tilde{a}^2(C-\tilde{\zeta}_0)^2}$$

$$+ (1+\tilde{\zeta}_0)e^{-\tilde{a}^2(1+\tilde{\zeta}_0)^2}$$

where to repeat $C \equiv (Y_{22}/wY_{11})$, and where $\tilde{A} = (A/a)$, $\tilde{a} = awY_{11}$, $\tilde{\zeta}_0 = (\zeta_0/wY_{11})$. A corresponding set applies to the other side of the mixing layer. If the left sides of these equations are known for a given point in space, the values of $\tilde{A}$, $\tilde{a}$, and $\tilde{\zeta}_0$ may be determined from Equations (15).

With these values known the conditioned expectation yielding $\bar{Y}_1$ can be carried out; we have

$$-w\bar{Y}_1 = \int_{-wY_{11}}^{0} P_\zeta \zeta d\zeta$$

which gives

$$\frac{\overline{Y}_1}{Y_{11}} = (1-\overline{I}) + \frac{\tilde{A}}{2\tilde{a}} \left(e^{-\tilde{a}^2 \tilde{\zeta}_0^2} - e^{-\tilde{a}^2(1+\tilde{\zeta}_0)^2} \right)$$

$$- \frac{1}{2}\tilde{A}\tilde{\zeta}_0 \pi^{\frac{1}{2}} \left[erf\left(\tilde{a}(1+\tilde{\zeta}_0)\right) - erf(a\tilde{\zeta}_0) \right]. \tag{16}$$

The solution of Equations (15) for $\tilde{A}$, $\tilde{a}$, and $\tilde{\zeta}_0$ must be carried out numerically. It is readily possible to eliminate $\tilde{A}$ and to deal with two equations for $\tilde{a}$ and $\tilde{\zeta}_0$. We have solved these by forming an error measure and by employing either a direct search procedure in two dimensions or a steepest descent; the latter can be extended to cases where several a_i coefficients are incorporated.

We turn now to the question of determining the quantities required to utilize this treatment; namely, $\overline{Z}_1$, $\overline{\zeta}$ (cf. Equation (8)) and $\overline{\zeta'^2}$. In general we might expect that the first two of these are calculated by numerical treatment of Equations (5a) and (7a) with some suitable model or additional equations for the two trans-port terms, $\overline{u_2' Z_1'}$ and $\overline{u_2' \zeta'}$. In addition, in general, we can follow the current developments in the new phenomenology of turbulent shear flows (cf., e.g., Hanjalic and Launder [15] for a widely available paper providing an appropriate entry into the extensive literature on these developments) in order to compute $\overline{\zeta'^2}$. The appropriate equation therefore is

$$\frac{1}{2} \frac{\partial}{\partial x_k} (\overline{u}_k \overline{\zeta'^2}) + \frac{1}{2} \frac{\partial}{\partial x_2} (\overline{u_2' \zeta'^2}) + \overline{u_2' \zeta'} \frac{\partial \overline{\zeta}}{\partial x_2} = -\nu \overline{\frac{\partial \zeta'}{\partial x_k} \frac{\partial \zeta'}{\partial x_k}}. \tag{17}$$

To close this equation modelling of the terms $\overline{u_2' \zeta'^2}$, $\overline{u_2' \zeta'}$ and of the dissipation term $\nu \overline{(\partial \zeta'/\partial x_k)\partial \zeta'/\partial x_k}$ is required. Previous experience in modelling provides a guide here. However, it appears to be accepted widely by those working in the phenomenology of turbulent shear flows that additional attention must be devoted to the treatment of scalars and in particular to the modelling needed to close Equation (17). This may be true to some extent even for the transport terms $\overline{u_2' Z_1'}$ and $\overline{u_2' \zeta'}$. The need for additional atten-tion to the modelling of scalars is related to the significance of intermittency in determining the unconditioned means appearing in the conservation equations; consider the case of a scalar θ which has the constant value, θ_c, in the external flow. Then in the external flow the "fluctuation" of θ has the constant value $(\overline{\theta}-\theta_c)$; consequently, the flux term $\overline{u_2' \theta'}$, which appears in the equivalent of Equation (14), would be the result of fluctuations in the turbu-lent fluid alone and would thus decrease at least at the rate of $\overline{I}$ at the outer edges of a shear layer. Whether the usual modelling properly accounts for such behavior appears doubtful.

Accordingly, in [9] a heuristic assumption for $\overline{\zeta'^2}$ is developed by adding to the intensity due to intermittency a contribution which results from balancing in Equation (14) the production term, $\overline{u_2'\zeta'}(\partial\overline{\zeta}/\partial x_2)$, and the dissipation term on the right side. Thus for the upper portions of the mixing layer

$$\frac{\overline{\zeta'^2}}{(w\overline{Y}_{11})^2} = \frac{1-\overline{I}}{\overline{I}} \left(\frac{\overline{\zeta}}{w\overline{Y}_{11}} + 1\right)^2 + \kappa\gamma U\left(\frac{1+C}{\gamma}\right)^2\left(-\frac{\overline{u_1'u_2'}}{U^2}\right)\left(\frac{1}{U}\frac{\partial\overline{u}_1}{\partial x_2}\right)\Big/\left(\frac{\partial\overline{u}_1}{\partial x_2}\right)_{x_2=0} \quad (18)$$

where κ is a constant and where the quotient $\gamma U/(\partial\overline{u}_1/\partial x_2)_{x_2=0}$ introduces a length scale required by the dissipation term. If in Equation (17) $\kappa = 0$, $\overline{\zeta'^2}$ is due only to intermittency, ζ having the value, $(-w\overline{Y}_{11})$, $(1-\overline{I})$ of the time and $(\overline{\zeta}+(1+\overline{I})w\overline{Y}_{11})/\overline{I}$, the remainder of the term, so as to yield the specified mean $\overline{\zeta}$.* On the other hand, if $\overline{I} = 1$, i.e., the flow is fully turbulent, Equation (18) corresponds to the modelling based on the local balance indicated above. Thus Equation (18) contains in a crude and probably inaccurate manner the requisite physics of the intensity of the fluctuations of ζ.

The applications of this theory in [9] are to the two-dimensional mixing layer since it provides a simple test flow. Distributions of both intermittency and mean velocity are assumed. Crocco-type relations connect the streamwise velocity and $\overline{\zeta}$ and $\overline{Z}_1$ and Equation (18) is employed to determine at a given value of the similarity variable $\eta = x_2/x_1$ the left sides of Equations (12). In Figures 8a and 8b we show typical results for $C = 0.5$, i.e., the case when the concentration of reactant M_1 in stream 1 is relatively larger than that of reactant M_2 in stream 2 so that the reaction zone tends to be on the lower portion of the mixing layer ($\eta < 0$) and for two velocity ratios in the two streams: $r = 0$, corresponding to one stream relatively quiescent, and $r = 0.3$. The regime in which mean values of both reactants coexist is the reaction zone; according to our previous discussion it is across this zone that the flame sheet interface oscillates.

Additional work needs to be devoted to the phenomenology of scalars before calculations of the sort presented here can be considered convincing. However, the present formulation, based as it is on a rather general, flexible description of the pdf of ζ, would appear to be capable of incorporating as much information on the fluctuations of ζ as we are likely to be able to predict.

---

* The "edge" of the reaction zone in this case is associated with the location in the flow where this second value equals zero; until this value is positive there is no reactant M_2 present.

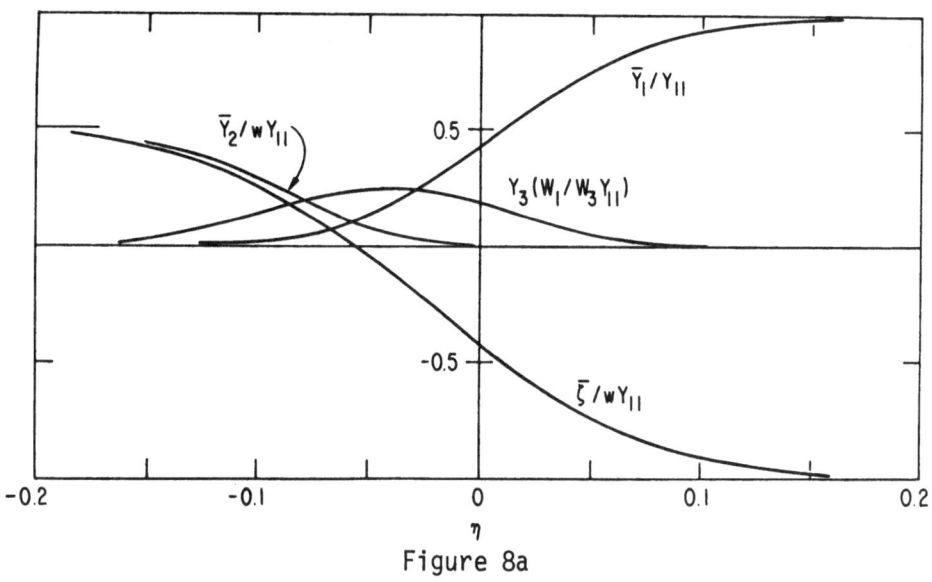

Figure 8a

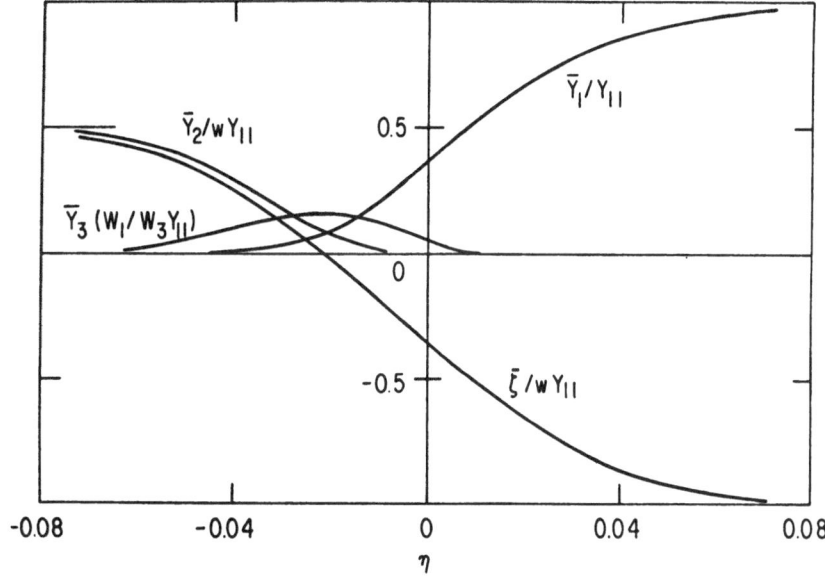

Figure 8b

Figure 8. Predicted Distributions of Mean Concentration in the Two Dimensional Mixing Layer (From Reference 9): $C = 0.5$, $\kappa = 0.1$

a) $r = 0$
b) $r = 0.3$

The accuracy of a given number of terms in the representation of P_ζ must be compared with experimental data on the pdf's of passive scalars. However, a formalism would appear to be available for the treatment of the rather special but basic case of turbulent reacting flows considered here.

5. CONCLUSIONS

We have discussed with primary emphasis in the applicable physical aspects a case of a simple chemical reaction, $M_1 + M_2 \rightarrow M_3$, occurring in a turbulent shear flow under conditions such that chemical kinetics are infinitely fast and the reactants highly diluted. For shear flows there is considered a flame sheet which oscillates across the reaction zone and which can be thought of as an interface with turbulent fluid on both sides.

The problem of describing analytically the mean composition and mean rate of creation at each point in these turbulent flows is shown to be related to the computation of a conditioned expectation of a passive scalar variable, ζ. Accordingly, recent experimental data on the probability density function of temperature in the wake of a heated cylinder is reviewed in order to emphasize the importance of intermittency; of the physical bounds on the permissible values of a variable such as concentration; and of highly skewed, non-Gaussian pdf's. Experimental data from turbulent flows of helium-air mixtures are shown to have direct relevance to reacting flows in providing pdf's of passive scalars and in providing information on the dynamics of, and the velocity field associated with, the oscillating "flame sheet."

A formal procedure permitting an a priori calculation of the mean composition and mean rate of creation in turbulent shear flows is suggested. To be exploited this procedure requires knowledge of the fluctuations of ζ with intermittency taken into account. Further work on the phenomenology of such fluctuations appears necessary before satisfactory calculations using this procedure can be considered to prevail.

There remains much additional experimental and theoretical work on turbulent reacting flows before even relatively simple cases can confidently be computed. Experimental data on the characteristics of interfaces separating two turbulent fluids distinguished by two scalar quantities would assist in the development of the theory. In addition flows involving reactants in sufficient concentrations so that significant density fluctuations arise must be treated theoretically.

6. REFERENCES

1. O'Brien, E. E., "Turbulent Diffusion of Rapidly Reacting Chemical Species," Vol. 18, Advances in Geophysics, Academic Press.

2. Lin, C. H. and O'Brien, E. E., "Turbulent Shear Flow Mixing and Rapid Chemical Reactions: An Analogy," (to appear).

3. Burke, S. P. and Schumann, T. E. W., "Diffusion Flames," Ind. and Engineering Chemistry, 20, 998-1004 (1928).

4. Libby, P. A., "Theoretical Analysis of Turbulent Mixing of Reactive Gases with Application to Supersonic Combustion of Hydrogen," ARS J., 32, 388-396 (1962).

5. Ferri, A., "Mixing Controlled Supersonic Combustion," Annual Review of Fluid Mechanics, Vol. 5, 1973, Annual Reviews, Inc., Palo Alto, California, pp. 301-338.

6. Gibson, C. H. and Libby, P. A., "On Turbulent Flows with Fast Chemical Reactions, Part II: The Distribution of Reactants and Products Near a Reacting Surface," Combustion Science and Technology, 6, 29-35 (1973).

7. Corrsin, S. and Kistler, A., "Free-Stream Boundaries of Turbulent Flows," NACA TR. 1244 (1955).

8. LaRue, J. and Libby, P. A., "Temperature Fluctuations in the Plane Turbulent Wake," Physics of Fluids (accepted).

9. Libby, P. A., "On Turbulent Flows with Fast Chemical Reactions, Part III: Two-Dimensional Mixing with Highly Dilute Reactants," Combustion Science and Technology (submitted).

10. Way, J. and Libby, P. A., "Application of Hot-Wire Anemometry and Digital Techniques to Measurements in a Turbulent Helium Jet," AIAA J., 9, 1567-1573 (1971).

11. Stanford, R. and Libby, P. A., "Further Applications of Hot-Wire Anemometry to Turbulence Measurements in Helium-Air Mixtures," Physics of Fluids (accepted).

12. Kovasznay, L. S. G., Kibens, V., and Blackwelder, R. F., "Large-Scale Motions in the Intermittent Region of a Turbulent Boundary Layer," J. Fluid Mech., 41, pt. 2, 283-325 (1970).

13. Lundgren, T. S., "Distribution Functions in the Statistical Theory of Turbulence," Physics of Fluids, Vol. 10, No. 5, 969-975 (1967).

14. Lundgren, T. S., "Model Equation for Nonhomogeneous Turbu-
     lence," Physics of Fluids, Vol. 12, No. 3, 485-497 (1969).

15. Hanjalic, K. and Launder, B. E., "A Reynolds Stress Model of
     Turbulence and Its Application to Thin Shear Flow," J. Fluid
     Mech., Vol. 52, Pt. 4, 609-638 (1972).

ACKNOWLEDGMENTS

The research reported here was supported by the Office of
Naval Research under Contract N00014-0226-005 (Subcontract No.
4965-26) as part of Project SQUID. The author gratefully acknow-
ledges the assistance of Dr. John La Rue and Mr. Richard Stanford.

APPENDIX I

ADDITIONAL EXPERIMENTAL RESULTS

We present here some additional experimental results related
to conditioned sampling based on specific, "gate" values of the
concentration of helium (cf. Figure 6a). A general picture of the
velocity field associated with oscillating surfaces of fixed concentration within our porous tube results from this sampling.
However, questions related to "overhangs" and other details of
surface geometry remain to be studied by the multiple probe experiment referred to earlier.

In Table A-1 we present these additional results for a "gate"
value, $c_0 = 0.001$. For practical purposes this may be considered
an interface between air and dilute helium. All data are at
$x/L = 0.75$, i.e., in a plane three-quarters up the tube from its
inlet end, but correspond to four radial positions. We present
first the unconditioned values of the two velocity components and
the concentration, of several of the moments of the fluctuations,
and of several significant correlations involving the fluctuation
of the radial velocity, v'. We next provide the corresponding
conditioned values based on the criterion

$$I = 1, \quad c > c_0$$

$$= 0, \quad c < c_0$$

Thus $\bar{I}$ is the percent of time that $c > c_0$. The notation $(\neg)_1$
corresponds to

$$(\neg)_1 = \lim_{T \to \infty} \frac{1}{T} \int^{T} (\ ) \, I \, dt$$

Note that this differs by a factor of $\bar{I}$ from that frequently used in research involving conditioned turbulence, an inessential difference. The average value of a quantity within the fluid with $c > c_0$ is $(\bar{\ })_1/\bar{I}$; the average value of a quantity within the fluid with $c < c_0$ is $((\bar{\ }) - (\bar{\ })_1)/(1 - \bar{I})$. The notation $(\ )\downarrow$ and $(\ )\uparrow$ is used to indicate respectively ensemble averages when the value of c drops from $c > c_0$ to $c < c_0$ and when c increases from $c < c_0$ to $c > c_0$. Finally, we give the crossing frequency, f_I, i.e., one half number of crossings corresponding to c dropping through c_0 divided by the total time, and the average time interval, $\bar{t}$, during which $c > c_0$, $I = 1$.

Analysis of these results supports the remarks which were made earlier and which were based on the limited data available earlier. Within the diluted helium the axial velocity is less and the radial velocity more toward the axis than that in the air. The radial velocities at the "fronts" when c drops through c_0 indicate a general rolling motion as discussed earlier. These features are shown graphically in Figure A-1, where the velocities are in

Table A-1

Experimental Results: x/L = 0.75

r/R	0.00	0.25	0.50	0.75
$\bar{u}$	8.825	8.884	8.754	8.335
$\bar{v}$	0	-0.525(-1)	-0.886(-1)	0.391(-1)
$\bar{c}$	-0.512(-3)	0.253(-3)	0.691(-2)	0.396(-1)
$\overline{u'^2}$	0.437(-1)	0.547(-1)	0.112	0.313
$\overline{v'^2}$	0.278(-1)	0.519(-1)	0.281	2.309
$\overline{c'^2}$	0.671(-6)	0.508(-5)	0.739(-4)	0.792(-3)
$\overline{v'u'}$	-0.272(-2)	0.785(-2)	-0.124(-1)	0.792(-3)
$\overline{v'c'}$	0.128(-4)	-0.837(-4)	0.310(-3)	0.891(-2)
$\overline{v'c'^2}$	0.443(-7)	-0.228(-6)	0.120(-4)	0.299(-3)
$\overline{u'^3}$	-0.558(-2)	-0.639(-2)	-0.131(-1)	-0.524(-1)
$\overline{v'^3}$	-0.809(-3)	-0.891(-2)	0.518(-1)	3.651
$\overline{c'^3}$	0.323(-8)	0.438(-7)	0.128(-5)	0.256(-4)
$\overline{u'^4}$	0.826(-2)	0.120(-1)	0.427(-1)	0.304
$\overline{v'^4}$	0.510(-2)	0.221(-1)	0.874(-1)	44.208
$\overline{c'^4}$	0.326(-10)	0.703(-9)	0.470(-7)	0.304(-5)

Table A-1   (Cont'd.)

r/R	0.00	0.25	0.50	0.75
		$c_0 = 0.001$		
$\bar{I}$	0.042	0.191	0.736	0.989
f_I, sec^{-1}	116.	304.	420.	70.2
$(\bar{t}), \text{sec}$	0.871(-3)	0.130(-2)	0.354(-2)	0.344(-1)
$(\bar{u})_1$	0.364	1.676	6.408	8.240
$(\bar{v})_1$	0.176(-2)	-0.270(-1)	-0.759(-1)	0.419(-1)
$(\bar{c})_1$	0.105(-3)	0.713(-3)	0.697(-2)	0.397(-1)
$(\overline{u'^2})_1$	0.403(-2)	0.177(-1)	0.911(-1)	0.308
$(\overline{v'^2})_1$	0.491(-2)	0.256(-1)	0.265	2.305
$(\overline{c'^2})_1$	0.519(-6)	0.422(-5)	0.603(-4)	0.774(-3)
$(\overline{v'u'})_1$	-0.528(-3)	0.387(-2)	-0.143(-1)	-0.217
$(\overline{v'c'})_1$	0.705(-5)	-0.578(-4)	0.391(-3)	0.878(-2)
$(\overline{v'c'^2})_1$	0.446(-7)	-0.256(-6)	0.113(-4)	0.304(-3)
$(\overline{u'^3})_1$	-0.141(-2)	-0.546(-2)	-0.191(-1)	-0.568(-1)
$(\overline{v'^3})_1$	0.567(-4)	-0.656(-2)	0.563(-1)	3.657
$(\overline{c'^3})_1$	0.320(-8)	0.447(-7)	0.138(-5)	0.263(-4)
$(\overline{u'^4})_1$	0.119(-2)	0.512(-2)	0.381(-1)	0.300
$(\overline{v'^4})_1$	0.213(-3)	0.177(-1)	0.865	44.197
$(\overline{c'^4})_1$	0.325(-10)	0.701(-9)	0.463(-7)	0.302(-5)
$\bar{u}\downarrow$	8.733	8.858	8.878	8.977
$\bar{v}\downarrow$	-0.820(-1)	-0.181	-0.150	-0.332
$(\overline{u'^2})\downarrow$	0.459(-1)	0.445(-1)	0.609(-1)	0.103
$(\overline{v'^2})\downarrow$	0.735(-1)	0.553(-1)	0.915(-1)	0.330
$(\overline{v'u'})\downarrow$	-0.107(-1)	-0.394(-2)	-0.604(-2)	-0.482(-1)
$\bar{u}\uparrow$	8.734	8.842	8.858	8.854
$\bar{v}\uparrow$	0.758(-1)	-0.471(-1)	0.138(-1)	0.417(-1)
$(\overline{u'^2})\uparrow$	0.807(-1)	0.795(-1)	0.845(-1)	0.934(-1)
$(\overline{v'^2})\uparrow$	0.861(-1)	0.852(-1)	0.132	0.409
$(\overline{v'u'})\downarrow$	-0.361(-2)	0.199(-1)	0.128(-1)	-0.482(-1)

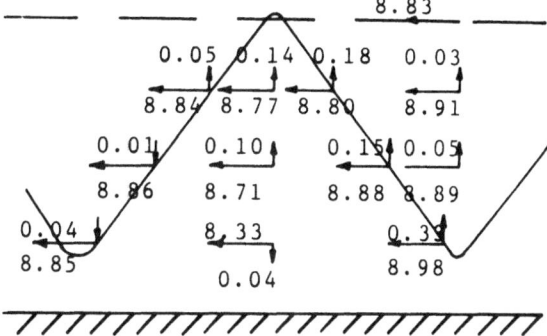

Figure A-1. The Mean Velocity Field Associated with the Contour
$c = c_0$

m/sec and where the contour $c = c_0$ is shown only schematically in
that there probably never exists a contour with the shape shown.

The intermittency expressed in terms of $\overline{I}$ and the crossing
frequency in terms of the ratio $f_I/f_{I,max}$ are shown graphically
in Figure A-2. Also shown are theoretical relations between these
quantities based on the assumption that the location of the contour
is normally distributed and on the resulting theory of Rice.
Clearly, more data are needed to establish firmly the relation
between intermittency and crossing frequency.

It appears that data of the sort presented here can supplement
that associated with the probability density functions of passive
scalars to provide a more complete physical picture of the behavior
of flame sheets in highly idealized turbulent reacting flows.
Additional data are needed.

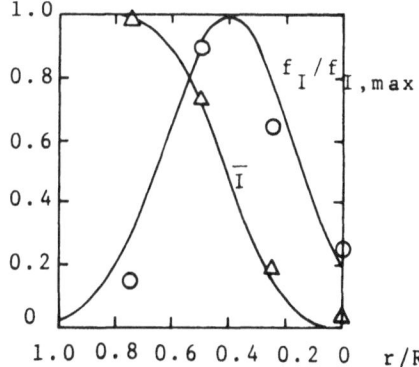

Figure A-2. The Radial Distribution of Intermittency and Crossing
Frequency

DISCUSSION

KOVASZNAY: (Johns Hopkins University, Baltimore, Maryland)

Regarding the folding interface, Paizis and Schwarz actually measured the foldedness. The paper is just out now.[1]

LIBBY:

La Rue and I found that on the downstream face in the heated wake, there were overhangs 40% of the time and overhangs on the upstream face only something like 8% of the time. But we were not able to tell the extent of the overhangs.

DOPAZO: (State University of New York, Stony Brook, New York)

Have you tried to exploit arguments on the self-preservation on the pdf? Is the pdf that is experimentally determined self-preserving?

ALBER: (TRW Systems, Redondo Beach, California)

The pdf, if you make the Gaussian assumption, depends only on the mean temperature and the rms, or the mean scalar and its rms. For the free shear layer, the rms is a constant along constant η-lines, which are simply lines of constant y/x.

LIBBY:

The same thing is true for the pdf's. Those x_1's, x_2's and x_3's that I had could be replaced by η's.

GRANT: (Shell Research Ltd., Chester, England)

We have done some work[2] on flame sheet motion using Langmuir probes, i.e. ion-probes, similar to the work that Karlovitz[3] did a long time ago, except he did pre-mixed flames. We just detect that

---

[1] Paizis, S. T. and W. H. Schwarz (1974) J. Fluid Mech. 63 2 p. 315-345.

[2] A. J. Grant and J. M. Jones, (1974), "Low-frequency diffusion flame oscillations." Proposed paper for publication in Combustion and Flame.

[3] B. Karlovitz, D. W. Denniston, D. H. Knapschaefer and F. E. Wells, 1953, Fourth Symposium (International) on Combustion, pp. 613-620. Williams and Wilkins. Baltimore.

motion of the sheet, an on-off signal, and the set of results that we get show that the turbulent diffusion flame sheet is dominated by low frequency waves. Along with the shadowgraph work we have done, we get the same sort of idea of the flamebrush that Williams talked about yesterday, that it is an ensemble of laminar diffusion flames.

O'BRIEN: (State University of New York, Stony Brook, New York)

I am not sure I got the point of your remark about intermittency. That information is contained in the probability density functions anyway, if they are determined properly.

LIBBY:

The point is that, certainly in that one example, 76% of the time the temperature was over the gate value. That is a fact of life. When we use those data in a theory, we put a delta function at the origin, whose integrated strength is 0.24. Of course, if you have all the elements in the pdf, then it has got to integrate to one. The only thing is you have to decide whether or not to integrate with or without the delta function.

O'BRIEN:

So, really your remarks are addressed to experimenters.

LIEPMANN: (California Institute of Technology, Pasadena, Calif.)

I am interested in your remark about Rice's zero-corssings.

LIBBY:

Well, Rice has a theory which says that if the location of a surface is Gaussianly distributed, one can deduce quite a few things about the crossing frequency and the like. What I did was simply to take the information on the intermittency distribution, from that deduce the constant in Rice's theory, and then compute the crossing frequency.

KOVASZNAY:

One must assume that the slope is distributed Gaussian with distribution too.

LIBBY:

I think that if you make a statement that the location is Gaussian, everything else follows. You do not have to make the additional statement about the slope.

LIEPMANN:

Rice's expression[4] for the number of zero crossings N_0 is quite easily obtained[5] from the joint probability $p(\xi,\eta)$ $d\xi d\eta$ for the position ξ and slope η of a stationary stochastic signal $I(t)$.

$$N_0 = \int_{-\infty}^{+\infty} p(0,\eta)\ |\eta|d\eta$$

if $p(\xi,\eta) = p_1(\xi)p_2(\eta)$ i.e., if ξ and η are statistically independent

$$N_0 = p_1(0)\ \overline{|\eta|}$$

or

$$N_0 = const\ \sqrt{\overline{\eta^2}} \sim \overline{\left(\frac{dI}{dt}\right)^2}$$

For a stationary, stochastic process ξ and η are always uncorrelated. They are statistically independent if p is Gaussian.

LIBBY:

I did that simply because it seems to me that there is an alternative way of looking at the reaction problem. One is simply that you happen to know what the mean rate of destruction of a reactant is at a crossing, and you know the crossing frequency; that is the end of the problem. That is quite a different approach from that band or the pdf. No one has worked it out as far as I know, because even if you have a theory for one of the elements in the product, you do not know what the other one is at the moment. But I just thought it was useful perhaps to pursue that in terms of some data.

CHEVRAY: (State University of New York, Stony Brook, New York)

I would like to make a comment on the measurements of the pdf of the temperature field. I agree that, far downstream, it will

---

[4]   Rice, S. O., The Bell System Technical Journal, 23, 82 (1944). 24, 46 (1945).

[5]   Liepmann, H. W., Die Andwendung eines Satzes über die Nullstellen Stochastischer Funktionen auf Turbulenzmessungen, Helvetica Physica Acta, v. 22 (1949) pp. 119-126.

become symmetrical and perhaps Gaussian. However, it can be shown
that the characteristics of this temperature pdf are related to an
accumulated entrainment. This is particularly noticeable in the
initial sections where the pdf is found to be highly skewed.
Although the effect persists downstream, it is blurred by molecular
diffusion which becomes a predominant factor so that the resulting
pdf becomes more symmetrical and compressed as one progresses downstream. Our recent measurements in a heated axisymmetric jet show
clearly this effect.

LIBBY:

Well, I was going to say that; I think the evidence that
La Rue and I developed with respect to what we call range-
conditioned-statistics, I think this is the same thing. Anyway,
what you do is to collect the data on temperatures where the
lengths are within a specified ϵ of the mean. Then you do salami
chops through those lengths. What we found is that the skewness
of the temperature was very high at the front and the back, but
almost zero in the middle. We interpreted that to mean that the
fronts and backs were virgin turbulence and the middle was old
turbulence. And I think that is along the same lines that your
comment is.

CHEVRAY:

I forgot to mention that my comment applies to the centerline;
outside of it, of course, the pdf is even more skewed as a direct
consequence of intermittency.

LIBBY:

Well, we are fully turbulent on the center and we have no
distinction of old and new on that line. It is all pretty much
Gaussian.

DIFFUSION-LIMITED FIRST AND SECOND ORDER CHEMICAL REACTIONS IN A TURBULENT SHEAR LAYER

I. E. Alber

Applied Technology Division, TRW, Inc.

Redondo Beach, California

I briefly wanted to introduce to the Workshop today, some recent experimental and analytical work performed at TRW Systems[1] by R. Batt and myself, on the properties of passive and reactive scalars in low speed turbulent shear layers.

I have sought to develop a simplified statistical model for predicting the effects of turbulent mixing on diffusion-limited, chemically reacting flows by examining two prototype problems:

1. First Order Reaction: The dissociation of highly diluted nitrogen tetroxide (N_2O_4) injected at low temperatures into the high speed portion of a turbulent free shear layer mixing with quiescent room temperature ambient air (Figure 1a) $N_2O_4 \rightleftarrows 2NO$.

2. Second Order Reaction: The turbulent diffusion flame formed by the reaction of initially highly diluted unmixed reactants A and B which are brought together in a turbulent shear layer to form the product C by the one step reaction (Figure 1b) $A + B \rightarrow C$.

The general method of solution that was adopted in this study, for such diffusion-limited reactions, can be termed to be a form of the turbulent ensemble or probability density function approach. In addition to my work in this area, several similar models have been developed by other researchers and applied to the shear layer reaction problem.[3,4,5]

The present study is unique in that the ensemble analysis has for the first time been used to study a particular reacting flow

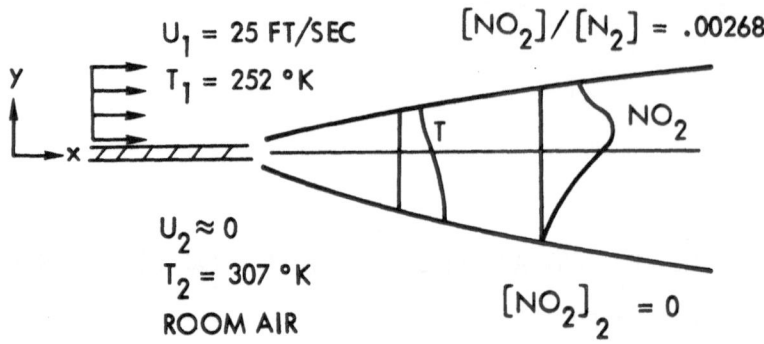

Figure 1a. First-Order Reaction Turbulent Mixing Problem Dissocia-
            tion of N_2O_4. $N_2O_4 \rightleftarrows 2NO_2$.

for which detailed measurements of the fluctuating chemical system
are available. The measurements referred to were made in the N_2O_4
dissociation mixing layer experiment by Dr. Batt[2] of TRW. In his
experiments, measurements were made of the fluctuating NO_2 concentration and temperature levels throughout a low speed turbulent
shear layer using light absorption and hot wire probes. From his
data, Batt was able to obtain a quantitative measure of the probability density function for the inert species turbulent mixing
field, a necessary ingredient of any statistical theory.

    Batt's measured temperature pdf profiles are shown plotted in
Figure 2, at various locations across the shear layer. For reference, the shear layer mean temperature ($\overline{T}$) profile is plotted in
similarity coordinates, $(T_2 - \overline{T})/(T_2 - T_1) = f\left(\frac{\sigma y}{x-x_0}\right)$. Note that the
pdf to the far left of the figure came from a region close to the
high speed edge of the jet and those to the right from the low
speed edge of the jet. These pdfs are quite skewed near the intermittent jet boundaries. Batt's data however, indicate that a broad

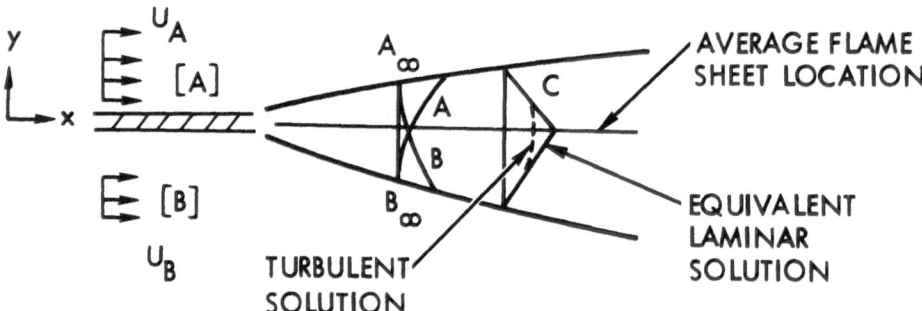

Figure 1b. Second-Order Reaction Turbulent Mixing Problem.
            $A + B \rightarrow C$.

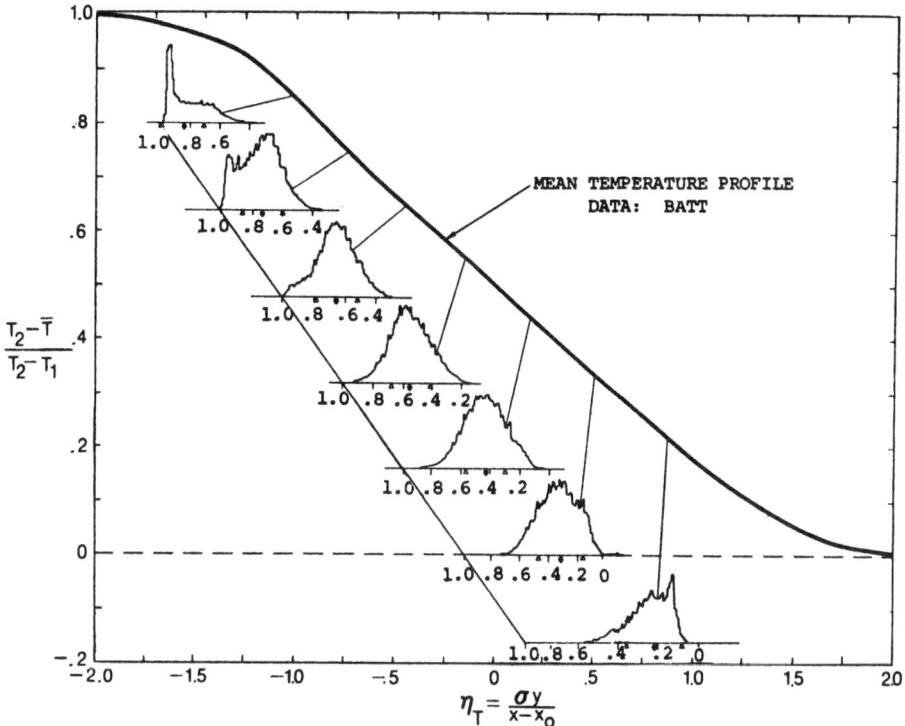

Figure 2. Measured Temperature Probability Distribution at Different Locations Across the Shear Layer. (◆ - Mean Temperature; Δ - Standard Deviation) Data: Batt.[2]

domain, approximately the mid-50% of the shear layer is fully turbulent and nearly Gaussian in its statistics. This is in contrast to the data that Prof. Libby has just shown us for the wake of a cylinder, where the scalar pdfs become quite skewed just above the wake axis.

For diffusion-limited chemical reactions occurring near the mid-portion of a shear layer, it therefore appears that one can approximate the pdf quite well by the Gaussian formula

$$f(T) = \frac{1}{\sqrt{2\pi\sigma_T^2}} \exp\left[-(T - \bar{T})^2/2\sigma_T^2\right] \tag{1}$$

To use equation (1) in a diffusion-limited analysis, all that is required is a shear layer turbulence model which predicts the mean scalar concentration or temperature $\bar{T}$, and its variance $\sigma_T = \overline{T'^2}$ across the shear layer. The mean temperature profile can be calculated by a fairly conventional eddy diffusivity theory (Pr_T = .5), and is found to be nearly linear across the jet. The square root of the variance, or the rms temperature $\sqrt{\overline{T'^2}}$, can be calculated by a combined set of turbulent energy and mean ($\overline{T}$) and fluctuating thermal ($\overline{T'^2}$) conservation equations. A typical calculated $\overline{T'^2}$ profile, determined by such a model is shown in Figure 3, along with a plot of Batt's rms temperature data. One notes a broad region, covering approximately the mid-75% of the jet, in which the calculated rms intensity is nearly uniform, with a magnitude between 12 and 15%. This trend is confirmed by Batt's data. A slight double hump in $\sqrt{\overline{T'^2}}$ not calculated by the present model, is apparent in the data.

The Gaussian pdf assumption for the temperature field can be used to determine analytically the ensemble mean and variance of the NO_2 concentration field for the N_2O_4 experiment of Batt. For the conditions of Batt's experiment (Figure 1a), the chemical times of dissociation and recombination are so much shorter than the typical turbulent diffusion times that one can consider that each temperature-N_2O_4 eddy is essentially in local chemical equilibrium. Thus, if one simply specifies the local temperature, and assumes that the N_2O_4 available for reaction has a minus one correlation with temperature, then the local instantaneous concentration of NO_2 can be determined exactly from the equilibrium formula,

$$[NO_2]/[N_2] = \frac{K_p(t)}{4P} \left\{ \left[1 + \frac{16\,K_p(T)}{P} \left(\frac{T_2 - T}{T_2 - T_1} \right) \frac{(N_2O_4)a_1}{(N_2)_1} \right]^{\frac{1}{2}} - 1 \right\} \quad (2)$$

$$K_p(T) = \exp\left[20.72 - \frac{6747}{T} \right] \text{ atm}$$

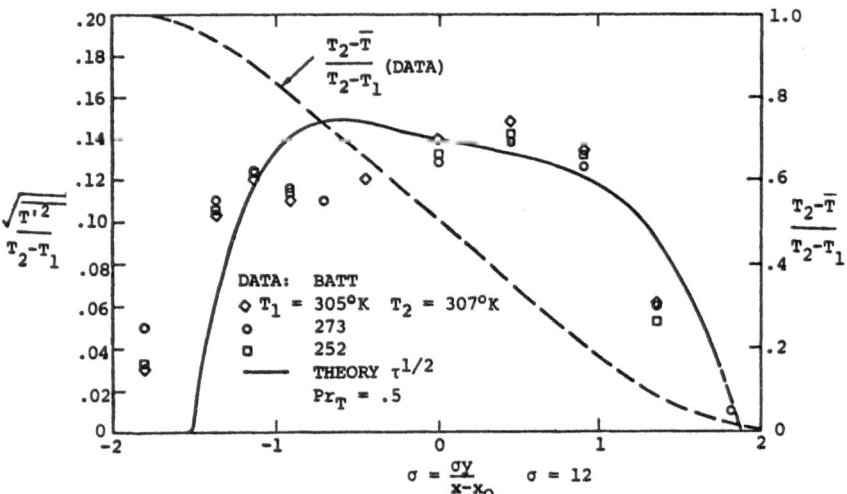

Figure 3. Temperature Intensity Profiles. Data: Batt
(u_1 = 23 fps, $X - X_0$ = 18.5 in)

The ensemble mean $\langle NO_2 \rangle / \langle N_2 \rangle$ can then be readily calculated from eqs (1) and (2), using the following equation for the first concentration moment of the probability density function $f(T)$,

$$\langle NO_2 \rangle / \langle N_2 \rangle = \int_{-\infty}^{\infty} [NO_2]/[N_2] \; f(T) \; dT \tag{3}$$

Figure 4 shows a comparison between the calculated ensemble mean concentration profile of NO_2 and the NO_2 data of Batt ($T_1 = 252°K$ case) plotted as a function of the normalized distance across the jet. The dashed line in Figure 4 represents the results of an equilibrium calculation based on the local mean temperature only. As seen in Figure 4, the peak value of $\langle NO_2 \rangle$ is reduced from the mean calculation by only about 5% at the $T_1 = 252°K$ condition. (Larger deviations are found for lower values of T_1). The uncertainty in the fiber optics probe data is between 10 and 15%. The agreement between the ensemble average calculation and the data is fair, but the comparison is not conclusive as to the significance of the ensemble procedure.

A much more definitive demonstration of the power and simplicity of the ensemble technique is the calculation of the rms concentration fluctuation $[NO_2]'$, shown in comparison with Batt's data in Figure 5.

The agreement between theory and data as to the magnitude and form of the concentration intensity profiles (Figure 5) is quite good over most of the jet with some disagreement near the low speed

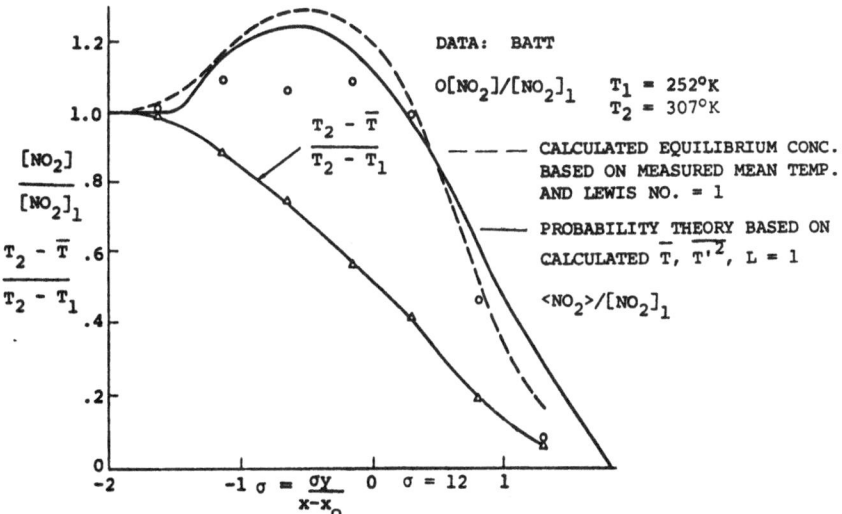

Figure 4.   Mean Mole Fraction Profiles of NO_2. Comparison of Mean Equilibrium and Ensemble Average Theory with Data of Batt ($T_1 = 252°K$).

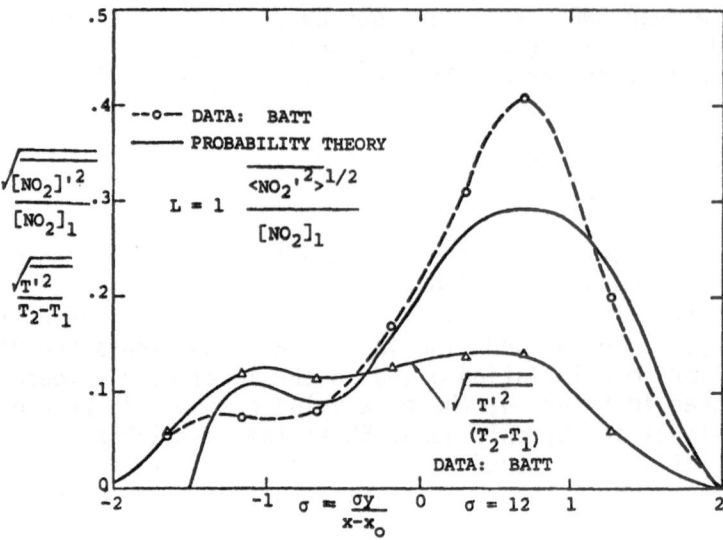

Figure 5. NO_2 Intensity Profiles.   Comparison of Probability
Theory with Data of Batt (T_1 = 252°K).   Also Includes
Measured Temperature Intensity Profiles.

edge, where $[NO_2]'$ peaks.   The measured rms level at this point is
about 40% while the calculated value is approximately 30%.   A
calculation of the rms intensity of NO_2 was also performed by
Batt,[1] using the equilibrium relation for $[NO_2]$, eq. (2); but he
replaced the present assumption on the temperature field with
direct hot wire readings of the temperature time traces.   Batt's
resultant calculated intensity traces were quite close to the
present theory, which assumes a normal temperature population.
The only noticeable difference between the two calculations is
near the peak $[NO_2]'$ location where Batt's local equilibrium
calculation gives an intensity value of 34%.   If one chooses to
think in terms of pdf models, the calculation of Batt using the
measured temperature time traces, employed the 'exact' probability
density function for the scalar field throughout the flow.   His
data analysis results thus showed little difference from the
present model.

The good results shown by the ensemble average theory for
first order quasi-equilibrium reactions, gives us some confidence
that the corresponding diffusion flame theory for the one step
second order reaction A + B → C, will show similar agreement with
measurements when data become available.

The same basic ensemble average analysis, developed for the
N_2O_4 problem, can be applied to the isothermal second-order

chemical shear layer problem, by replacing the temperature pdf with one for the equivalent inert species $\psi \equiv A + C$ (the pdf having variance σ_ψ). One finds (for the stoichiometric case) that a wide turbulent diffusion flame region is formed in the central one-third of the jet with the ensemble average of <A> being non-zero in regions where <B> is non-zero and vice versa (see Figure 6). This effect is characteristic of the experimental measurements of Hawthorne, et al.,[6] for hydrogen-oxygen turbulent diffusion flames.

The implications of these results for turbulent chemical reactions in a shear layer is that peak levels of the product or reaction, <C>, are reduced on the order of 20 to 30% (and inte-grated fluxes reduced 4.3%) for a normal turbulent shear flow over what would have been predicted from an equivalent laminar flame sheet theory. The analysis also indicates that the beneficial performance effects of increased shear layer spreading rates will be reduced to some degree by increased scalar intensity fluctua-tions generated in the mixing region. RMS fluctuation distribu-tions of C across the jet are found to be double humped with a minimum value of $\sqrt{C'^2}$ at the average flame sheet location (see Figure 7).

An application of the present theoretical model to actual combustion problems, requires an extension of the analysis to include compressible non-stoichiometric flow conditions with a further generalization to include more general flow geometries. I suggest that a verification of the second-order analysis should be made by new experimental measurements of the mean and fluctuat-ing properties of a two species turbulent reaction shear layer.

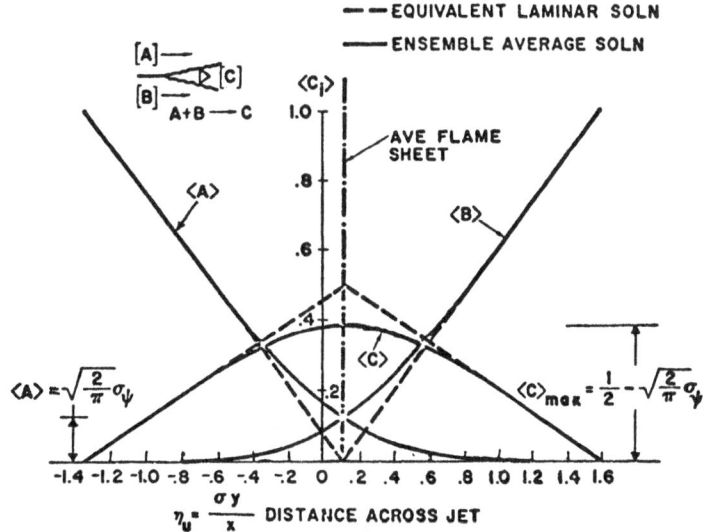

Figure 6. Ensemble Average of Reactants and Products
$(B_\infty/A_\infty = 1)$ $\sigma_\psi = .15$

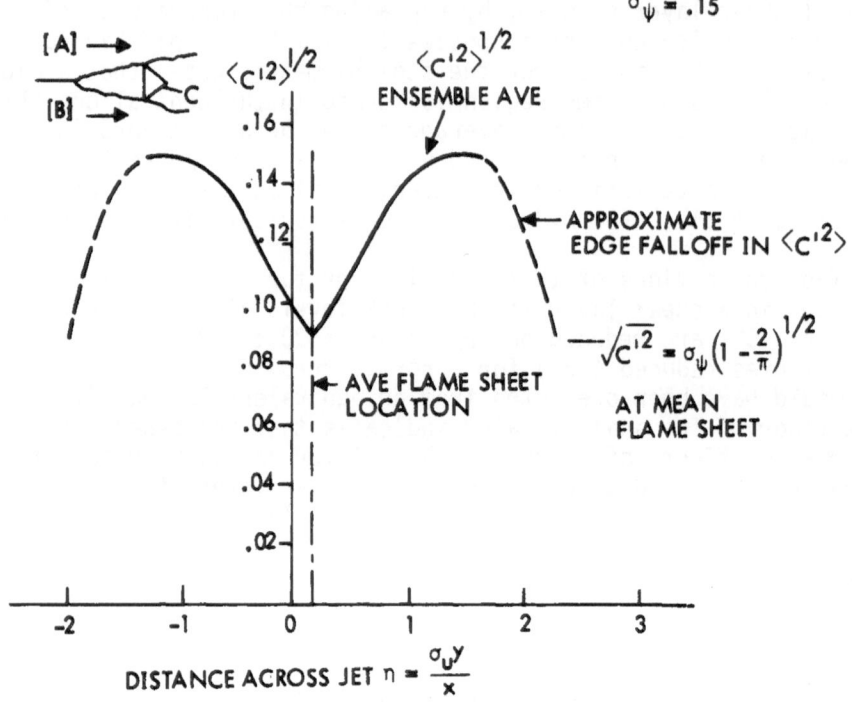

Figure 7.   RMS Product Fluctuation Levels

$$\frac{\langle C'^2\rangle^{\frac{1}{2}}}{A_\infty}, \quad (B_\infty/A_\infty = 1)$$

REFERENCES

1.   Alber, I. E., and R. G. Batt, "An Analysis of Diffusion-
     Limited First and Second Order Chemical Reactions in a Turbu-
     lent Shear Layer" AIAA Paper No. 74-593, presented at the
     AIAA Fluid and Plasma Dynamics Conference, Palo Alto,
     19 June 1974.

2.   Batt, R. G., "Turbulent Chemical Kinetics Investigation,"
     P.S.S. III, Task 4.11 Final Report, SAMSO TR 74-62. TRW
     Report 18117-6023-RU-00, February 1974. See also Batt, R. G.,
     T. Kubota and J. Laufer, "Experimental Investigation of the
     Effect of Shear Flow Turbulence on a Chemical Reaction," AIAA
     Paper No. 70-721, presented at the AIAA Reacting Turbulent
     Flows Conference, San Diego, June 1970.

3.   O'Brien, E. E., "Turbulent Mixing of Two Rapidly Reacting
     Chemical Species," Physics of Fluids, Vol. 14, pp. 1326-1331,
     July 1971.

4. Toor, H. L., "Mass Transfer in Dilute Turbulent and Non-Turbulent Systems with Rapid Irreversible Reactions and Equal Diffusivities," American Institute of Chemical Engineering Journal, Vol. 8, pp. 70-78, March 1962.

5. Libby, P. A., "On Turbulent Flows with Fast Chemical Reactions Part III - Two Dimensional Mixing with High Dilute Reactants," University of California, San Diego Paper, to be published October 1973.

6. Hawthorne, W. R., D. S. Wedell and H. C. Hottel, "Mixing and Combustion in Turbulent Gas Jets," Third Symposium on Combustion Flame and Explosion Phenomena, pp. 267-300, 1949.

ON TURBULENCE STRUCTURE AND MIXING MECHANISM IN FREE TURBULENT

SHEAR FLOWS

H. E. Fiedler

Institut für Stromungstechnik

Berlin, Germany

Some results of recent experimental investigations into the turbulence structure of simple shear flows including conditional sampling techniques are reported and discussed, with special emphasis on common features.

One outstanding feature, most prominent in the two-dimensional shearlayer, is the existence of distinctly coherent large scale (vortical) motions. To visualize this structural peculiarity more clearly and at the same time endeavour an explanation for the scalar mixing mechanism the temperature field in a plane turbulent shearlayer between still ambient air and a heated stream of air at low speed was looked into. The measurements included characteristics of mean temperature and temperature fluctuations obtained by conventional averaging and by averaging separately over the turbulent and non-turbulent parts of the flow. In the light of the obtained results questions concerning structure, transport (mixing) mechanism, entrainment etc. are discussed.

1. INTRODUCTION

In 1969 Wygnanski and Fiedler published an experimental investigation into the round jet. In 1970 the same authors did a similar investigation--with then however improved experimental equipment (conditional sampling)--into the two-dimensional shearlayer. In 1973 Champagne, Pao and Wygnanski looked once again into the same shearlayer (with on line computer processing of the signals) and Gutmark and Wygnanski 1974 investigated the two-dimensional turbulent jet. With the completion of the latter work a

body of experimental evidence was worked out for three different
simple shear flows with essentially the same equipment.

In the following I shall try to give a summarizing view of
some results of these investigations before going into some greater
detail with the investigation of the heated shearlayer which myself
and Messrs. Korschelt and Reitebuch have been working on for some
time.

2.   SOME STRUCTURAL FEATURES

2.1.   Self Similarity and Spread Rates

Those shear flows which by principle are self similar or
approximately self similar attain this state of similarity only
after a certain time. In assuming this time to have a characteristic value for most shear flows (Townsend), one can obtain certain
ratios for the "development lengths" in those flows. In contrast--
as it seems--to these considerations, which propose a longer
development length for a two-dimensional jet than for the round
jet the following has been observed.

In the round jet total self preservation is attained at
approximately $x/d = 70$. For the longitudinal fluctuations the
self preserving state is, however, reached at $x/d = 40$ already
(Wygnanski and Fiedler (1969)). In case of the two-dimensional
jet Gutmark and Wygnanski (1974) claim that the approach to self
preservation occurs much earlier: According to their measurements
the normalized turbulent intensities on the center-plane of the jet
attain their self preserving state at about 30 slot widths downstream of the nozzle. In contrast, however, to the findings in the
axisymmetrical jet there is no indication here that the transverse
and lateral components of the velocity fluctuations attain self
preservation later than the longitudinal fluctuations (Figure 1).
If one looks at the dissipative structure it is, however, obvious
that self similarity is not (even at $x/d = 100$) reached, since then
it must be $\lambda \sim x^{3/4}$, which is not the case throughout the region
of observation.

In comparing the measurements in the two-dimensional jet with
others one finds certain differences in the spread, the decay, the
hypothetical origin and the distribution of the turbulent intensities. It seems most plausible, that these differences may be
attributed to different initial conditions. This assumption is
strongly backed by the investigation of Champagne et al. (1973)
into the two-dimensional shearlayer. One of the main points in
this investigation had arisen from the disturbingly different flow

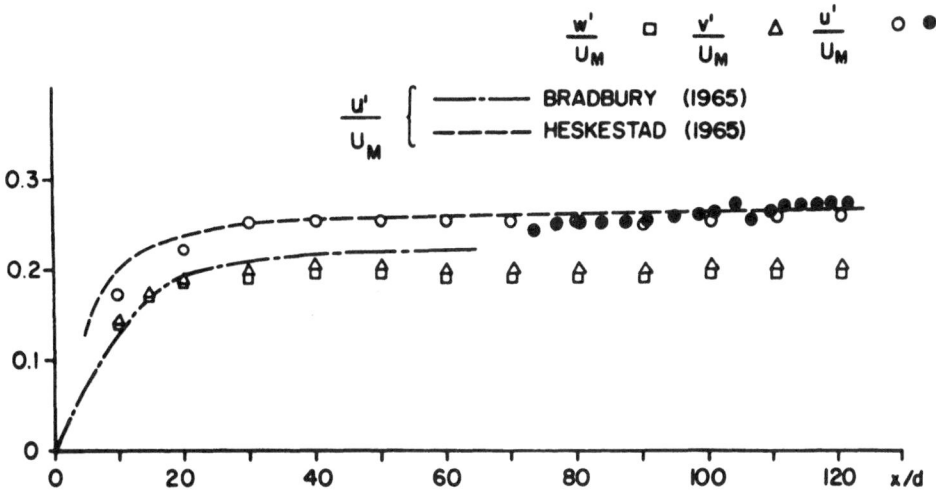

Figure 1.   Variation of Turbulent Intensities along the Center of
the Two-dimensional Jet (Gutmark and Wygnanski 1974).

Author	x_0, cm	U_E, m/s	$\dfrac{y_{0.5} - y_{Plate}}{x - x_0}$	$\dfrac{y_{0.95} - y_{0.1}}{x - x_0}$
Liepmann & Laufer (1947)	0.0	18	0.031	0.17
Wygnanski & Fiedler (1970)	-1.9[+]	12	0.048[+]	0.22[+]
Batt, Kubota, Laufer (1970)	6.0 / -4.3[+]	15.2 / 15.2	0.06 / 0.07[+]	0.17 / 0.22[+]
Spencer & Jones (1971)	0.0	30.5	0.04	0.188*
Sunyach (1971)	0.0	18	0.046	0.187
Present study	6.0	8	0.036	0.185

[+] Trip wire used

* Obtained by assuming an error function profile with $\sigma = 11$ as
the authors did not show an experimental velocity profile for
this case.

Figure 2.   Spread Ratios as Measured by Different Authors
(Champagne, Pao and Wygnanski 1973)

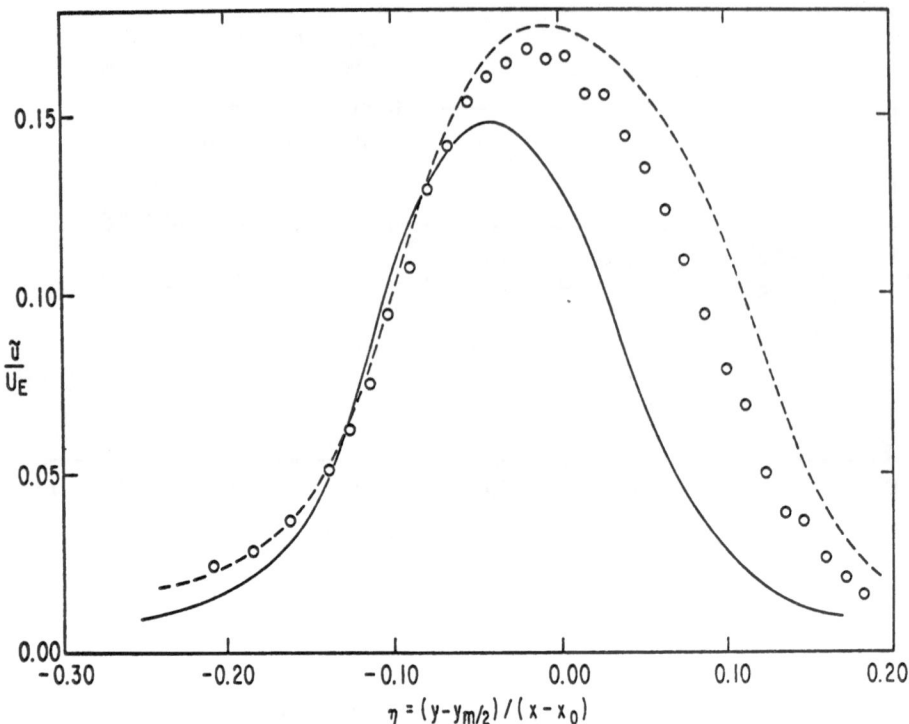

Figure 3. Comparison of Longitudinal Turbulence Intensity Distri-
butions. 0, present; ————, Liepmann and Laufer;
------, Wygnanski and Fiedler. (Champagne, Pao and
Wygnanski 1973)

Author	U_e m/s	Max x, cm	$Re_{x, max}$
Liepmann & Laufer	18	75	9×10^5
Wygnanski & Fiedler	12	60	5×10^5
Batt et al.	15.2	64	6×10^5
Spencer & Jones	30.5	56	1×10^6
Present Study	8	80	4×10^5

Figure 4. Reynolds-Numbers of Different Investigations (Champagne,
Pao and Wygnanski 1973)

development as observed by Liepmann and Laufer (1947) and by
Wygnanski and Fiedler (1970). In comparing these two publications
with others and their own investigation they find clear correlation
between spreading rate and initial condition (trip-wire), where a
stronger spreading, as caused by a trip-wire, occurs primarily on
the low velocity side of the flow (u'-fluctuations) as can be seen
from Figures 2 and 3.

Two explanations for this behaviour are discussed:

1. A disturbance (non existence) of the "regular vortex
   mechanism" by a disturbed initial condition and

2. An insufficiently high Reynolds-Number of the flow.
   (Figure 4).

--I shall comment on No. 1 later--.

2.2. Some Typical Characteristics of Mean and Turbulent
     Features in the Overall and in the Turbulent
     Regime (Conditional Sampling)

In the two-dimensional shearlayer as well as in the plane jet
the difference between the overall mean velocity and the mean
velocity in the turbulent regime is not very strong (Figure 5).
It is, however, evident, that for equal velocity difference the
turbulent velocity at the interface of the mixing layer is higher
than the turbulent velocity at the interface of the jet (Figure 6).

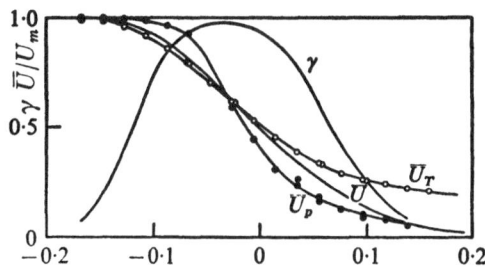

Figure 5. Conventional and Turbulent-Zone-Averaged Velocities in
           the Plane Shearlayer (Wygnanski and Fiedler 1970)

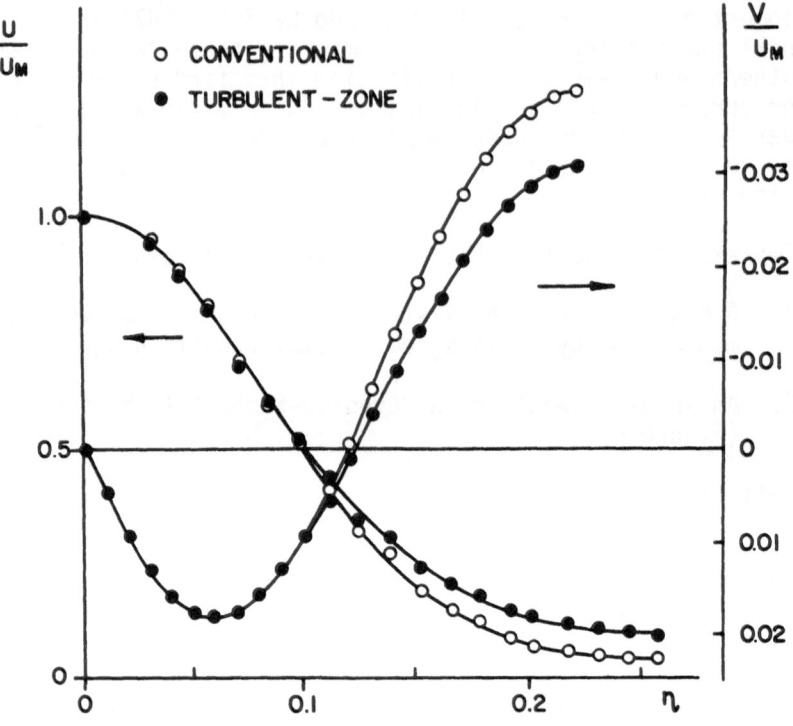

Figure 6.   Conventional and Turbulent-Zone-Averaged Velocities in
            the Plane Jet (Gutmark and Wygnanski 1974)

        With respect to the turbulent intensities it is found that no
isotropy--also not at places where the shear stress is zero--
exists. It is everywhere

$$\overline{u'^2} > \overline{v'^2} \approx \overline{w'^2} \qquad\qquad (\text{Factor} \approx 2)$$

The turbulent zone-averaged intensities show a similar behaviour
in the plane jet and in the plane shearlayer: they attain rela-
tively high values in regions where $\partial \bar{u}/\partial y \approx 0$[†] (Figures 7 and 8).[†]
This is especially true for the turbulent shear stress distribu-
tion, where one may assume negative shear stress in the irrotational regimes (Kibens and Kovasznay (1969)). These turbulent intensities are, however, still far from being homogeneous in the turbulent
domain (also no local isotropy).

_____

[†]   i.e. at the edges of the flow.

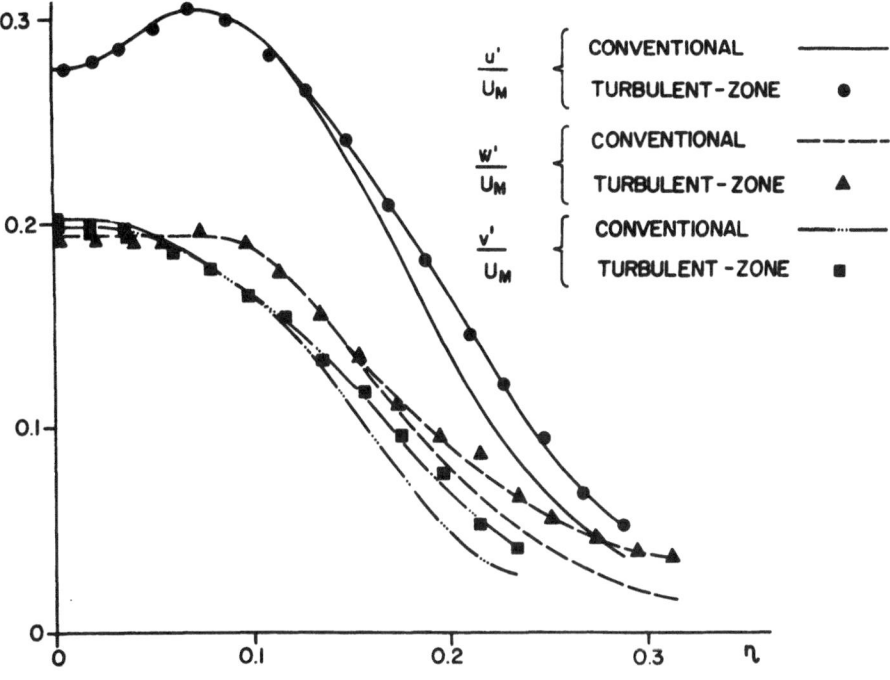

Figure 7.   The Conventional and Turbulent-Zone-Averaged Distribu-
            tions of the Three Components of Turbulent Intensities
            in the Plane Jet (Gutmark and Wygnanski 1974)

In comparing the energy-balances in the turbulent zones of the
shearlayer and the plane jet we find certain similarities especially
on the low velocity sides. While, however, the dissipation shows a
very homogeneous distribution over the whole flow region in both
cases and also the production appears very similar for both flows,
we find the diffusion dominated by different mechanisms: On the
low velocity side of the shearlayer diffusion is largely by velocity fluctuations, while in the plane jet the pressure transport is
dominating. This may be an erroneous result, since the flow is not
self similar in the dissipative region (see above).

One observation, unfortunately however only available for the
shearlayer, is of particular interest: the point averaged velocity
profiles which from their straightlinedness clearly suggest the
existence of large vortices in the flow field as can be seen from
Figure 9. This interpretation is in good agreement with the peculiar shape of the v'-signal, the remarkably constant values of the
eddy viscosity and eddy diffusivity terms in the turbulent regime
as well as with the shape of the two point lateral correlations,

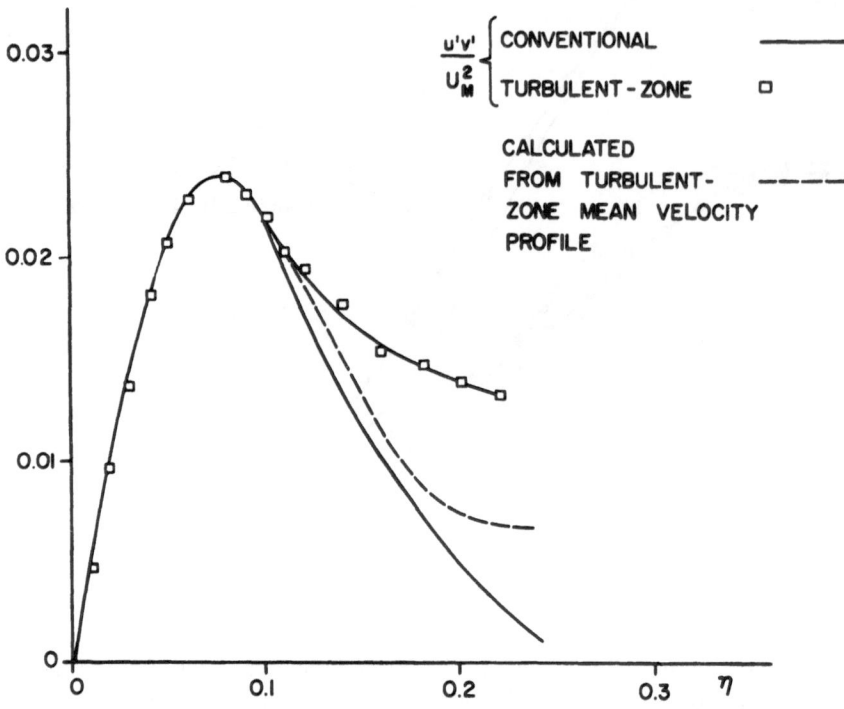

Figure 8.   The Conventional and Turbulent-Zone Distribution of the
            Turbulent Shear Stress in the Plane Jet (Gutmark and
            Wygnanski 1974)

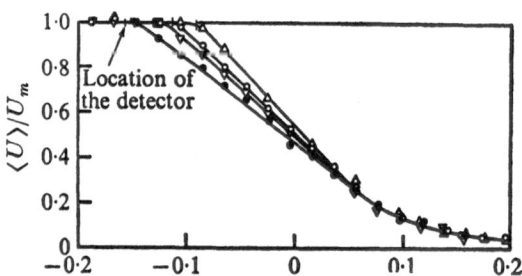

Figure 9.   Point Averaged Velocity Profiles in the Plane Shearlayer
            (Wygnanski and Fiedler 1970)

particularly in the turbulent jets. There, however, a marked difference shows between the round and the plane jet. The lateral
correlation in the plane jet has a very prominent negative dip--
considerably stronger than the round jet--indicating a stronger
coherence of the large eddies.

In all three shear flows the ratio $\Lambda g/\sigma \approx 1$ for $y = \bar{y}$, indicating that the interface is primarily contorted by the large
eddies.

With respect to the two interfaces in the shearlayer the
statement from Wygnanski and Fiedler's (1970) paper claiming
independence is in contradiction with the above observation and
with the idea of the existence of structural coherence. It can no
longer be considered valid: Recent measurements by the author of
the correlation of δ_1' with δ_2' show a clear interdependence with an
expected maximum at an angle of approximately 50° (Figure 10).

In this context the interesting measurements of flatness
factors in the plane jet (Gutmark and Wygnanski 1974) ought to be
mentioned, which clearly characterize the difficulty of obtaining
intermittency distributions (being a problem in itself) from flat-
ness-factors (Figure 11), and which, apparently, caused the authors
to use the sum $u'^2 + k(\partial u'/\partial t)^2$ as their intermittency indicating
signal.

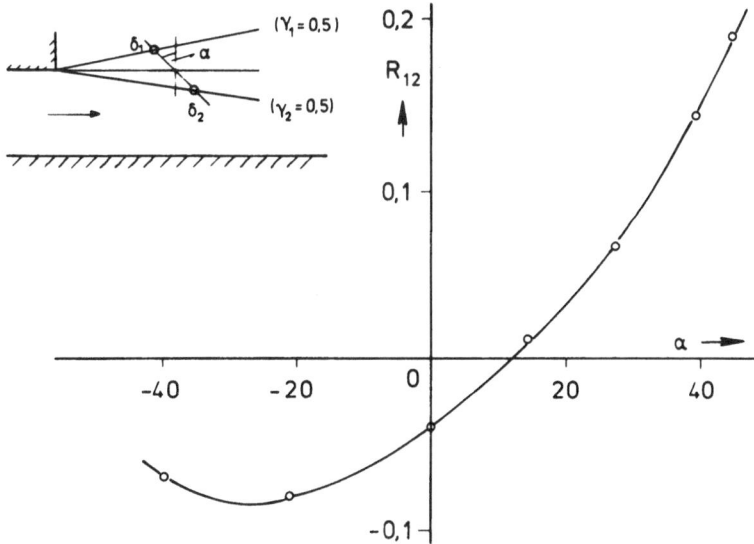

Figure 10.   Correlation of δ-signal (Intermittency) in Plane Shear-
            layer

Let us look here in some more detail at the question of structural coherence: Since all conceivable free shear flows are spreading and--at the same time--decaying, any existing large scale structure (e.g. a vortex) can only have a limited life-time if it is compatible with the general similarity behaviour of the flow.

A "coherent" large vortex may be characterized by its frequency:

$$f_V \sim \frac{\Delta u_V}{b_V} \sim x^{-n-m}, \qquad \text{with} \qquad u_V \sim x^{-n}$$

$$\text{and} \qquad b_V \sim x^m$$

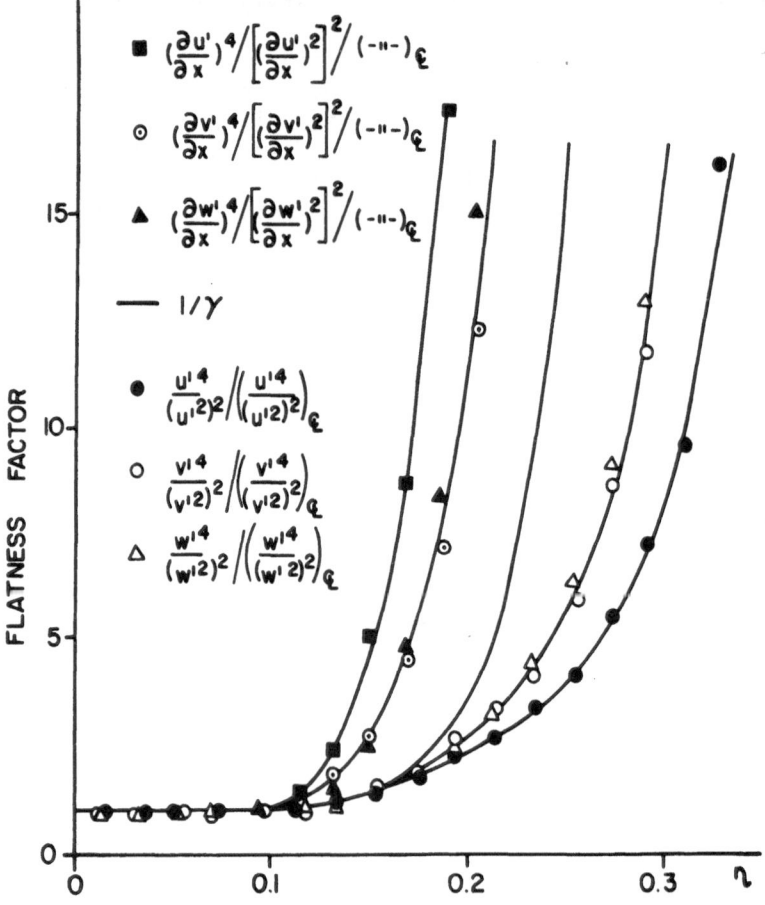

Figure 11.   Different Flatness Factors in the Plane Jet (Gutmark and Wygnanski 1974)

For some typical flows therefore:

shearlayer: $f_V \sim x^{-1}$

plane jet: $f_V \sim x^{-1.5}$

round jet: $f_V \sim x^{-2}$

plane wake: $f_V \sim x^{-1}$

round wake: $f_V \sim x^{-1}$

Perfect coherence would of course require $f_V \sim x^0 = const.$
Thus we should expect the most prominent coherence in the shearlayer as well as in both wake configurations. The least coherence
is, on the other hand, to be expected in the round jet--which
agrees qualitatively with the observation reported above. As a
simple parameter to characterize the coherence in a flow a normalized length may be used:

$$L_c = \frac{x(f/2) - x(f)}{x(f)} = 2^{1/m+n} - 1$$

We obtain:

total coherence: $L_c = \infty$

shearlayer: $L_c = 1$

plane wake: $L_c = 1$

round wake: $L_c = 1$

plane jet: $L_c = 0.59$

round jet: $L_c = 0.44$

no coherence: $L_c = 0$

From this it is obvious, that for a study of coherent features
the plane shearlayer should be the best choice.

3.   HEAT TRANSFER IN THE SHEARLAYER

3.1.   Objectives and Motivations

This experimental investigation which will be published in
greater detail elsewhere (Fiedler 1974) was primarily aimed at

finding a model for the transport mechanism of a scalar quantity in a typical turbulent shear flow. As a first configuration the two-dimensional shearlayer between a heated stream of air and cold air at rest was chosen for the following reasons.

   a.   Its velocity structure is well known.

   b.   Self-preservation is attained at an early stage.

   c.   Heat-transfer measurements have not been reported previously.

   d.   It is the simplest conceivable configuration.

   A second objective of this investigation was to obtain better information about the turbulence structure by observing and analyzing not the flow field itself, but the temperature field and its characteristics.

3.2. Experimental Arrangement

   A schematic of the test section is given in Figure 12. All measurements reported were obtained with a velocity of 8 m/sec and a temperature difference of 26° C. Mean and fluctuating temperature characteristics were measured with the use of a resistor probe with a frequency response of better than 2000 Hz.

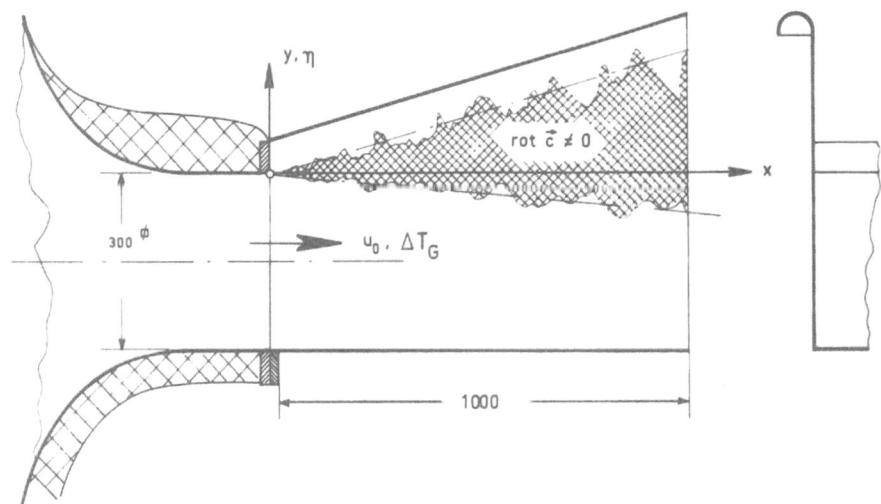

Figure 12. Schematic of Test Section for Shearlayer

3.3. Experimental Results

From a number of preliminary measurements of $\bar{T}$ and T' and $\partial T'/\partial t$ the self preserving region of the temperature field was found for values $x > 500$ mm. The following measurements were then restricted to the self preserving region ($Re_x \geq 3 \cdot 10^5$). It is interesting to note here, that in this flow the expected proportionality from similarity considerations:

$$\overline{\left(\frac{\partial T'}{\partial t}\right)^2} \sim \frac{\overline{T'^2}}{\lambda_T^2} \sim x^{-1}$$

holds perfectly.

The distribution of the mean velocity was measured for $\Delta T_G = 0$ and $\Delta T_G = 26°$ C. For the cold flow the spread parameter was $\sigma = 11$. In the heated flow case the spread, especially in the outer i.e. the low velocity region of the shearlayer, is somewhat larger. The lateral distribution of the mean temperature exhibits a strikingly different character when compared with the mean velocity--Figure 13. The velocity distribution curve has a single inflexion point indicating an ordinary gradient diffusion mechanism for the momentum. In contrast to this the mean temperature distribution curve has three inflexion points with a somewhat flat region at large values of γ. This shape of the curve suggests a transport mechanism dominated by large scale motion.

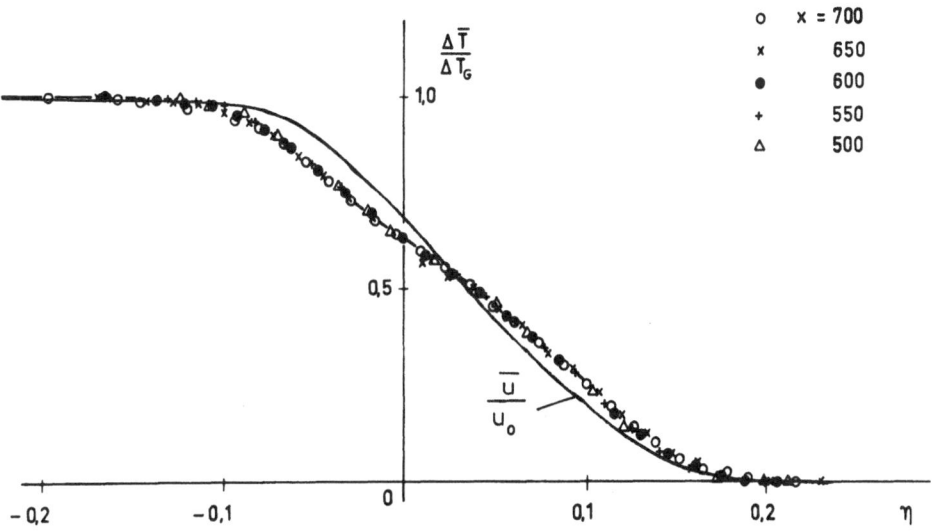

Figure 13. Distribution of Mean Temperature and Mean Velocity in Plane Shearlayer

The distribution of the fluctuating intensity across the flow
is shown in Figure 15. It has two maxima and one minimum, all of
which are approximately located at the position of the inflexion
points of the mean temperature curve. The maximum value of the
fluctuating variance (at η = .07) is .044 which is considerably
larger than the maximum velocity fluctuation variance which in the
same kind of flow is approximately .03. The existence of three
extrema can be explained from the typical shape of the T'-fluctua-
tions (sawtooth). For further characterization of the fluctuating
signals skewness and flatness distributions were measured. Further
measurements included spectra of T' at different locations of η,
integral spectra and auto-correlations of the T' signal.

Measurements of the mean temperature in the turbulent and non-
turbulent regions are presented in Figure 14. The mean temperature
distribution in the turbulent zone is remarkably flat and homogeneous, which is in contrast to the mean velocity characteristic
(Figure 5).

The mean variance of the temperature fluctuation in the turbulent domain is shown in Figure 15.

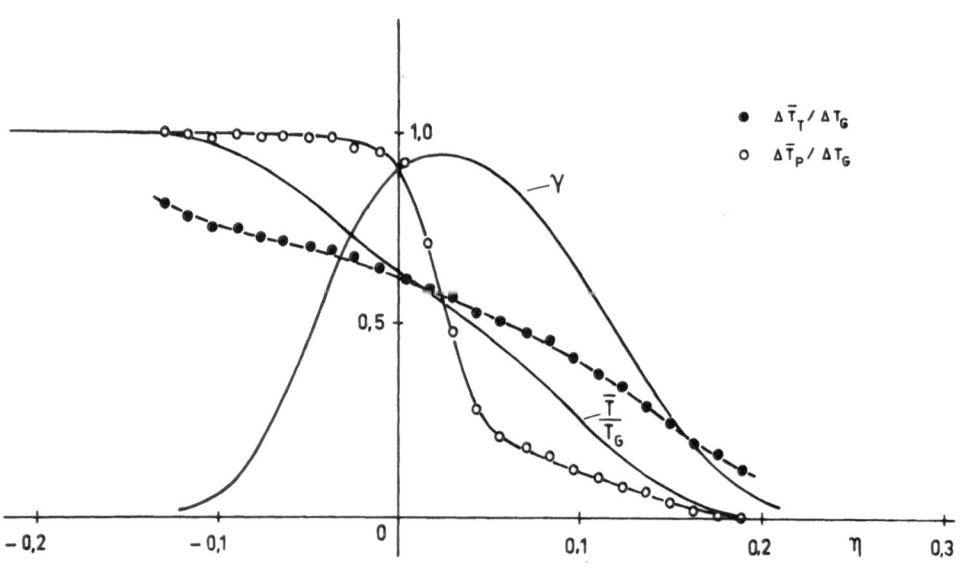

Figure 14.   Zone Averages of the Mean Temperature in Plane Shear-
             layer

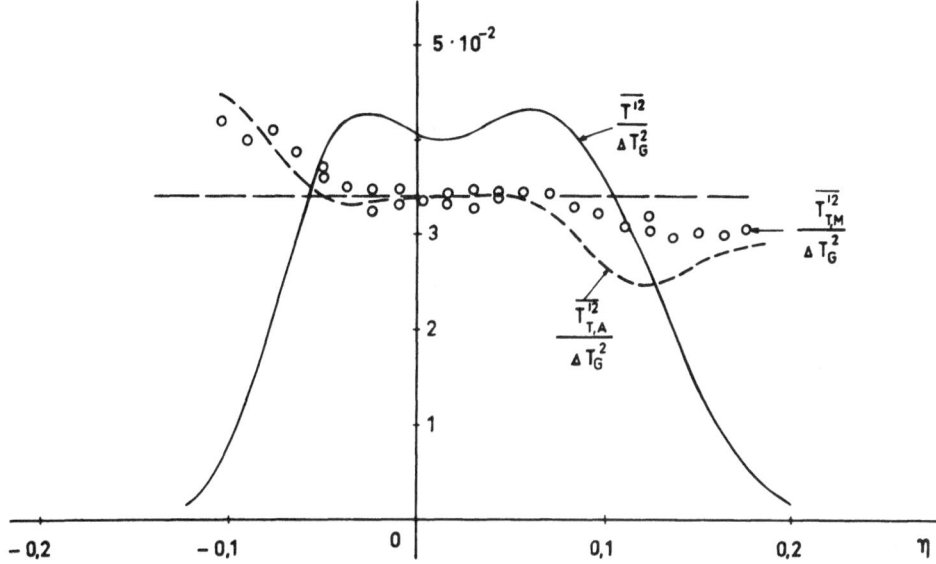

Figure 15. Temperature Fluctuation Variance in the Turbulent
            Regime of Plane Shearlayer

3.4. Further Observations and Discussion

The reported observations may be summarized as follows:

a. From the shape of the temperature fluctuations as well as
    from the mean quantity distributions it can be assumed
    that the transport mechanism is largely due to a large
    scale coherent vortical motion (see also Brown and Roshko
    1971 and Winant and Browand 1974).

b. The mean temperature distribution in the turbulent domain
    is largely homogeneous.

c. The temperature across a single vortex is approximately
    linear in the mean.

d. The temperature fluctuation variance in the turbulent
    domain is approximately constant.

Apparently the essence of the observation is the existence of
large scale coherent structures and their strong domination of the

scalar transfer mechanism. These large scale coherences and organized motions in fully turbulent flows have been reported lately by
various authors and have consequently become the basis of various
theoretical considerations (e.g. aeroacoustics etc.). The observation of coherence within the experiments reported here was, however,
primarily by deduction from averaged measurements.

　　　To obtain more direct identification we felt that a simultaneous view of the total cross-section of the flow--i.e. not only an
averaged measurement (even a conditional one)--might give us a
somewhat better understanding.

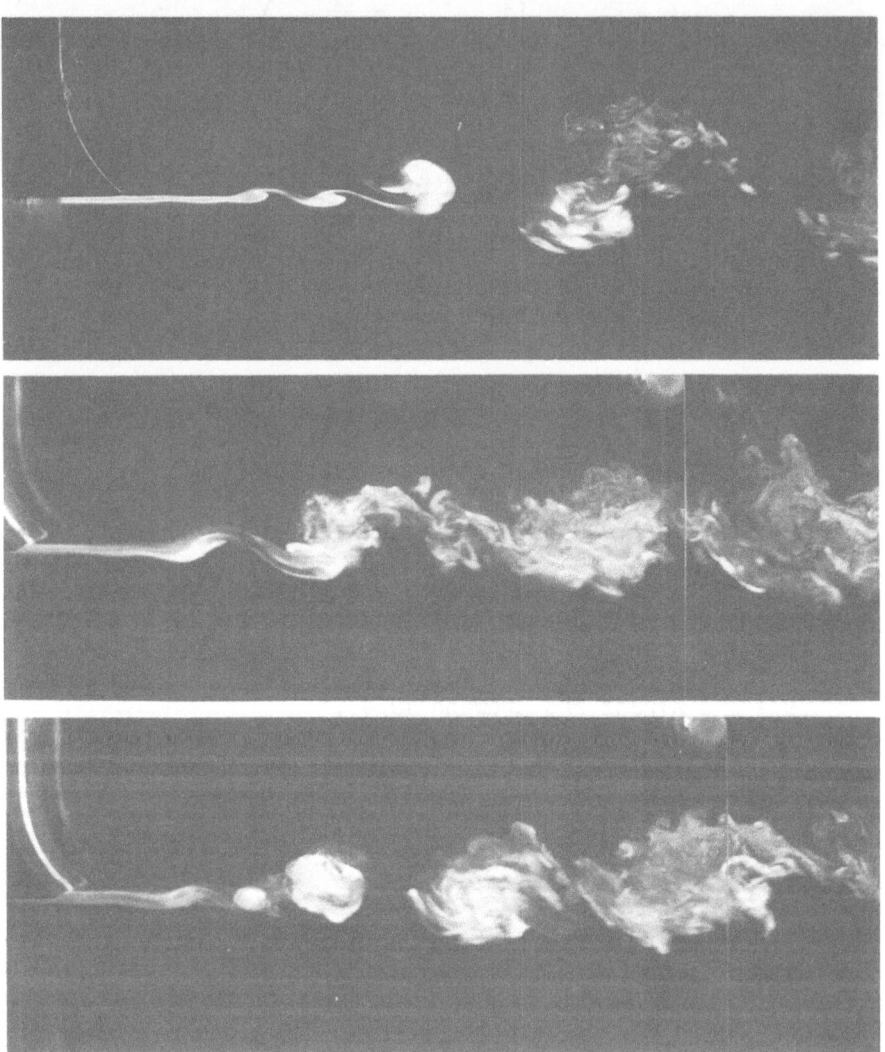

Figures 16, 17 and 18.   Vortical Structure in Plane Shearlayer

One possibility is of course the method of visualization (Roshko), and therefore a great amount of smoke photographs with laser illumination and some high speed camera films were taken (Figures 16, 17 and 18). We were, however, never able to obtain those strikingly coherent patterns as reported by Roshko. This may have something to do with the fact that in our case the densities of both streams were virtually equal.

A more immediate method, however, which was applied in a similar application already by Kovasznay et al. (1962), consists in the use of an array of probes arranged evenly spaced across the flow. The following slides show temperature traces from 10 points across the heated shearlayer. These slides show very clearly the assumed large scale features and allow a good interpretation of the structure. The very sharp cuts across the flow--dividing as it seems two vortical lumps--are remarkably straight with an angle of 49° which agrees well with that obtained by Jones et al. (1973) from correlation measurements. Their existence is independent of the initial conditions (see Champagne et al. (1973)). This is evident from Figures 19 to 21.

Figure 22 shows an interpretation of the temperature traces.

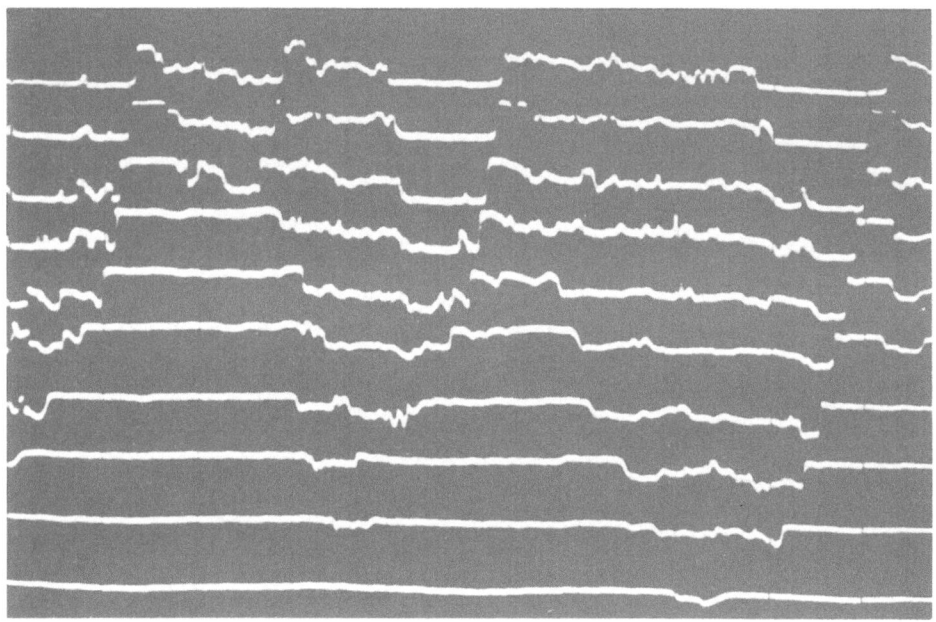

Figure 19.   Temperature Traces from Rake-probe across Flow--No Trip Wire

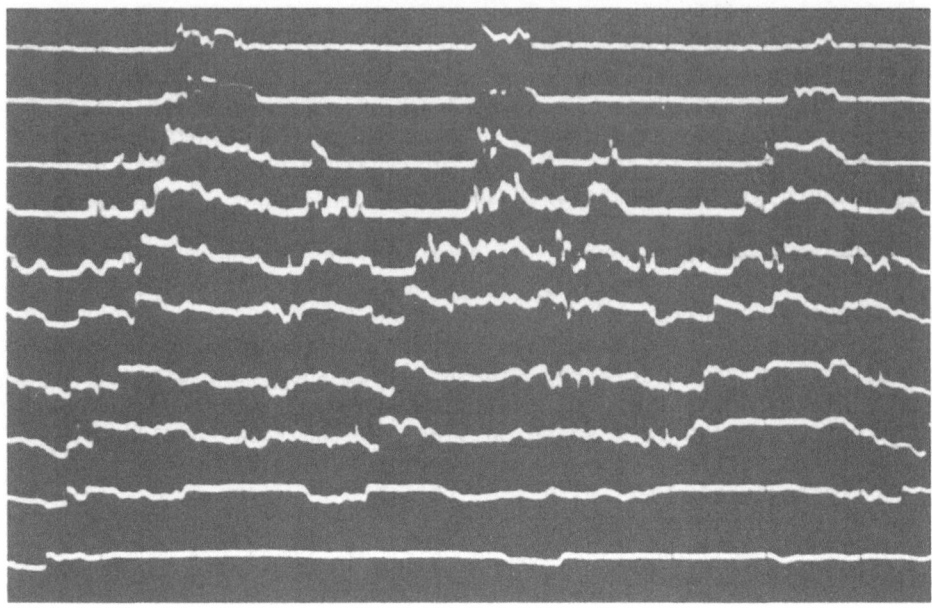

Figure 20. Temperature Traces from Rake-probe across Flow--Small Trip Wire

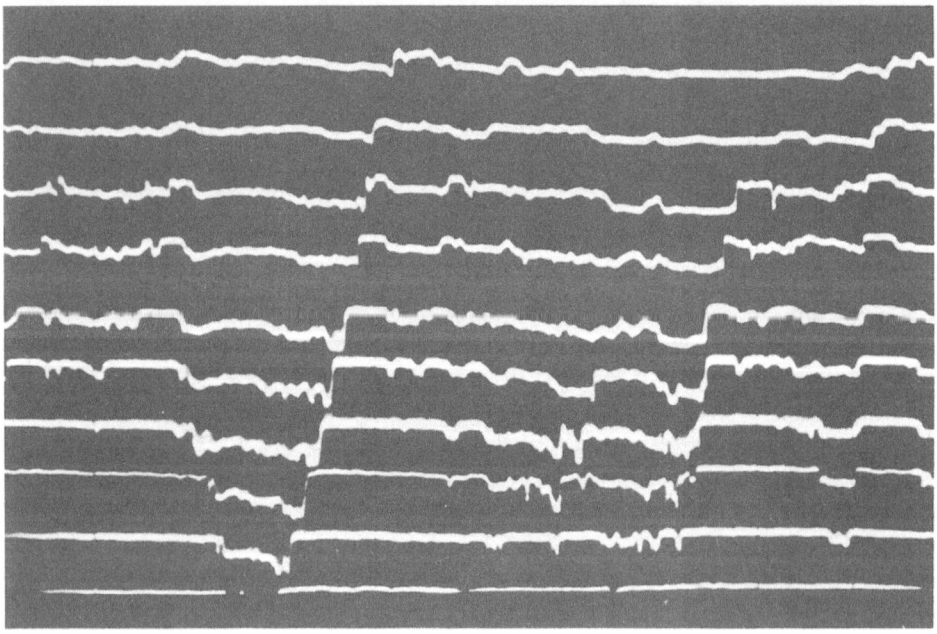

Figure 21. Temperature Traces from Rake-probe across Flow--Large Trip Barrier (Spoiler)

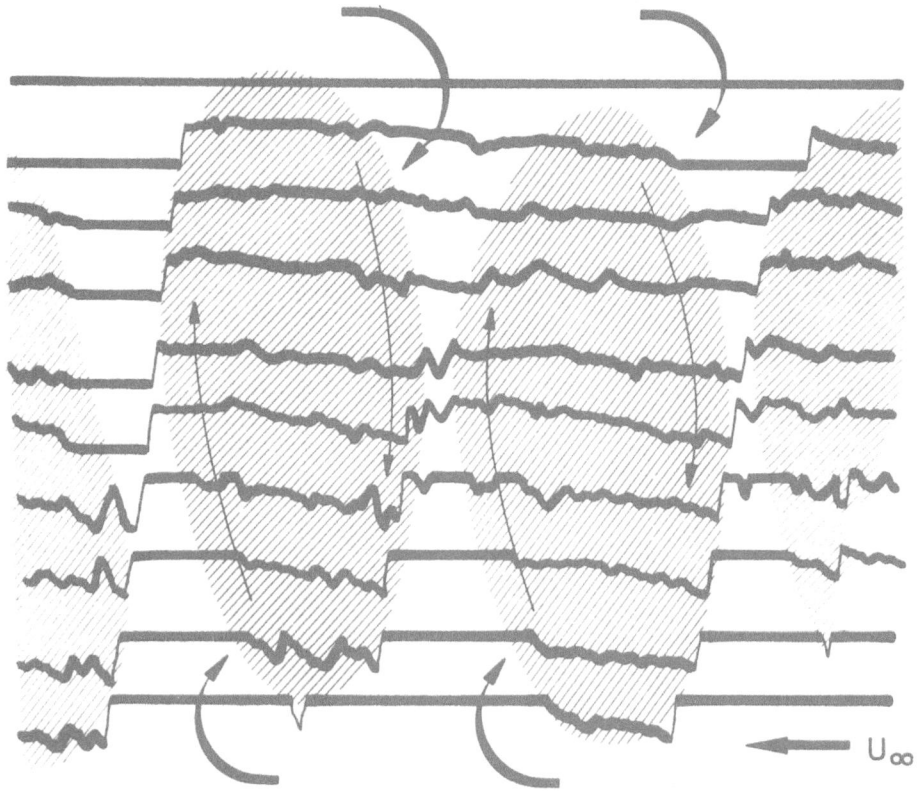

Figure 22.   Interpretation of Temperature Traces

The following are obvious.

    a. The turbulent region consists of clearly separated lumps, which have vortex-like characters.

    b. Entrainment is present only on the "lee-side" of the vortices.

    c. Within the "vortices" the "mean" temperature distribution is essentially linear with an essentially homogeneous fluctuation intensity.

Figure 23 shows temperature traces from the multi-probe in horizontal (z) position. From this it is clear that the coherence is, to a certain degree, two-dimensional, at least over a width corresponding to the "vortex diameter" ($d_V \approx \overline{y}_1 - \overline{y}_2$).

From a flow film, which cannot be shown here, it seems apparent (as from some of the still slides already shown) that the continuity of the vortices is established by a pairing mechanism (Winant and Browand 1974). For the time of its existence the convection velocity of the single vortex has to accelerate while the vortex itself grows, without, however, growing into its neighbours. This might be the consequence of a slow motion of the vortex center towards the high velocity before and after pairing (Figures 16, 17, 18 and 24), while the mean convection velocity as obtained from the motion of the cuts is of the order of 0.5 u_∞ (but) with strong variations and a (reasonable) tendency to be smaller at the low velocity and larger at the high velocity side.

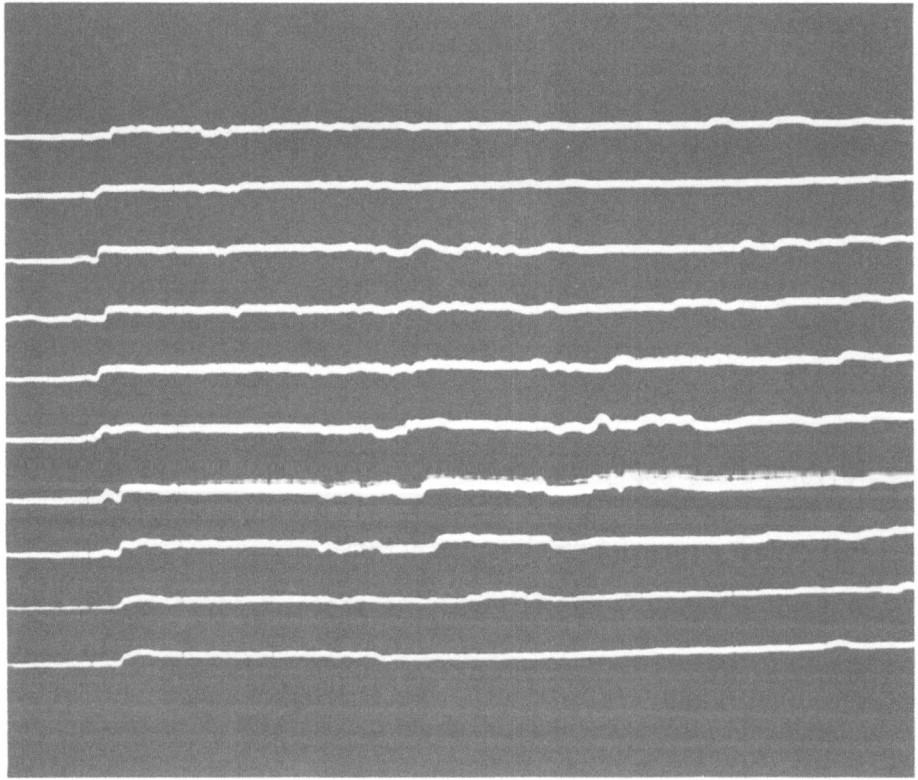

Figure 23.   Temperature Traces at Constant η-position

Figure 24. Pairing Mechanism of Laminar Vortices (Michalke and
Freymuth 1966)

In closing this lecture I would like to discuss briefly two
questions which emerged from the reported observation:

1. What scalar transfer mechanism is suggested from this
observed structure?

The transfer mechanism is assumed to be mixed (Townsend) where
within a vortex, i.e. in the turbulent regime (Figures 25 and 26):

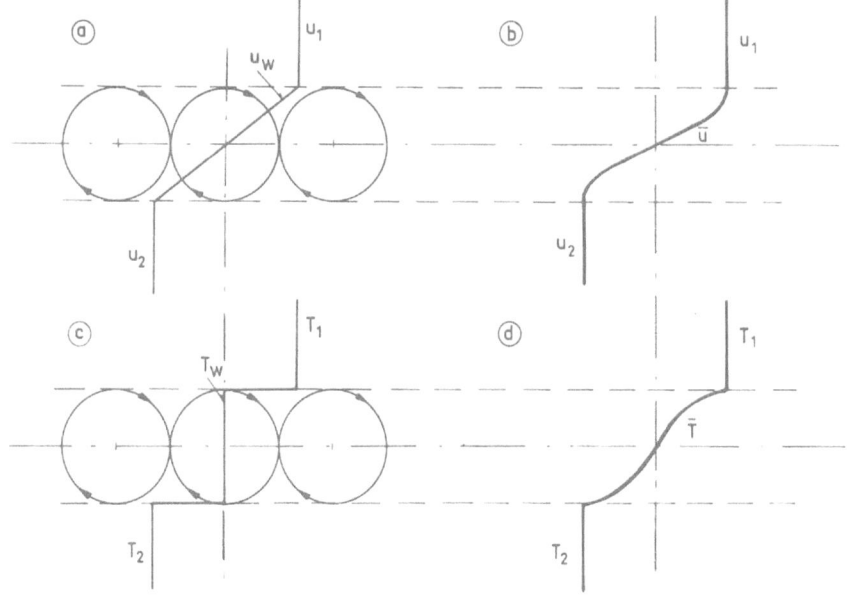

Figure 25. Velocity and Temperature Distribution in Idealized
Vortex Sheet

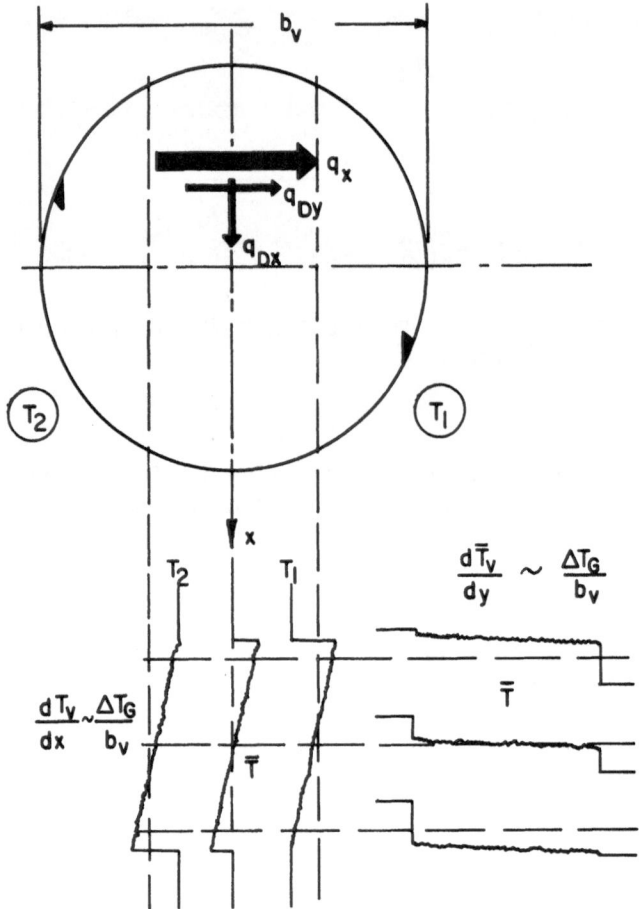

Figure 26.   Model of Heat Transport in Single Vortex

$$\overline{(v'T')}_V \sim \Delta\bar{u}_V \cdot \Delta T_{G_V} - \varepsilon_V \left.\frac{d\Delta\bar{T}}{dy}\right|_V$$

and on the basis of the reported observations:

$$\overline{(v'T')}_{V(T)} = const, \qquad\qquad \overline{(v'T')}_P = 0$$

$$\overline{v'T'} \sim \gamma$$

This model agrees well with the measurements as can be seen from Figure 27.

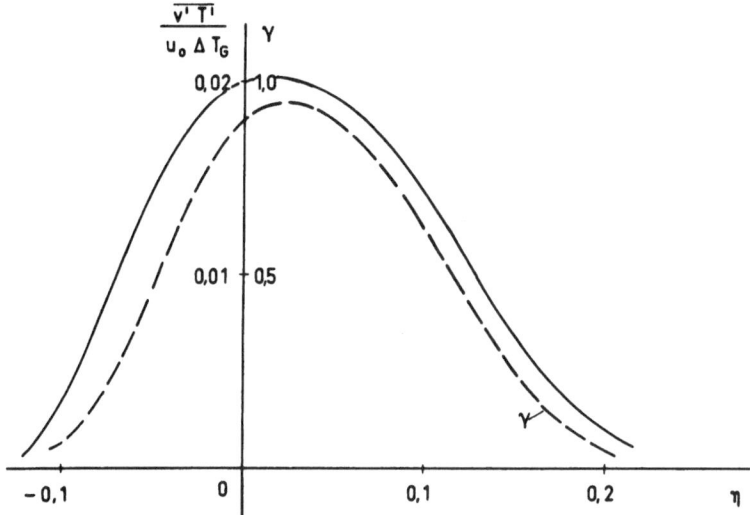

Figure 27.   The Turbulent Diffusion Term, Evaluated from $\Delta\bar{T}/\Delta T_G(\eta)$
             and $\bar{u}/u_0(\eta)$ and Compared to the γ-distribution in
             Plane Shearlayer

2.   From the foregoing one may expect a general relationship
     for those large vortices characterized by the statistics
     of the turbulent front and the diffusivity of the small
     scale motion within these vortices.

     Corrsin (1957) has already shown such a relationship,
     without, however, having been able to look at it in
     greater detail. In the following a correlation is
     attempted, based on a great number of published data on
     simple shear flows, which were evaluated in cooperation
     with I. Wygnanski.

It is a plausible assumption to put:

$$\varepsilon \sim \sqrt{\overline{v'^2}} \cdot \Lambda_{\text{Lagrange}} \sim \overline{v'^2} \cdot \theta$$

where θ = Euler-Auto-Correlation-length in moving frame.

Corrsin and Kistler (1954) have shown, that

$$\sigma_{intermittency} \sim \sqrt{\theta \, \overline{v'^2} \, t}$$

i.e. with $t \sim x/\bar{u}(\bar{y})$

$$\sigma \sim \sqrt{\theta \, \overline{v'^2} \, x/\bar{u}}$$

hence: $\varepsilon \sim \sigma^2 \, \bar{u}/x$

$$\varepsilon \sim \left| \underbrace{\left(\tfrac{\sigma}{x}\right) \sigma \, U_0}_{A} - \underbrace{\left(\tfrac{\sigma}{x}\right) \sigma \, \Delta\bar{U}(\bar{y})}_{B} \right|$$

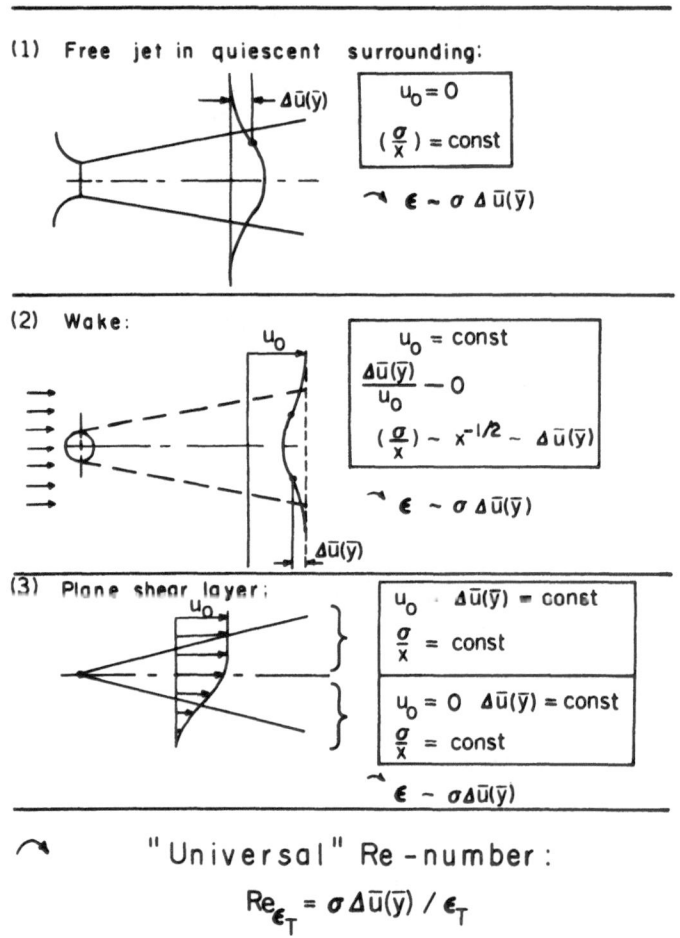

(1) Free jet in quiescent surrounding:

$U_0 = 0$

$\left(\tfrac{\sigma}{x}\right) = $ const

$\varepsilon \sim \sigma \, \Delta\bar{u}(\bar{y})$

(2) Wake:

$U_0 = $ const

$\dfrac{\Delta\bar{u}(\bar{y})}{U_0} - 0$

$\left(\tfrac{\sigma}{x}\right) \sim x^{-1/2} \sim \Delta\bar{u}(\bar{y})$

$\varepsilon \sim \sigma \, \Delta\bar{u}(\bar{y})$

(3) Plane shear layer:

$U_0 \cdot \Delta\bar{u}(\bar{y}) = $ const

$\dfrac{\sigma}{x} = $ const

$U_0 = 0 \quad \Delta\bar{u}(\bar{y}) = $ const

$\dfrac{\sigma}{x} = $ const

$\varepsilon \sim \sigma \Delta\bar{u}(\bar{y})$

"Universal" Re -number:

$$Re_{\varepsilon_T} = \sigma \, \Delta\bar{u}(\bar{y}) \, / \, \varepsilon_T$$

Figure 28.   "Eddy Viscosity" for Different Flow Configurations.

It is then found, that for a great number of simple flows

$$Re = \left|\frac{\sigma \, \Delta u}{\epsilon}\right|_T = \left.\frac{u_0 - \bar{u}}{\bar{u}} \frac{x}{\sigma}\right|_T = const., \qquad \text{(see Figure 28)}$$

having a "universal" value of

$$Re \approx 14 \qquad \text{as is shown in Figure 29.}$$

This obviously is a stability Reynolds-Number, relevant to the large scale motion in a shear flow.

Some further aims of this investigation are:

1. Explanation of the entrainment mechanism and the establishment of a general correlation with characteristics of the turbulent front.

2. Further identification of the "natural" (Liepmann 1972) vortex with eduction measurements triggered by the "cut".

3. Correlation of temperature and flow field in the turbulent domain.

4. Similar measurements in different flow configurations, e.g.:

   shearlayer with heated lip

   plane heated jet

   plane jet with heat transfer across flow
   (different air temperatures on both sides of jet) etc.

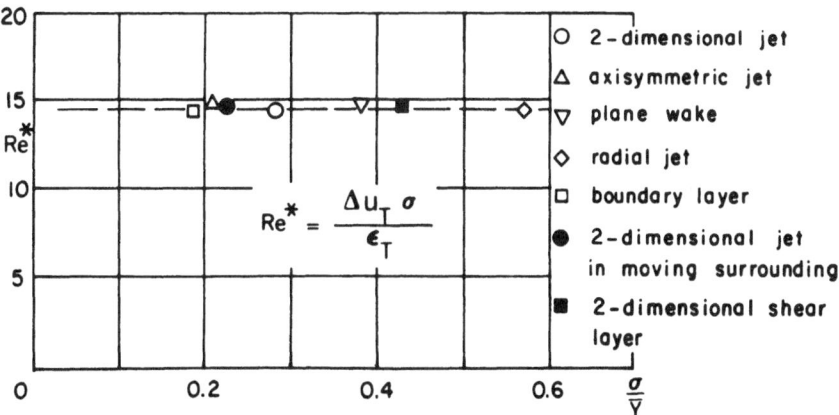

Figure 29. "Universal" Reynolds-Number for Different Flow Configurations

REFERENCES

1. Brown, G. & Roshko, A. 1971 The effect of density non-uniformity on the turbulent mixing layer. AGARD-CP-93.

2. Champagne, F. H., Pao, Y. H. & Wygnanski, I. 1973 On the two-dimensional mixing region. Internal Rep.

3. Corrsin, S. 1957 Some current problems in turbulent shear flows. Naval Hydrodynmcs., Pbl. 515.

4. Corrsin, S. & Kistler, A. L. 1954 Free stream boundaries of turbulent flows. NACA TN 3133, Johns Hopkins Univ., Rep. 1244.

5. Fiedler, H. E. 1974 Transport of heat across a plane turbulent mixing layer. Advances in Geophysics 18.

6. Gutmark, E. & Wygnanski, I. 1974 On the two-dimensional turbulent jet. T.A.E. Rep. No. 201, Haifa, Israel.

7. Jones, B. G. et al. 1973 The turbulence structure in the plane two-stream mixing layer. Nuclear Engineering Program, University of Illinois at Urbana-Champaign, Urbana, Illinois 61801.

8. Kibens, V. & Kovasznay, L. S. G. 1969 The intermittent region of a turbulent boundary layer. Johns Hopkins University, DA-31-124-ARO-D-313.

9. Kovasznay, L. S. G. et al. 1962 Proc. Heat Transfer and Fluid Mech. Institute p. 1. Stanford University Press.

10. Liepmann, H. W. 1972 Experimental fluid mechanics: the impact of modern instrumentation. Theoretical and Applied Mechanics Proceedings Moscow University.

11. Liepmann, H. W. & Laufer, J. 1947 Investigations of free turbulent mixing. NACA No. 1257.

12. Michalke, A. & Freymuth, P. 1966 The instability and the formation of vortices in a free boundary layer. AGARD CP. No. 4 Part II.

13. Winant, C. D. & Browand, F. K. 1974 Vortex pairing: the mechanism of turbulent mixing layer growth at moderate Reynolds number. J. Fluid Mech., 63, 2 pp. 237-256.

14. Wygnanski, I. & Fiedler, H. E. 1969 Some measurements in
     the self preserving jet. J. Fluid Mech., <u>38</u>, 3 pp. 577-612.

15. Wygnanski, I. & Fiedler, H. E. 1970 The two-dimensional
     mixing region. J. Fluid Mech., <u>41</u>, 2 pp. 327-361.

DISCUSSION

GOLDSCHMIDT: (Purdue University, West Lafayette, Indiana)

The first thing is I am glad you showed us these pictures by
Wygnanski et al. on intermittencies on opposite sides of a jet.
Bradshaw and I[1] have reported on a definite flapping in jets,
which I think is related to what you have shown.

Secondly, we have not reported on this yet, but at the same
time we also looked at the effect of screens in the plenum chamber
(upstream of the nozzle) on the widening rate. The results showed
an increase of the s-widening rate with an increase of the turbulence intensity at the mouth of the jet. I am wondering if you
want to comment on that.

FIEDLER:

It is well possible that in the investigation of Gutmark and
Wygnanski jet flapping of the kind you are referring to was present.
The authors however do not comment on this possibility. According to their observation, however, the very large scattering of
peripheral intermittency (i.e. standard deviation) was largely
reduced by encasing the test section with screens, eliminating thus
any room drafts etc.

As to the second question this problem has not been looked
into in the investigations discussed herein. Your observation
seems, however, quite plausible to me and is in agreement with
e.g. Vagt's[2] observations in the round jet.

KOVASZNAY: (Johns Hopkins University, Baltimore, Maryland)

Did you reach similarity, new similarity, with respect to the
temperature values?

---

[1] "Flapping of a Plane Jet," Physics of Fluids, Vol. 16, No. 3,
     March 1973, pp. 354-355.

[2] J. Vagt: Untersuchungen zur Turbulenzstruktur von Freistrahlen.
     HFI Internal Report 1969.

FIEDLER:

Similarity was reached for all distributions, including the temperature characteristics, for x-values larger than 500 mm, which corresponds to a Re-number of

$$Re_x = \frac{x_1 \, u_0}{\nu} \gtrsim 3 \cdot 10^5$$

For the temperature field the similarity behaviour was especially checked by also measuring the time derivative of the fluctuating signals, the axial distribution of which appeared to be in perfect agreement with similarity theory.

NAGIB: (Illinois Institute of Technology, Chicago, Illinois)

Did you say that when they put the screens around the jet, it stopped flapping, or did I misunderstand you?

FIEDLER:

I could not say anything about flapping of the jet. Its intermittency spread, however, became definitely smaller in the presence of screens.

BEVILAQUA: (Aerospace Research Labs, Dayton, Ohio)

Your observation that the thermal discontinuity at the boundary is relatively sharper than the velocity discontinuity is for me especially significant. It is consistent with the observations of Professor Kovasznay and his students[3] that there is a distinct interface for the turbulent vorticity, but that the mean and rms velocities vary smoothly across the boundary. The thermal interface is made sharp by the rotation of the large eddies, which brings hot fluid to the surface from the opposite side of the mixing region. Since the turbulent vorticity is also a scalar quantity, it seems possible that the turbulent interface is produced by this same large eddy mechanism, rather than through the process of gradient steepening supposed by the superlayer hypothesis.

FIEDLER:

This may well be so, but you will still have to have something like a superlayer, i.e. some viscosity dominated turbulent front, since this seems to be the only way by which the entrained fluid

---

[3] Kovasznay, L. S. G.; Kibens, V. R.; and Blackwelder, R. F. 1970 Large scale motion in the intermittent region of a turbulent boundary layer. J. Fluid Mech. 41, 283-325.

(engulfed on the lee-side of the turbulent outbursts) may obtain
vorticity. But I agree that vorticity should behave similar to
temperature in this respect, which should be most obvious and
analogous for the case of a heated nozzle lip configuration
(experiment planned).

ALBER: (TRW Systems, Redondo Beach, California)

Your experiment seems to be very well done, and it looks
something like the study that Sunyach and Mathieu[4] performed
several years back. I was noticing, since I have been working
with these rms temperature fluctuations, that yours seems to be
somewhat higher than theirs. Your T' intensity seems to be like
about 20% compared to their 14%, something like that. Have you
compared your work with theirs?

FIEDLER:

Sunyach and Mathieu's experiments were made at a Re-number at
which my experiments did not show similarity yet ($Re_x \approx 1.5 \cdot 10^5$).
On the other hand they were using a 2 μ wire operated with constant
current for measuring their temperature fluctuations and I have not
read anything about their frequency response.

From my observation it was obvious that e.g. also the thermal
inertia of the prongs has to be compensated for since otherwise
this may cause erroneous results especially due to the steep
temperature jumps at the turbulent fronts. So in considering these
points differences of approximately 20% in the RMS-values as
observed might well be expected.

---

[4] Sunyach, M. and Mathieu, J., 1969, "Zone de Melange d'un Jet
     Plan, Fluctuations Induites dans le cone a potential-intermit-
     tence," Int. J. Heat-Mass Transfer, Vol. 12, pp. 1679-16-97.
     See also: Sunaych, M., 1971, "Contribution a l'etude des
     Frontieres d'Ecoulements Turbulents Libres," Sc. D. Thesis,
     L'Université Claud Bernard de Lyon.

CONDITIONAL (POINT-AVERAGED) TEMPERATURE AND VELOCITIES IN A HEATED TURBULENT PLANE[†]

P. E. Jenkins   and   V. W. Goldschmidt

Northern Arizona University   Purdue University

Flagstaff, Arizona   West Lafayette, Indiana

In continuation of some earlier work on the measurement of particle transport coefficients, we have undertaken conditional measurements in a heated jet. Data have been obtained for the measure of the conditional point average $u\theta$ and uv correlations.

This report is thus concerned with the point averages of the velocity and temperature fields. These are the values at a (x,ξ) coordinate taken while the interface is at a preselected coordinate (x,y_d). ξ is the distance from y_d to the point of measurements, (see Figure 1). These measurements give a temperature and velocity profile measured with respect to the moving interface. From it, transport coefficients within the fully turbulent region can be derived.

A signal proportional to $\frac{\partial^2 u}{\partial + \partial y}$ was used to detect the passing interface. The signal is formed by,

$$\frac{u_2 - u_1}{\Delta y} = \frac{\Delta u}{\Delta y} = \frac{\partial u}{\partial y}$$

where Δy = .095 inches or approximately one-half the microscale for this region of the flow field. Figure 1 shows a schematic of the detecting set-up relative to the jet.

[†] A more complete account has been submitted for publication elsewhere.

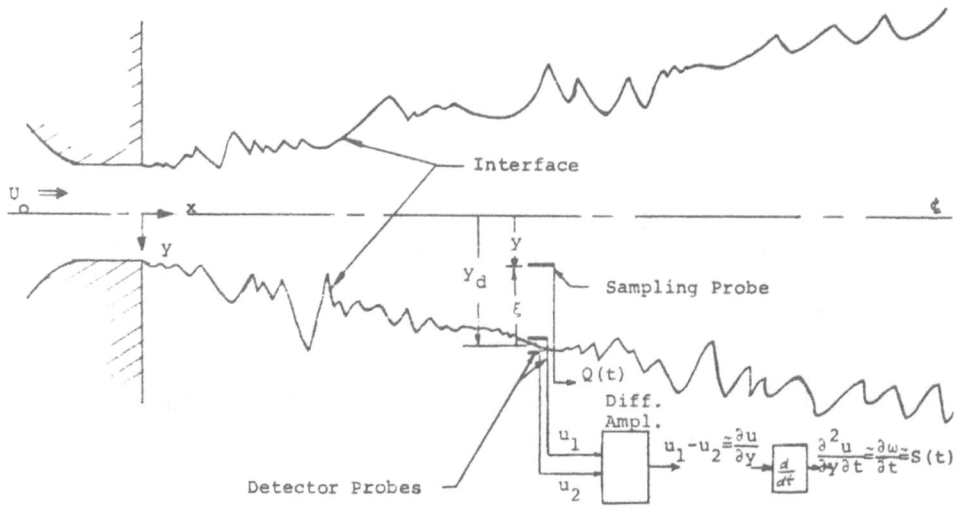

Figure 1.   Schematic of Conditional Sampling Flow Field Set-Up

In the detection measuring process, the detector probes were operated in the velocity mode, with the sampling probe operating either in the velocity or constant current mode, depending on the property being measured.

The experimental set-up consisted of a plane jet of air heated by a 4 Kw. mesh wire heating element located in a 56 × 51 cm. plenum chamber. A gradual contraction led to a smaller (15.2 × 30.4 cm.) plenum chamber with flow straightening elements. The flow then discharged through a 1.27 × 30.4 cm. vertical slot on a 91.4 × 30.4 cm. wall. Two confining horizontal walls (91.4 × 121.0 cm. long) were installed to maintain the two dimensionality of the jet. The jet Reynolds number (based on the jet slot width) was held constant at 1.43×10^4. The measurements were taken at $X/D = 35$, 45, and 50, at a temperature excess of 35°C above ambient. The mean velocity and temperature profiles were measured and reported earlier, Jenkins and Goldschmidt (1973).

In measuring the conditional temperature and velocity profiles, the relative position of the respective temperature and velocity interfaces were required for detecting purposes. To determine the relative interface location, the detector probe set-up as shown in Figure 1 was used. By comparing the output traces from a Honeywell visicorder (with a 5 khz. frequency response) for temperature and velocity, the two interfaces were noted to coincide. To further confirm the coincidence of the interfaces, the intermittencies, γ_u and γ_θ, and the interface

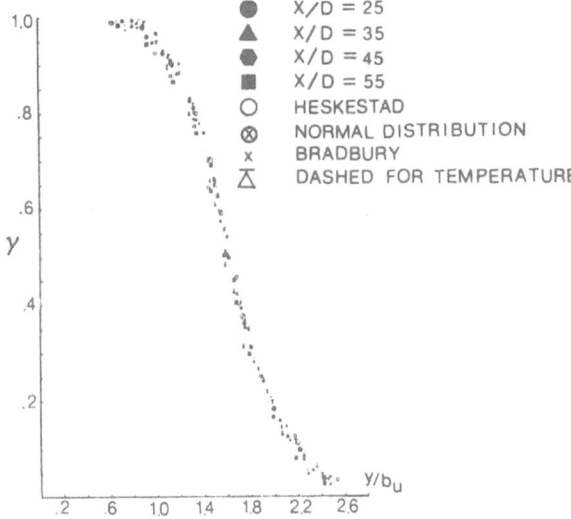

Figure 2.   Intermittency vs. y/b_u

crossing frequencies, f_{γ_u} and f_{γ_θ}, were measured for both the velocity and temperature, respectively.   Figures 2 and 3 show the results. $f_{\gamma max}$ is in the order of 50 to 30 Hz.   The coincidence of the intermittencies and crossing frequencies implies that respective interfaces are also coincident.

The point averaged velocity and temperature profiles can then be measured with respect to the moving interface.   The detector

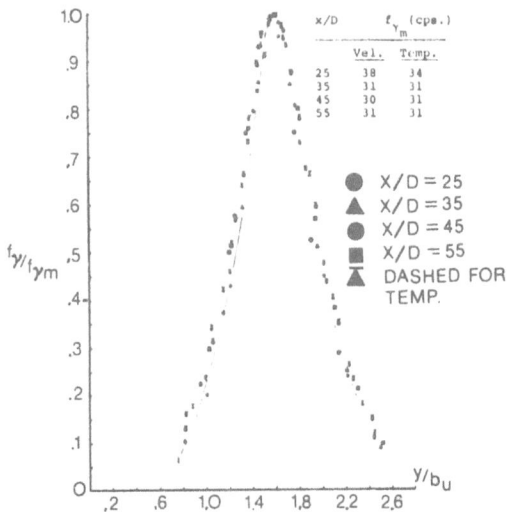

Figure 3.   Crossing Frequency vs. y/b_u

(which was set on a velocity model), is capable of determining
when the interface is crossing at a preselected y_d station. This
station can be identified by the corresponding intermittency,
$\gamma(y_d)$. The coincidence of the temperature and velocity interfaces permits the measurement of $\hat{\theta}$ and of $\hat{U}$ by operating the
"sampling" probe $(Q(t))$ on either a temperature or velocity mode.
A correction for the temperature contamination on the velocity
signal (and vice versa) was not made as this was noted to be a
small effect when operating the sampling probe at extreme overheat
values.

The point average properties were taken at three X/D stations
and five different γ (or y_d) locations. The plots of Figure 4
show the values at X/D = 35 and 45. (Similar trends were noted at
X/D = 50). The data points in Figure 4 correspond to profiles
measured relative to the interface as it crosses five different
"γ - locations." The similarity of the point averaged profiles
is gratifying. Averages were taken independently for the interface moving inwards and outwards and their differences, if any,
were within the experimental scatter. It appears then that the
point average profiles, with respect to the moving interface, have
a consistent shape.

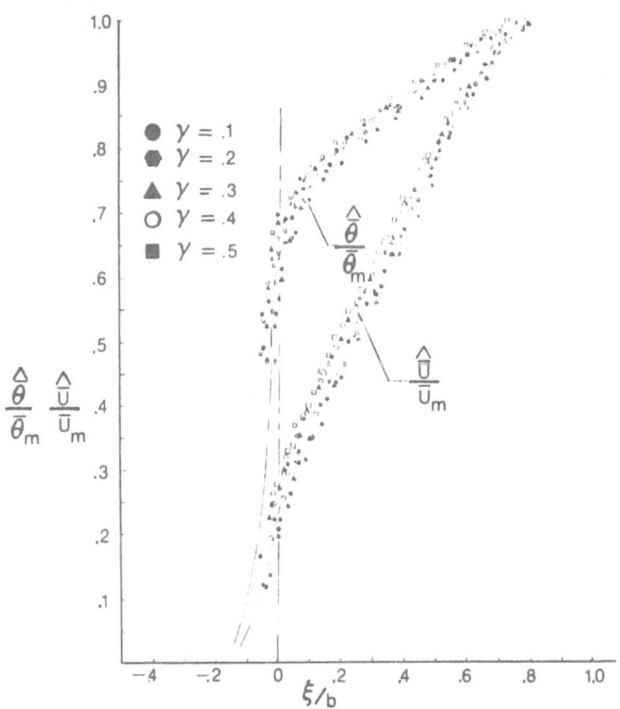

Figure 4.   Point Average Temperature and Velocity Profiles
            X/D = 35, 45; θ_0 = 35°C

A comparison between the turbulent transport of heat and momentum acquires more significance when based on point average properties (instead of the conventional mean quantities). The flatter temperature profile and the coincidence of the interfaces shows that heat transports no further than momentum (although it looks like such on the time averages*). In addition, within the turbulent zone it seems as if the transport properties for $\hat{\theta}$ are indeed different from those for $\hat{U}$.

ACKNOWLEDGEMENTS

The reported work was part of a project carried out at Purdue University under the sponsorship of the NSF. The authors acknowledge the support received through Grant GK 19317.

REFERENCES

1. Bradbury, L. S. G., "The Structures of a Self-Preserving Turbulent Plane Jet," JFM, 23, Part I, pp. 31-64, 1965.

2. V. W. Goldschmidt, M. K. Householder, G. Ahmadi and S. C. Chung; "Turbulent Diffusion of Small Particles Suspended in Turbulent Jets," Progress in Heat and Mass Transfer, Vol. 6, Pergamon Press, 1972.

3. Heskestad, G., "Hot Wire Measurements in a Plane Turbulent Jet," JFM, pp. 721-729, Dec. 1965.

4. Jenkins, P. E. and Goldschmidt, V. W., "Mean Temperature and Velocity in a Plane Turbulent Jet," ASME, JFE, Dec. 1973, pp. 581-584.

DISCUSSION

LIBBY: (University of California, La Jolla, California)

When you measured the velocity, did you trigger on the temperature? That is, did you discriminate on the temperature to get the velocity? In my view, the thing to do is to trigger on the temperature and measure both temperature and velocity. The temperature discrimination is so much less ambiguous.

---

*   This possibility was suggested as early as May 10, 1968, by L. S. G. Kovasznay during a seminar at Purdue University.

GOLDSCHMIDT:

We triggered on velocity on the cold jet. On the heated jet, we triggered on both, although we used the velocity as the trigger for most of the reported values and reported here results from that.

MORKOVIN: (Illinois Institute of Technology, Chicago, Illinois)

How much of a change in your width did you have?

GOLDSCHMIDT:

I'm guessing, in the order of ten to twenty percent (see Jenkins and Goldschmidt, JFE, December 1973, pp. 581-584).

DIFFUSION FROM A PERIODICALLY HEATED LINE-SOURCE SEGMENT AND ITS APPLICATION TO MEASUREMENTS IN TURBULENT FLOWS*

Roald A. Wigeland and Hassan M. Nagib

Mechanics, Mechanical and Aerospace Engineering Dept.

Illinois Institute of Technology, Chicago, Illinois

A study of the diffusion of heat from a line-source segment was performed in three test flow conditions with different free-stream turbulence intensities. The line-source segment, or heated spot, was 0.001 inch in diameter and about 0.1 inch long. It was constructed using a length of 0.001 inch diameter platinum-10%-rhodium wire stretched across the test section of the wind tunnel (approximately 3 inches in diameter). The wire was first plated with copper to a diameter of 0.0015 inch everywhere except for the 0.1 inch spot located near the center of the test section. Since the resistivity of the platinum-10%-rhodium wire is much greater than that of copper, the plated sections have much less resistance per unit length than the unplated spot. When current is applied, the temperature of the plated part is much less than that of the unplated part.

In order to evaluate the size of the thermal wake of the heated spot at different downstream locations and to determine the decay rate of the wake centerline temperature, surveys of the wake were done with a cold-wire probe, properly oriented so as to minimize averaging effects across the length of the measuring cold-wire.

Temperature profiles in the wake were measured at various distances downstream of the spot, where x is downstream distance

* Supported under AFOSR OSR Grant No. AFOSR-73-2509. The details of the study are available in an AFOSR Scientific report with the same title which is based on the Master's thesis of the first author.

measured from the heated segment. The non-dimensional temperature
profiles in the wake are Gaussian in the direction perpendicular
to the wire, i.e., the y direction, and almost Gaussian in the
direction parallel to it, i.e., the z direction, as is shown in
Figure 1. The slightly non-Gaussian behavior is due to the temperature of the plated sections of the wire, and is less than 10%
of the peak centerline temperature.

The cross-section of this thermal wake takes on an elliptical
shape at short distances downstream from the heated spot, and tends
toward a circular shape with increasing downstream distance.
Figures 2 and 3 demonstrate this behavior, where Figure 2 displays
the temperature contours of the wake close to the heated spot, and
Figure 3 displays similar contours far downstream of the spot.

The spreading of the temperature wake in both the y and z
directions is shown in Figure 4. The straight lines connecting
each set of data points are indicative of the spreading rate of
the wake for that test flow condition, i.e., turbulence level. It
can be seen that for lower turbulence levels, i.e., flow condition
"G" having approximately 1% turbulence intensity, the spreading
rates are lower than for the high turbulence levels, such as flow
condition "C-3" with approximately 4% turbulence intensity. The
results of the study demonstrate that the wake spreading rate in

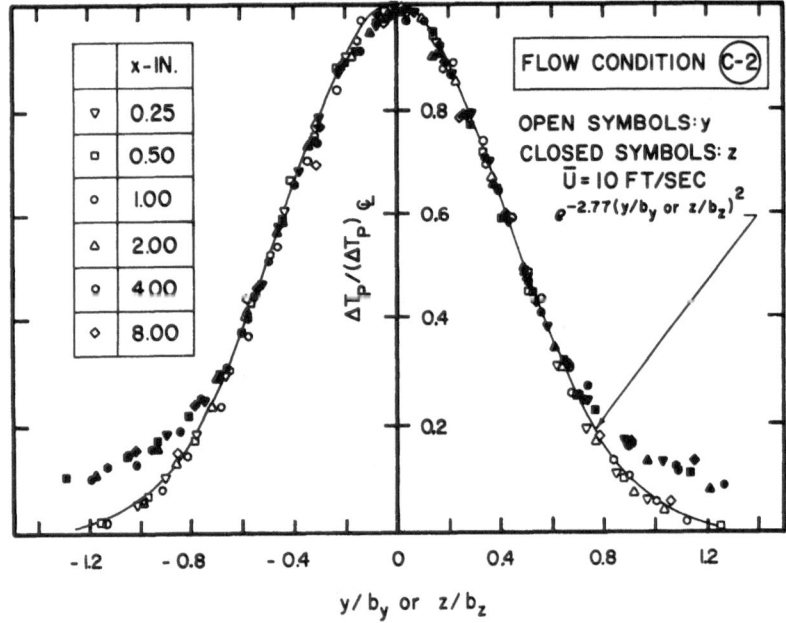

Figure 1.   Normalized Peak Temperature Profiles along y and z in
            Wake of Heated Spot at Different Axial Locations in
            Test Flow Condition "C-2"

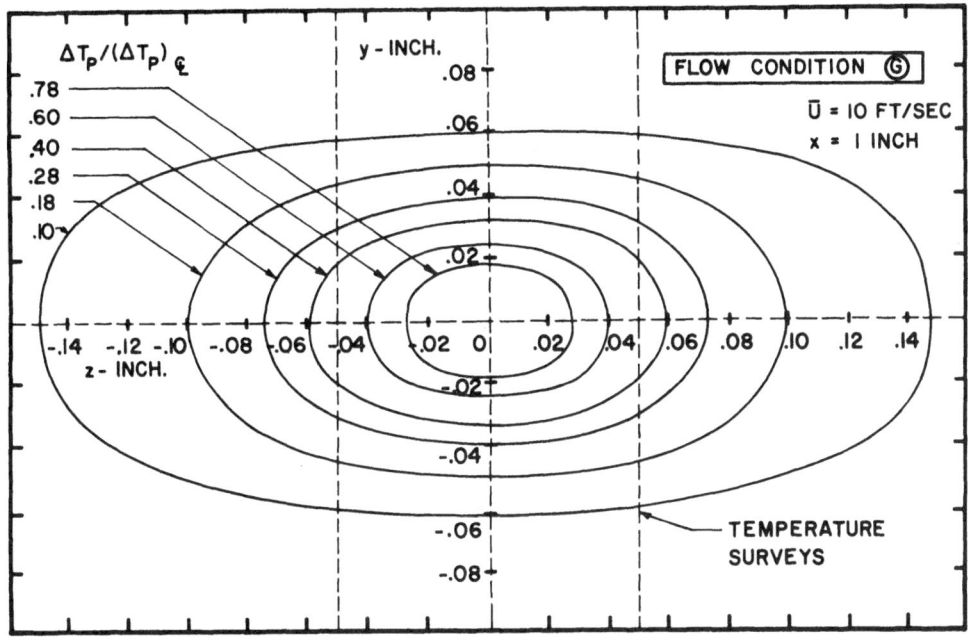

Figure 2. Contours of Constant Peak Temperature in Wake of Heated Spot at x = 1 Inch in Test Flow Condition "G"

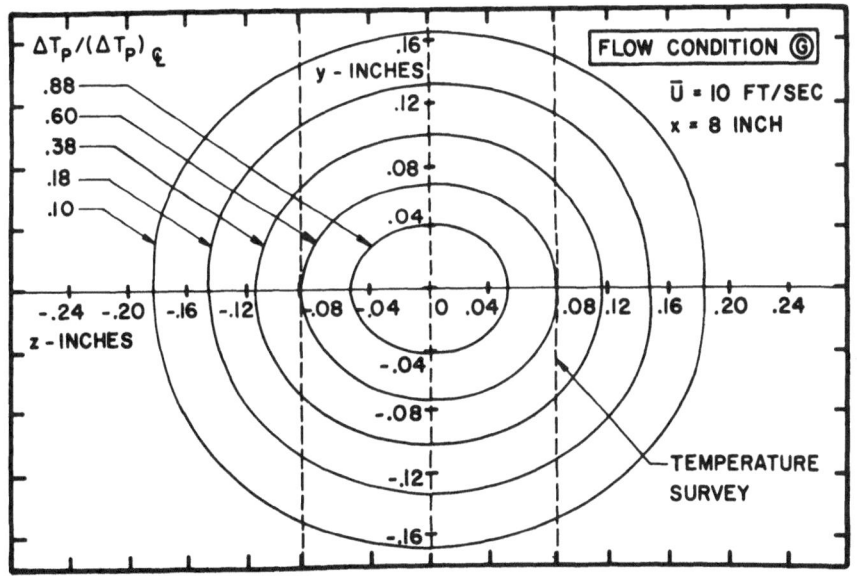

Figure 3. Contours of Constant Peak Temperature in Wake of Heated Spot at x = 8 Inches in Test Flow Condition "G"

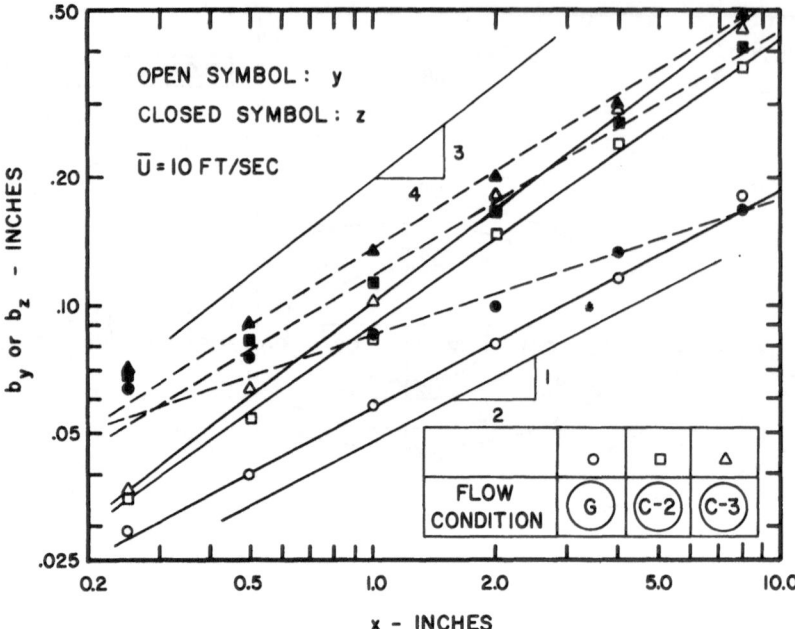

Figure 4. Half-Width of Temperature Wake along y and z at Different Axial Locations in Test Flow Conditions "G", "C-2",
and "C-3"

the direction perpendicular to the wire does not differ from the
laminar case until the free-stream turbulence intensity exceeds
approximately 1%; beyond this point it increases with the intensity.
The spreading rate in the direction parallel to the wire, however,
increases even for very small turbulence levels.

The centerline temperature decay is shown in Figure 5. The
slopes of the straight lines of this figure are indicative of the
decay rates for each flow condition. The flow condition with the
lowest turbulence intensity has the lowest temperature decay rate;
the rate increases with increasing turbulence level. The deviations of the curves at small distances from the wire are due to
averaging along the cold-wire sensing probe. The centerline temperature decay rate also increases from that of the laminar case
even for small turbulence intensities.

The heated spot used for the diffusion studies was also used
as a heat tracer for determination of mean streak lines and average
convection velocities over long distances in turbulent flow fields.
The system is made up of a periodically heated line-source segment,
a temperature sensing cold-wire probe, and the associated signal

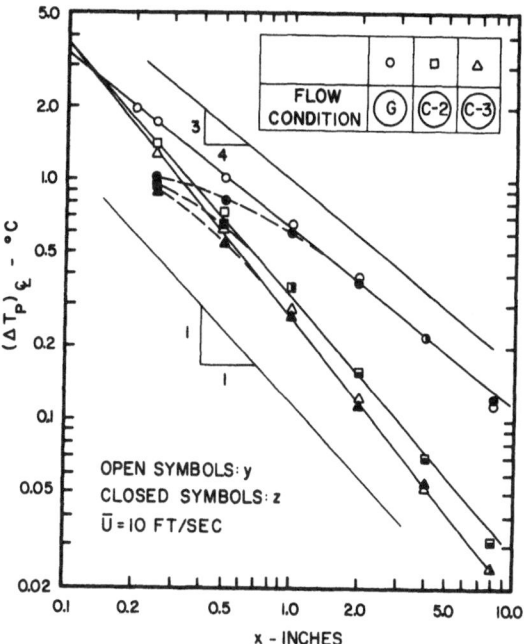

Figure 5.   Decay of Centerline Peak Temperature with Downstream
Distance in Test Flow Condition "G", "C-2", and "C-3"

processing equipment. Periodic sampling techniques in the form of
cross-correlations between the heating signal and the cold-wire
output enhance the resolution of the technique and permit the mean
velocity measurements.

Figure 6 shows the change in amplitude and phase shift of the
cross-correlation between the heated spot input and the cold-wire
output for different separation distances between the two. The
heated-spot technique depends on many parameters, and was optimized
with respect to these parameters whenever possible. For average
free-stream turbulence intensities of up to 3%, both the mean
streak lines and the average convection velocity of particles from
the heated spot to the cold-wire probe can be measured for a distance of at least 8 inches. This technique is useful for velocities up to 25 feet per second or higher.

The method was also evaluated for very high turbulence conditions in the shear layer of a free jet and the results are
encouraging. Separation distances of up to 4 inches were used
between the heat marker and the temperature sensing probe, both
being in the shear layer of the jet.

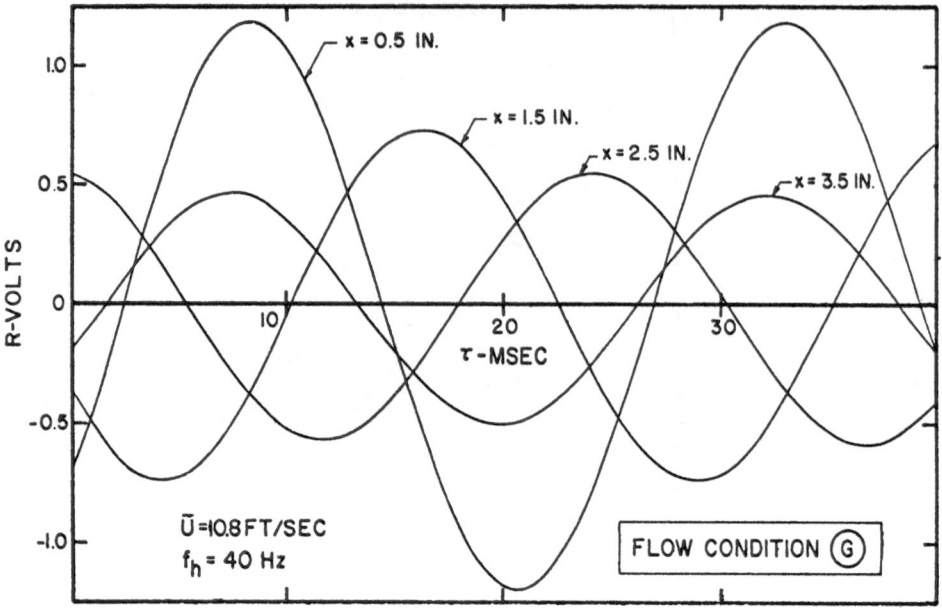

Figure 6.   Typical Cross-Correlation Functions at Different Axial
            Locations in Test Flow Condition "G"

DISCUSSION

GOLDSCHMIDT: (Purdue University)

    Do you have any kind of a feeling why it would do that?

NAGIB:

    If you are referring to the spreading of the thermal wake,
several mechanisms are involved. The spreading rate of the temperature wake in the direction parallel to the wire, for a free
stream which is laminar or has very low turbulence level, is
governed by molecular diffusion only. In the other (perpendicular)
direction, the additional mechanism based on the spreading of the
momentum wake is dominant. For higher turbulence intensities, a
mechanism which includes turbulent diffusion and the "flagging" of
the thermal wake from side to side by the large eddies is the
dominant one for the spreading in both directions.

    As far as the increase of the spreading rate with increasing
turbulence intensity is concerned, we do not have a complete

understanding of it yet. However, since the increased turbulence level leads to more transport by the turbulent motions and more energetic large eddies, then the spreading rate of the thermal wake should increase with higher turbulence intensities.

GENERAL DISCUSSION

Panelists

    H. W. Liepmann (Caltech) - Chairman

    K. N. C. Bray (University of Southampton)
    D. E. Coles (Caltech)
    S. J. Kline (Stanford University)
    M. V. Morkovin (Illinois Institute of Technology)
    M. Summerfield (Princeton University)

S. N. B. MURTHY: (Purdue University)

    The panel discussion is intended to survey the prospects for
research in turbulent mixing with special emphasis on taking molecular chemical action into account. I also hope that there will be
some discussion on how the modelling schemes--which have become
developed as far as to model the conditionally weighted pdf data--
may become influenced by the understanding of the large scale
coherent structures.

    It is my privilege to ask Professor Liepmann to chair and lead
the panel.

K. N. C. BRAY: (University of Southampton)

    I have some personal impressions and I will try to be very
brief. I will confine myself to the turbulent combustion aspects
of this meeting, and there are only three things that I want to say.

    The first concerns the probability density function (pdf),
which has obviously become so fashionable recently in the modelling
of reaction processes in turbulent flows. We have seen many different forms of the pdf over the last couple of days, varying from a
selection of Dirac delta functions through beta functions, Gaussians,

segments of straight lines, this and that. Everybody seems to have
got the right answer for their own case, and I am left wondering
how sensitive are all our predictions to the particular form of pdf
that is chosen. Is not there a need somewhere to test the sensitivity of our models to this assumption which may or may not prove
to be a sensitive one. If predicted results should turn out to be
strongly dependent on the form of the pdf, then this whole approach
may not be very useful, because the means available for predicting
the pdf appear to be unsatisfactory.

Secondly, we have talked a great deal about the coupling effect
of turbulence on combustion particularly through the pdf. However,
the other aspect of coupling, namely the effect of combustion on the
turbulence, does not seem to have been mentioned. If one looks at
the full horror of the complete equations of reacting turbulent
flow, refs. 1, 2, one finds many terms, which are absent without
combustion. In particular, one finds terms containing the divergence
of velocity, and terms related to the time-average covariance of
density with velocity which classical turbulence people have been
very happy to ignore, of which perhaps some are important in the
combustion situation. What I was particularly happy to see at this
meeting was that some of these terms have been modelled, I believe
for the first time, by Donaldson. I have argued at another meeting
(ref. 3) that some of these terms can be important, particularly in
the balance equation for turbulence kinetic energy, where it appears
that the interaction of combustion on the turbulence is somewhat
complicated in that the combustion can both remove the turbulence
energy and generate it. I will not go into details here. But I
think that it should be frankly admitted that the effect of the
combustion on the turbulence is something that we do not properly
understand. If one uses as a dimensionless measure of a turbulent
shear structure, for example, the ratio of Reynolds stress to the
turbulence kinetic energy, the variation across an inert jet is very
well known and tends towards a universal curve. I have made similar
calculations from published experimental data obtained in free jet
flames and have obtained curves of similar shape but with magnitudes
four or five times greater than in the cold flows. Now, is this an
influence of the combustion on turbulence? How can it be explained
in terms of present day modelling?

My third point is to stress the urgent need for more experimental data in turbulent combustion. If those of us who try to do
experiments aim to keep the models as honest as possible--and it is
just an impossible task--we need to measure the quantities which
they are trying to model. We have heard in the last couple of days,
that they are modelling intensities, spectra, probability density
functions, length scales, etc., for both vector and scalar fields.
We need to go to turbulent flames and measure these things to see
if they are being modelled realistically, and this is an enormously

difficult task. One aspect of it that can be done--we are trying
to do it at Southampton and other people are too--is the use of
laser-Doppler anemometry on the turbulent velocity field in combustion systems. We are measuring time-resolved velocities and
obtaining spectra, length scales, probability densities, correlations and so on, of the velocity field. But there is an even more
urgent need for a comparable local, time-resolved measurement of
the concentration field, and as I see it, no one can now make such
measurements. That is a challenge to all of us: how can one
measure the spectra, the covariances, the probability density
function, etc., of the composition field, not in a cold flow, but
in a real turbulent flame?

References

1. K. N. C. Bray: "Equations of turbulent combustion I Funda-
    mental equations of reacting turbulent flow." University of
    Southampton AASU Report No. 330, 1973.

2. K. N. C. Bray: "Equations of turbulent combustion II Bound-
    ary layer approximation." University of Southampton AASU
    Report No. 331, 1973.

3. K. N. C. Bray: "Kinetic energy of turbulence in flames," paper
    presented at AGARD Propulsion and Energetics Panel Specialists
    Meeting, Liege, Belgium, April 1974. To be published by AGARD.

D. COLES: (California Institute of Technology)

I have an opinion which I have expressed in private and which
I want to make public. It did not necessarily arise at this meeting, but it was reinforced here. I would like to go back a few
years to the Stanford conference. We have three of the organizers
of the Stanford contest here at the table, and several of the con-
testants in the audience, and so some of you are familiar with that
event. It is not much of an oversimplification to say that that
meeting was an attempt to eliminate in a more or less humane way
the weaker methods among the then-existing methods for calculating
turbulent flow. It did succeed in a modest way in fulfilling that
objective. It also had another effect, associated with a secondary
objective. We hoped that after we cast our net there would be one
big fish left, which was obviously the way to go and which had been
overlooked or underestimated or whatever. That did not quite happen
either. What did happen was that the technique of integrating the
equations of mean motion with a big computer and with rate equations
became not only respectable but fashionable. I have sometimes
thought of this situation as an arms race to see who can mobilize
the most impressive array of rate equations for Reynolds stresses

and length scales and so forth, and then make the whole thing run
on a computer. Now, that was the theme of yesterday's session.
This morning we heard from an entirely different bunch of people,
the eddy chasers. Yesterday was modeling, today was eddy chasers.
I submit that the divergence of interest is not between the people
who know something about turbulent shear flow and the people who
know something about chemical kinetics. The divergence of interest
is between the modelers and the eddy chasers. These two groups are
going 180 degrees apart and the gap is opening rapidly. How are
you going to manage any kind of technology transfer between what
people were saying yesterday and what people were saying this
morning? The eddy chasers begin with intermittency. The modelers,
I think, cope very badly with intermittency, if they take account
of it at all. The eddy chasers have thrown away their spectral
analyzers. They do not draw bars over the terms in the equations
anymore; they draw hats and tildes and funny brackets, and they
mean something entirely different. They are trying to save the
phase information which is thrown away in the classical kind of
Reynolds averaging. I do not think that escalating the order to
which you do Reynolds averaging is going to clarify the physics of
turbulence. I think that eddy chasers, if they are lucky and if
they interact with each other, have a much better chance of doing
that. We are a kind of a club, I guess, all the people who draw
the hats and the tildes. The lights are going on all over the
world, at Johns Hopkins and Stanford and USC and CalTech and Ohio
State and the Indian Institute of Science and Imperial College and
what have you. We can see, I think, that we do not yet agree in
detail in any respect. But we have already begun the third generation of experimental research on these problems. The third generation is the computer-assisted generation, and the fruit of this
could be that the notion of averaging will change permanently. I
think that this could very well happen in twenty or thirty years--
everything happens very slowly in this business. What people will
mean by averaging is a probability average over a set of operators
whose properties we are all trying to establish and measure. I
think this will have nothing at all to do with correlations and
spectra and the whole machinery of generalized harmonic analysis
which has so far been mobilized on the analytic side. And my point
at the end is that something should be done on the analytical side
to develop this new possibility. The experimenters will do it if
the theoreticians do not.

S. J. KLINE: (Stanford University)

Professor Coles has, not surprisingly, said some of the things
I had planned to say.

The progress in turbulence, if you look at it over a short
time, does not seem to be very substantial, but then one has to

remember there are only two speeds in research: slow and reverse.
If one looks at knowledge about turbulence over a ten-year period,
then it is quite clear that we have made quite a bit of progress
in the understanding of the <u>physical</u> phenomena in a number of
exemplary flow fields. What disturbs me is the same thing that
seems to be bothering Don Coles, that there is still an appreciable
gap between that physical knowledge and what 'mathematical modelers'
are able to actually use. I agree with him, as most of you know;
I am myself an eddy-chaser, in that terminology. Also, I agree
with him that long-term averages of fluctuation quantities are mis-
leading--often seriously misleading. We have seen several examples
of this in the current meeting, perhaps the most notable being the
work of Roshko and his colleagues, which, I think, is a very important piece of work in itself, and also shows very clearly how
long-term averages can mislead understanding of the flow. Erik
Mollo-Christensen,* in his inimitable way, has pointed out the ease
with which one can create figments of statistics by telling us that
if we stand on the side of a road and measure the number of wheels
on a bunch of trucks, inter-weaving through a bunch of motorcycles,
we will pretty soon discern, for example, that there are 2.7985
wheels on each vehicle, or some other equivalent bit of nonsense.
Some of the measurements we have been making in turbulence are of
just that sort.

The other thing that I wanted to say on that same point is
that the 'animal' which Coles is asking us to describe may not be
just one kind of function. We may have to use the ideas that
Townsend[1] and Bradshaw[2] and Lahey,[3] among others, have discussed,
that there may be more than one part to the fluctuations, for
example, the flow might be decomposed into an active and inactive
part[4]; it may not expand appropriately in just one uniform set of
functions. We need to keep that in mind.

Another point I want to make is that the "surprises" are still
appearing in fluid mechanics. Every year or two somebody finds
something which we had not seen before. The history of the subject
as a science is by no means yet "closed." The Roshko pictures of
the jet are an example; the pictures that Coles did a few years
ago with concentric rotating cylinders are an example; some of the
data that Bradshaw showed us yesterday about 'difficult cases' are

---

*   AIAA Annual Lecture, 1973.
[1] Townsend, A. A., FJM, <u>11</u>, 97 (1961).
[2] Bradshaw, P., JFM, <u>29</u>, 625-645 (1967).
[3] Lahey, R. L., and Kline, S. J., Rept. MD- , M.E. Dept., Stanford
    Univ.
[4] It is not clear that these "decompositions" all refer to the same
    things.

examples. Things happen in these flows we would not have predicted, even qualitatively. The data which I cited in the discussion of Prof. Spalding's paper by Johnston, Halleen and Lezius for rotating channel flows is still another example of recent origin. One would have to be very shortsighted to believe that the 'surprises' are all finished, that no more will be found.

Now what does the continued existence of 'surprises' tell me about the conference and about what might be done with the existing methods? Let me, for the moment, take some wholly arbitrary numbers, for purposes of discussion. Let us say that ten to forty percent is a rough engineering prediction and that a few percent is a "fine-tuned" prediction. Then we ask when can we expect to do either in turbulent flows. History suggests--the 1968 Conference on Computation of Turbulent Boundary Layers particularly--that we can do fine-tuning only when we have what Jim Eggers asked for in a discussion yesterday--a set of systematic data over wide enough ranges so that we can do accurate interpolation in the computer. When we move beyond that, to other classes of flows, our record, to put it politely, is not very good. This record is something about which we would do well to be more explicit not only to warn the uninitiated but also to hold it clearly in our own consciousness. If we are going to get beyond this past record, to a better ability to extrapolate, then we are either going to have to extend the data base, which is slow and expensive, or we are going to have to do better mathematical modeling than we have done so far.[1] In fact, even when one goes to the kind of modeling that Professor Spalding is using--and I admire his willingness to take on the tough, significant engineering problems which somebody has to tackle--when one gets to that broad a scale, then one does not often get fine-tuned predictions; it becomes difficult. We again encounter, head-on, our inability to extrapolate. Let me consider still another example--the current work of W. M. Kays and R. J. Moffat. This is still ongoing research. Let me give you the variables for which they are mapping data. They are doing a whole series of flows (I think they have done five dissertations so far). Each case involves flow over a plate, and they are studying many kinds of variations of the following parameters: (i) pressure as a function of streamwise coordinate, x; (ii) variations of wall temperature (or heat flux) as a function of x; (iii) wall blowing or suction as functions of x; (iv) various kinds of surface roughness. They are building a broad data base of apparently high accuracy. However, to get predictions which are fine-tuned, they once again have to put significant modifications into the existing mathematical

---

[1]   See also the objections, on wholly mathematical grounds, by W. C. Reynolds (paper for AIAA meeting in June 1974).

models for boundary layer flows. In particular, they have to refine the inner-layer approximations because the heat transfer is very sensitive to those layers, and they also have to put in corrections not only to represent roughness but also to represent the blowing (or suction) and to represent pressure gradient. These things can and have been done, but only <u>after</u> the data bank began to accumulate sufficiently; in this sense the gross history is consistent.

What does all this suggest? It suggests two or three things to me. First of all, if we are going to understand the structures of turbulent flows and improve the mathematical models of them, then it has to be in terms that Don Coles described. I think it will have to be in terms of iterative models. Coles put this very well yesterday when he said that you can not find the structures until we have the instruments, the measurements, and we can not make the measurements until we have the structures. That obviously suggests iteration of the model, making measurements like flow-visualization studies--which give an overview of the complete structure. (I agree, of course, with Peter Bradshaw that flow-visualization measurements in general are coarse; they are not the finest measurements, but nevertheless they often do suggest the kinds of measurements that might give you better information.) Given that information, one can iterate the process and also, hopefully, iterate improvements in the mathematical models. There are some interesting possibilities.

Second, the history suggests a possible lack of uniqueness at least at the order of existing models. There has been the implicit idea, appearing several times in the conference, that we <u>should</u>, or we <u>ought</u>, to look for a set of <u>unique universal constants</u> which close Reynolds equations (or equivalent) at second order, or at second order plus a transport equation, or at higher levels. Coles has called this "the arms race"; W. C. Reynolds called it "the closure hierarchy" in his article on morphology in the 1968 conference proceedings. I do not know why this uniqueness is necessary. Certainly uniqueness of the constants is desirable, but if we are after engineering prediction, a unique set of constants is not essential, particularly if one is going to use the computer anyway. One could say, "If for a well-defined class of flows, say Class A, the constants that make the equations work are the following, but for the flow classes B, C, etc., they are instead another set of constants, the number of classes could be as large as several dozen without requiring an excessive body of data or undue difficulties in computation." Such a scheme is less elegant and less appealing, fundamentally, but it may be much more efficient in getting us to the kinds of fine-tuned engineering predictions we want than continuing the search for a single, unique, universal set of constants.

Finally, there is one piece of bad news which Professor Liepmann specifically asked me to report, so I will add it to the
general remarks above. A number of us have been talking about
bursts as a quasi-coherent structure in relation to turbulence production in boundary layers. But various workers have used different methods of detection. My recent student, G. R. Offen,[1] has
measured all the kinds of detection schemes we were aware that
various people have been using, plus several more Dr. Offen invented himself. Offen then compared these with the visual methods
and with each other. All of them agree with each other at roughly
twice what one would get by chance, but only about half what one
wants to get to say that there is good clear agreement, that is,
that each is measuring the same thing most of the time. We are
all measuring the same sort of thing, but apparently we do not yet
have good, consistent methods for picking out the coherent structures which Don Coles spoke about which are independent of the
technique and/or the observer.

MARK V. MORKOVIN: (Illinois Institute of Technology)

I spent some thirteen years in the industry with steady pressure for early useable results and I always ask myself what do I
want to use any set of tools for. For preliminary design? For
finer development? Or for deeper understanding of the underlying
phenomenon? As much as I now stress the latter, I cannot take
purely the eddy-chasers' attitude. With all the extra presentations here I am going to spare you more eddy-chasing, a set of
slides[2] dealing with conditional sampling of one of the simpler
complex flows of Bradshaw, namely the merging of two cylinder wakes.
With one wake heated just enough to provide conditional temperature
discrimination we were able to chase kinematically the interface
between the turbulent warm and turbulent cold fluid. There are
some surprises in it. There are some features which would lead you
to believe that the existing prediction methods can do a fairly
good job in that complex field.

---

[1] Offen, G. R., and Kline, S. J., Report MD-31, Dept. of Mech.
     Engrg. Stanford University (1973). See also forthcoming Proc.
     of Rolla Conference on Turbulence (1973).

[2] These are based on G. Fabris' dissertation: "Conditionally
     sampled turbulent thermal and velocity fields in the wake of a
     warm cylinder and its interaction with an equal cool wake,"
     Illinois Institute of Technology, May 1974; available from
     Xerox University Microfilms, Ann Arbor, Mich.

In other words, I would feel that there are a large number of non-reactive turbulent fields which for many very useful purposes can be dealt with rather well with the type of predictive methods we started on yesterday morning. Even with non-conditional Reynolds averaging, the basic conservation laws and asymptotically valid scaling laws such as the law of the wall provide very powerful constraints on the rich families of functions allowed as solutions by the different closure schemes. For many useful outputs we in effect use these function families for sophisticated interpolation between experimental data--provided enough data are available and provided we are extra careful in extrapolating to previously untested ranges of parameters. But I am only talking about non-reactive fields-- remember this is my virginal exposition to reaction processes in turbulence and my experience is with respect to the kinetic energetic aspects of turbulent fields, not so much with the finest scale detail mixing.

Even for non-reactive flows there were obviously two flaws that we have seen here and that have to be looked at. One concerns the pressure-velocity field correlations and its sticky modeling. Barclay Jones told us that pretty soon we will know something more about them experimentally so that the modeling could become more physical. The other one is one of the bad news of Bradshaw, the one about length-scale modeling inherent in the dissipation equation. I do not know what to say about it, it does not sound good. Maybe the decomposition of the pressure field might help. Nevertheless, those are basic flaws at the present time in the predictive methods. However, despite that, they seem to be of substantial utility for non-reactive problems.

Now, for reactive problems the extra important difference is the degree of mixedness. People have distinguished with emphasis between laminar and turbulent mixing. Not much has been said that there is a hybrid mixing process associated with nonlinear stability (such as for Browand's roll-ups and pairings). In connection with the instantaneous vorticity equation

$$\frac{D\vec{\omega}}{Dt} = \vec{\omega} \cdot \vec{\nabla}\vec{V} + \nu\nabla^2\vec{\omega}$$

I have been taught from way back that the first term on the right is the essence of turbulence because it characterized the three dimensional vorticity interaction and growth source. But if you have sharp local shear layers then you get instability phenomena which do not depend on this term at all as long as they remain

quasi-two-dimensional. So even without $\vec{\omega} \cdot \vec{\nabla}\vec{V}$ you get amplification, fast rise, and roll-ups which lead to large-scale laminar non-turbulent mixing. Some of this basically two-dimensional mixing you can see for sharp shear layers even with small scale turbulence on top for instance when you put in burner-like obstacles such as

in Fig. 1.[1] Here the spirals of the rolled-up shear layers are no
longer distinguishable because their small scale turbulence diffuses rapidly across the entrained irrotational spiraling tongues.
For tight laminar spirals of initially laminar shear layers molecular mixing should also be effective even though slower and more
spotty in completing the process. At the end I am going to stick
my neck out about a probably naive experiment along these lines
that I would like to propose to the reactive people.

Anyway then, when one wants to bring about extra mixedness,
I think one should also keep an eye on this hybrid instability
mechanism present in locally laminar flows. Many of our techno-
logically important flows are at only moderate Reynolds numbers
where the asymptotic turbulent ideas may be marginally applicable.
To what extent is the flow downstream of a mixing grid truly turbulent and isotropic or made up of a buffeted unsteady ensemble of
laminar sheets and filaments? We have to keep in mind that our
measuring devices are Eulerian and that we can get pseudosignals
from meandering shear layers not truly describing the dynamic
structure. I am just mentioning this twilight zone[2] between lami-
narity and turbulence which has a bearing on the degree of mixedness
for reactive flows.

With respect to this mixedness, I would like to return to the
question for what purposes do we want to know something, say the
degree of corrugation of the interface that Bradshaw cited as disturbing. Would it make a difference between extinction or non-
extinction of a reaction if there is a degree of uncertainty there?
I do now know. LaRue[3] and ourselves,[4] we have looked at the interface with a temperature wire, where I think you have more discrimination. LaRue in particular studied the sensitivity of the intermittency and crossing frequency to hold-time and threshold criteria,

---

[1] This slide, courtesy of H. Thomann of ETH, Zürich, was referred
to but not shown because of time pressure on the Panel. It is
a schlieren view of a Mach 0.56 field downstream of a 15 cm-long
9.5° wedge with a 4-cm full-span splitter plate and turbulent
separated shear layers. (The diagonal lines are wire supports
for the splitter plate at the windows--they do not influence the
base flow.)

[2] D. Küchemann in "Turbulence" and Vortex Motions (Zeit. f.
Flugwissen., 19, 8/9, 305, 1971) elaborates on this.

[3] John Ch. LaRue: "Temperature Characteristics in the Turbulence
Wake of a Heated Rod," Ph.D. Dissertation, University of California at San Diego, 1973.

[4] G. Fabris' dissertation, first footnote.

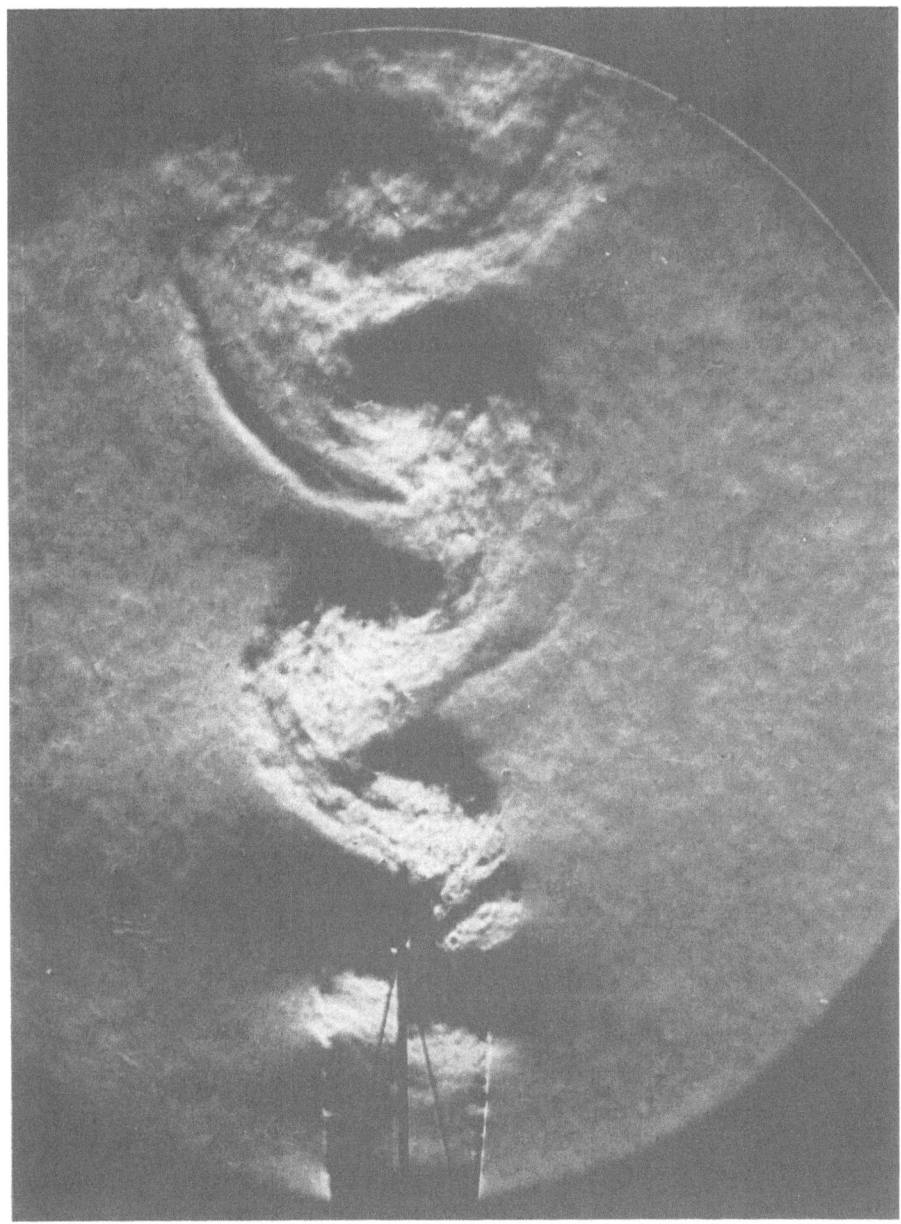

Figure 1

a very desirable work to look at. I cannot see that the uncertainty he showed would make much difference to any practical considerations perhaps other than the degree of mixedness. Would it make the difference between extinction and non-extinction?

For many purposes I think that Kovasznay's earlier statement on the interface is correct. I think that Bradshaw, unfortunately, is somewhat misled by the low Reynolds number of the smoke pictures which emphasize the large-scale irrotational enclaves. I think we have all been brainwashed by the low-Reynolds number smoke pictures. There is now evidence, Falco[1] at Cambridge University who has smoke pictures of a boundary layer at higher momentum Reynolds numbers (5800). It turns out that at these higher Reynolds numbers there are not these tremendous excursions,[2] that the structure is very much more fine-scale right at the interface. There does not seem to be quite as much uncertainty as we go to the higher Reynolds numbers but at low Reynolds numbers that is a different matter.

Let me end up with my probably naive suggestion for an experiment which might possibly give some guide to the question of mixedness. Here[3] in Figs. 2a and 2b are two successive shadowgraphs of the development of a tight laminar spiral roll-up of a separated shear layer caused by a partial barrier in a shock tube. A shock tube has got, of course, a lot of advantages for studying chemical processes. From the time lapsed between the instant the shock passed the barrier and the first picture and from the number of visible striations the rotation rate exceeded a million rpm for this medium shock strength with mean flow Mach number of about 0.5 --a very fast roll-up. Now, the viscous spirals and the irrotational tongues are very tight and molecular diffusion should rapidly increase the degree of mixedness. However, at the higher Reynolds numbers (which can be manipulated in these shock tube experiments) a secondary shear-layer instability can develop with its own vortical mixing as in Fig. 2b,[4] which increases the local mixedness.

---

[1]   F.g. Figs. 4 and 7 of R. E. Falco's "Some Comment on Turbulent Boundary Layer Structure Inferred from the Movements of a Passive Contaminant," AIAA Paper 74-99, January 1974.

[2]   E.g. in the movie of Fiedler and Head also from Cambridge University in connection with their paper in J. Fluid Mech., 25, 719, 1966.

[3]   Only a blackboard sketch was shown during the Panel presentations. The photos were taken by R. E. Duff and R. N. Hollyer in the Univ. of Michigan Physics Department's shock tube.

[4]   See also analogous primary and secondary roll-ups behind a flat plate accelerated from rest of D. Pierce. J. Fluid Mech., 11, p. 460, 1961; also reproduced in Batchelor's Introduction to Fluid Mechanics, Cambridge University Press.

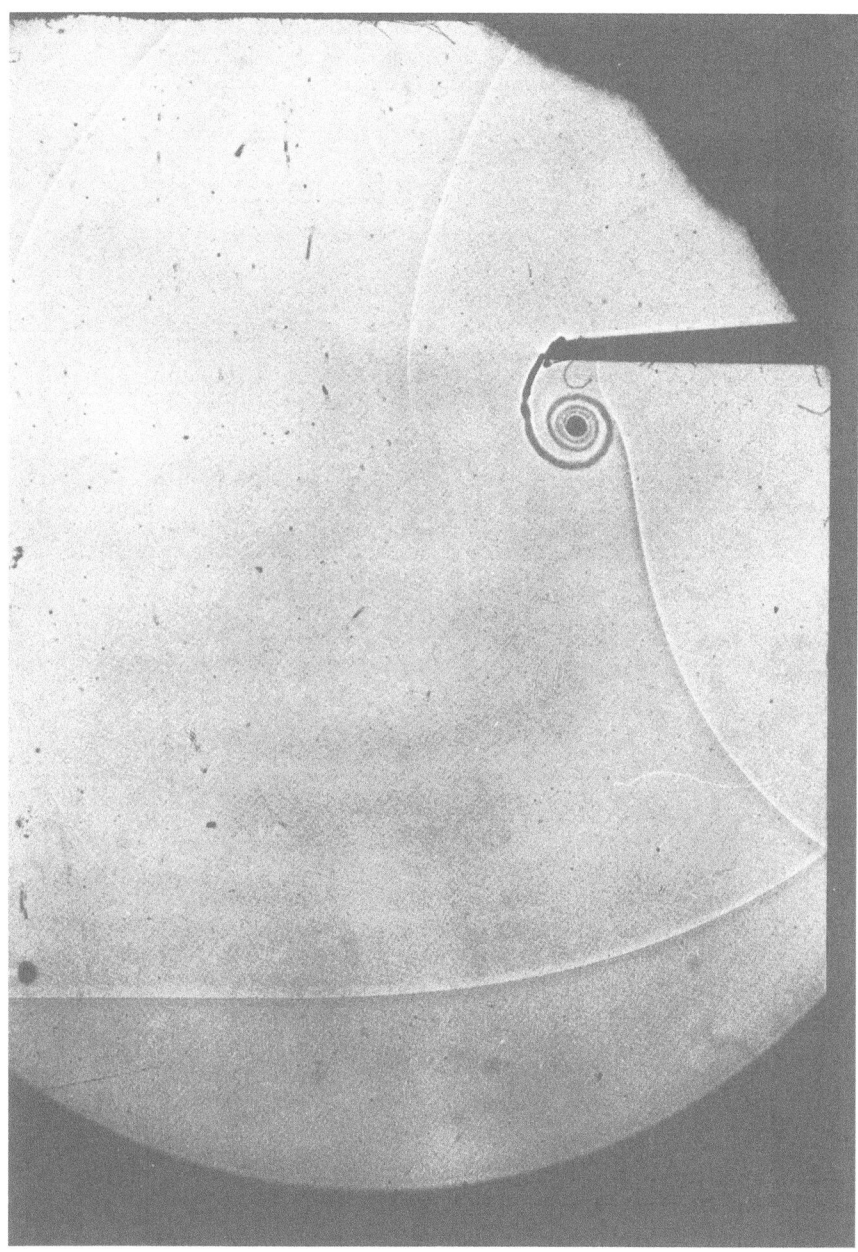

Figure 2a

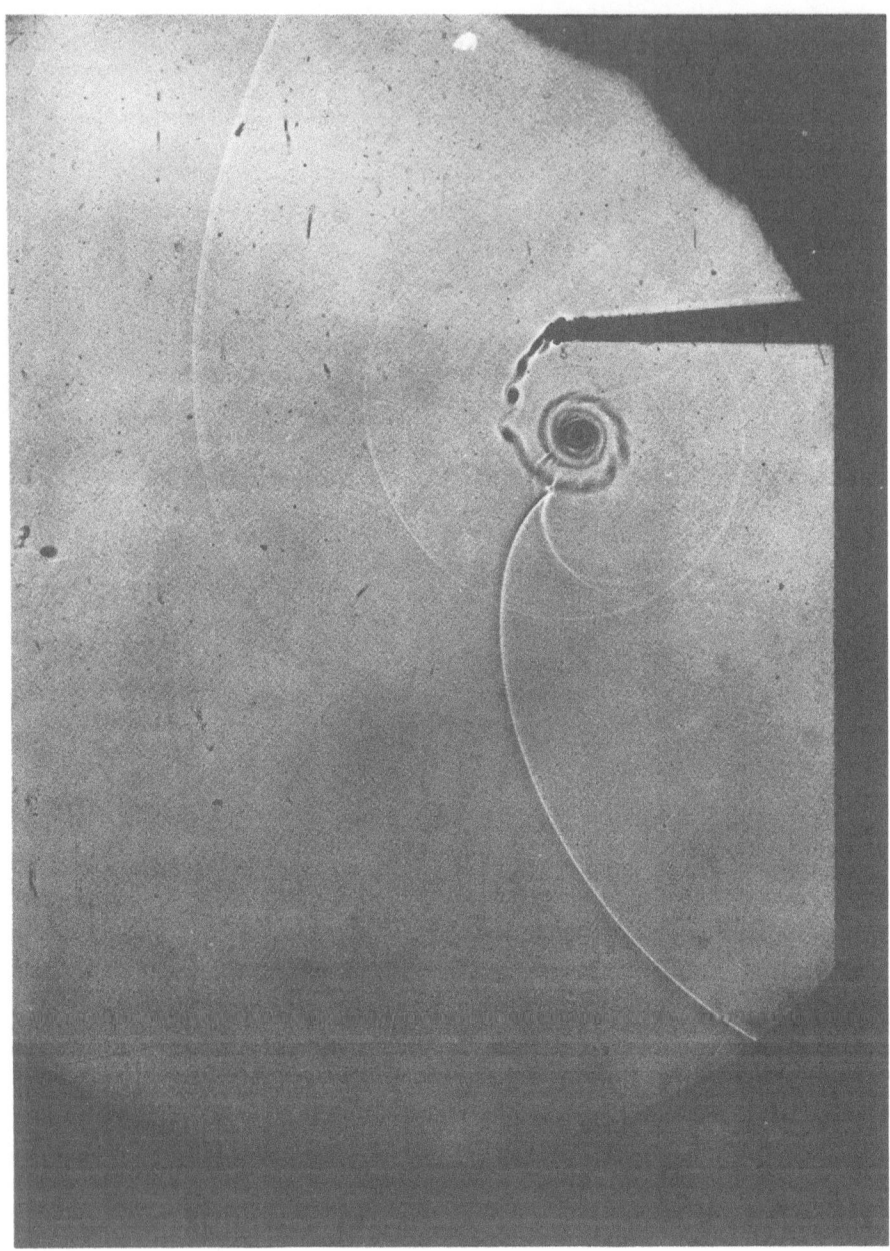

Figure 2b

With or without the secondary instability, I could arrange to have
only unmixed fuel to the left of the barrier by gently withdrawing
this partition from a fully-closed position to the desired partial
barrier position just before initiating the shock. I presume that
at some stage of mixedness we could ignite the diffusing mixture
locally with a laser. I think the experiment could potentially
tell us when there is sufficient mixedness in the spirals. It
offers high temporal and spatial resolution for better understanding of the degree of mixedness needed in reactive flows. Or am I
all mixed up?

M. SUMMERFIELD: (Princeton University)*

A laminar flame speed is an accepted idea, for example in
Professor Spalding's contributions on premixed flames. One of the
challenges in turbulent combustion is the ability to predict a
turbulent flame velocity. Perhaps this can be done by appealing
to the analogy between molecular diffusion and turbulent diffusion
and deducing the reaction rate by including the turbulent diffusion
into the little region where premixing arises.

WOLFSON: (A. F. Office of Scientific Research, Arlington, Virginia)

In modeling turbulent flames, it seems to me that there are two
very complex situations, one relating to turbulence and the other to
complex chemistry. I just feel that we are trying to lick two problems at the same time and really not being able to discriminate
which way we are going.

SUMMERFIELD:

Well, I would say that your misgivings are justified probably.
We do not know enough about the fluid mechanical situation and the
modeling is intended to make use of the little bit we do know in
order to interpolate or extrapolate. And then we do make use of
the injection rate mechanisms which were borrowed from quiescent,
molecularly mixed mixtures tested by some chemists. I suppose, in
defense of at least my point, you might say that the reaction that
finally does arise occurs in the microscale down at the molecular
impact level and if the gradients are not too great, in that region,
perhaps the reaction rate still follows the same laws. From then
on, if we do believe that latter point, the averaging is in the
hands of the kind of discussion that you heard here.

But your misgivings, I suppose, are well founded.

---

*   This is a summary of Professor Summerfield's remarks, which was
    prepared directly from the transcript of his recorded speech.

WOLFSON:

   May I amplify my feeling? As I have listened here these last
two days, there is a tremendous amount of attention being paid to
the fluid mechanical aspects of vorticity--turbulence iteration,
and very little being paid to the chemical reaction aspects of this
problem.

LIEPMANN:

   In order to get a chemical reaction at all, the species have
to be mixed on the molecular level. However complicated the resulting reaction is the molecules have to interact. We here have concentrated our attention mainly on the way this mixing on the
molecular level can be achieved in turbulent flow. At this stage
the reaction chemistry is of secondary importance. Of course, once
the mixing has been achieved, one deals with a chemically reacting
mixture and reaction kinetics becomes crucial.

WOLFSON:

   Even though they are mixed we still have to collide them in
order to get reactions.

LIEPMANN:

   That is exactly what we mean by being mixed on the molecular
scale.

STREHLOW: (University of Illinois, Urbana, Illinois)

   I would like to take issue with Professor Summerfield on one
or two accounts.

   First of all, your discussion of premixed laminar and premixed
turbulent flames implies, at least to me, that there is such a
thing as a turbulent burning velocity, and I am not sure that it
has been proved that there is such a thing in premixed systems.
You have to distinguish between low intensity turbulent flames and
high intensity turbulent flames, they have completely different
behaviors. In the low intensity turbulent flame, the type that is
on a burner, the flame is a continuous sheet that is held in a
laminar sublayer on the rim. The flame itself is blowing off
towards the tip at all times, as it propagates only perpendicular
to its surface and the vector velocity relations are such that an
element of the flame starts at the rim and ends up at the tip at
some later time. This flame is a transitional flame, it is responding to the turbulent fluctuation, and it is progressively getting
more wrinkled, as photographs have shown, for example those taken
by Wohl et al.,[1] Karlovitz[2] and by a number of other people.

Because of the transitional nature of this flame, the variety of velocities that are measured by an averaging technique, which has been used by a number of investigators, are apparatus-dependent. I do not think there is such a thing as a mean burning velocity. We performed some measurements, using an expanding spherical flame and we found that even at a diameter of 8 inches, which was when the flame burned through the core, in ordinary air, that the burning velocity as measured by the area versus volume ratio, was still increasing, even though the surface of the flame had a wrinkled behavior which was not changing. I do not know if there was a leveling-off point or not. We did not get a very high average burning velocity, maybe by factor of two compared to the laminar burning velocity.

One other point, in this particular flame sheet, even though we varied the scale, the average scale of the turbulence, by varying the grid size ahead of the flame, we found that the turbulent surface roughness of the flame had the same scale and this was essentially the normal cellular scale of that flame, the cellular stability scale. It was independent of the turbulent scale, and I think that the flame was responding slowly to the turbulent fluctuation, and responding in such a way that the cutoffs became cellular size bumps on the flame.

So, this is my main comment, that there is no such thing as a unique turbulent flame velocity.

SUMMERFIELD:

You do bring in some experiments that had to do with low levels of turbulence. And I think your remarks applied for those, for example, the expanding flame experiment, where turbulent flame speed was of the order of two times the laminar flame speed. Indeed that is probably a low level turbulence experiment. These diffusivities of ten are rather high levels for turbulent flows compared to, I think, the kind of experiment you are describing. The appearance of these natural cellular instabilities, for example, invariant to, perhaps, the initial perturbation of the flow would also indicate that we have a dealing in those experiments with low levels of turbulence. I would agree with you that this is not a uniform flame. I do not know what you mean by transitional flame. Of course, transitional with a lift is one thing and a core another. That may

---

[1] K. Wohl, L. Shore, H. von Rosenberg and C. W. Weil: "The Burning Velocity of Turbulent Flames," 4th International Symposium on Combustion, Williams and Wilkins (1953) pp. 620-635.

[2] B. Karlovitz. Open Turbulent Flames, ibid. pp. 60-67.

be what you meant by transitional, there is no uniform region, and
every criticism that you level against this kind of experiment is
valid, but I do not know any better measure than turbulent flame
speed.

There is such a thing as a turbulent combustion range, let us
put it that way. And I can report the numbers in terms of turbulent combustion rate.

STREHLOW:

What do you mean by turbulent combustion rate?

SUMMERFIELD:

Well, take the volume flow rate and divide it by the mean flame
area.

STREHLOW:

That is an apparatus-dependent number!

What I meant by transitional is as follows: let us consider a
laminar flame; let it be an oblique flame sheet, sitting on the
burner, and the flame is only propagating normal to itself, it is
only a wave. Therefore, as the flame propagates normal to itself,
the velocity vector crossing the normal to the flame translates at
some velocity along the flame. So the flame has to be seated at
the rim, it is constantly blowing off toward the tip, and it is
fighting through a turbulent region as it does this. And it starts
to become slightly wrinkled, more wrinkled in response. Response
is slow. I mean the flame has to travel many, many thicknesses,
just hundreds of thicknesses before it responds to these turbulent
fluctuations. So you have got nothing like an equilibrium turbulent
flame velocity. You have got an average over something that is very
hot at the tip, very small at the edge. If you have a bigger burner
you get a bigger number, for a smaller burner, you get a small number. There is nothing like an average for the burning velocity.
It is not a unique number, it is dependent on a piece of apparatus
that you are using to measure. That is my only point.

LIEPMANN:

I think the question does not seem to be settled at this point.

LIBBY: (University of California, La Jolla, California)

Since I gave a talk on experimental work this morning, I can
be excused for discussing some theoretical work this afternoon.

I do so because of the suggestion of Coles about the gap between the experimenter and the theoretician. This is a gap which, I guess, is widely recognized. I have tried to make a small step to close it in a manuscript being considered for publication in the Journal of Fluid Mechanics.[1] What I have tried to do is to develop a model equation for a scalar with a creation term. The scalar is either zero or one and the creation term is due to the crossing of the interforce at the point in question, so that creation turn is proportioned to the crossing frequency. What I have done is assumed that I know the unconditioned flow and then have calculated the two conditioned velocity components and the intermittency. I was bold enough to try to fit some data, and in some cases, it agreed pretty well, in other cases it did not agree too well. I believe this is the first calculation of this sort; naturally more work needs to be done but a modest step to close the gap discussed above has been made.

COLES:

Well, I am delighted to meet anybody who does not compute everything in terms of correlations.

I took an extreme position deliberately and willingly, and I want to say something in defense of that. Let me go back to the Stanford meeting. The beneficiaries of that meeting were supposed to be the people who do the world's work, not the people who addressed the meeting. And I thought enough data were available for this purpose. I spent several months of my life looking at other people's data; it can get to be quite a chore. I do not want to leave the impression that I think all the people who are doing the world's work should shut up shop and wait for the eddy chasers to provide some way of getting the job done.

SWITHENBANK: (Sheffield University, Sheffield, England)

I thought I might draw your attention to the utilization of the turbulence, because this is an important factor in practical combustion systems. I am a little depressed, at the end of this meeting, to find that the fundamental theories can not help us as yet with the analysis of reacting flow such as we need for designing combustion systems. The techniques which we are using at the present time, which are stirred reactor models, still appear to be about the only engineering solution. Consider the diagram in Figure 1. It shows that any practical combustion system consists of reacting material which is stirred together during which we have some combustion-ignition. At the end of the stirring process we must go into a plug-flow reaction section in order to get burn-

---

[1] "On the Prediction of Intermittent Turbulent Flows"

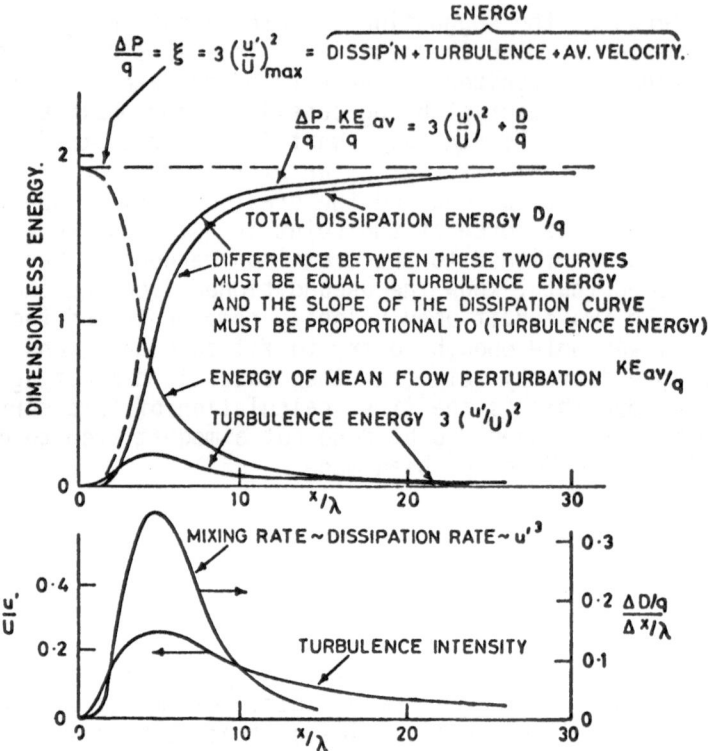

Figure 1.   Energy, Turbulence and Dissipation behind Baffle

out.   The model postulates a turbulent flow field, in which, when
all that turbulence is decayed away, leaves a residual unmixedness;
thus the system is not uniformly mixed and there is residual
unreactedness.   If this were a flowing system, the first stirred
section would be the partially stirred reactor that we use in the
model of the combustion systems.   Following the decay of the turbulence, the system changes to a plug flow reactor.   The next figure,
Figure 2, draws attention to this mixing limited combustion efficiency in terms of the parameters that we have been seeing already,
that is the correlation functions.   Here we have a binary mixture,
which, on the top diagram, is completely unmixed.   As we go down
the diagram, they become gradually more and more mixed.   That
happens in the stirred reactor.   When we get into the plug-flow
reactor, that is the reactor section in which no further mixing
takes place, the kinetics completes the combustion as far as permitted by the unmixedness.   We must, therefore, in our turbulence
studies take interest in the residual unmixedness, as illustrated
on the bottom of Figure 2.   I would suggest that using a technique
whereby we measure the correlation parameter, it will start at -1
as we go into our reactor, approach +1 as we go out of the stirred

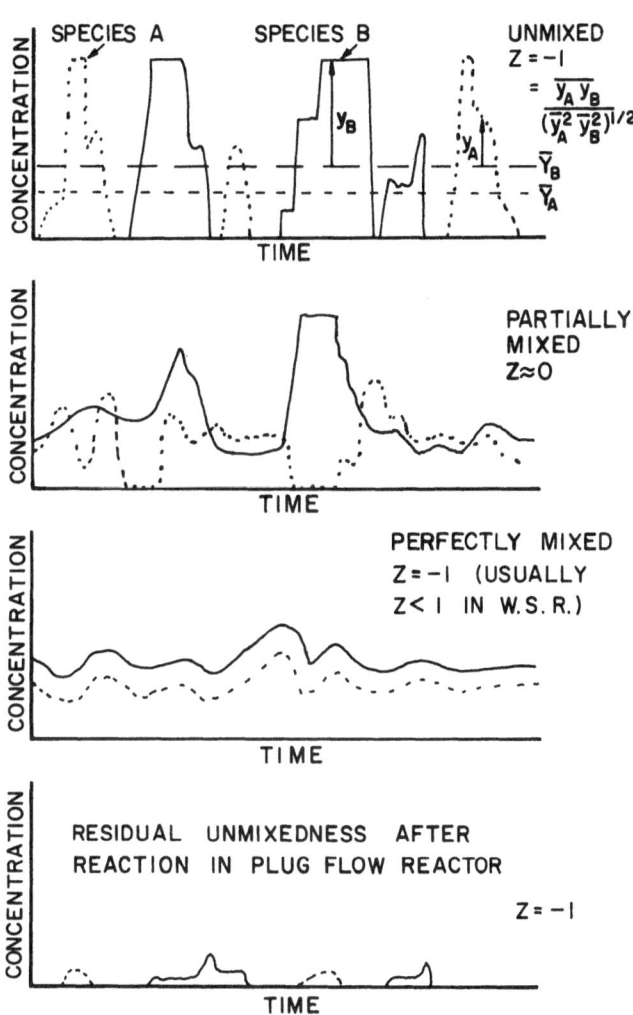

Figure 2.   Concentration and Mixedness Factor Z

section and will then revert to -1 as we complete the reaction. This parameter will therefore be very useful to us in order to follow the progress of reactions through a system. The final figure, Figure 3, demonstrates experimentally those points that I am making here from the turbulence point of view. This shows a gas analysis traverse across a 400 mega-watt boiler at the entrance to the economizer. The point I would draw to your attention is that we have a lot of residual unmixedness, since the reactant (oxygen) concentration values vary over a factor of six, i.e. from 0.5 percent to more than 3 percent. Were I to plot the carbon monoxide on that same diagram, it would move pretty well in opposition, showing us that we have, indeed, got a lot of unmixedness. I would appeal to fundamental turbulence workers to try to help we combustion engineers to sort out this serious problem of residual unmixedness. Just to give you an idea of how much this is worth; in terms of power station boilers, if we could overcome the problem that is shown here, it would save about a thousand million dollars per year in fuel cost on U. S. power station boilers alone. It follows that we can justify research in this area quite easily!

LIEPMANN:

   I think that is a very appropriate remark. I do hope the questions raised will lead to some stimulating research.

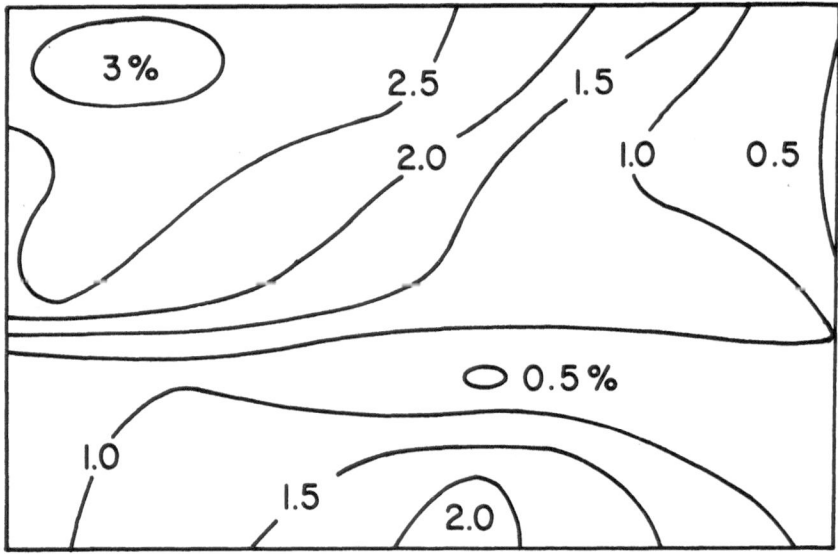

Figure 3.   Oxygen Concentration at Boiler Economizer Inlet Level –
            400 MW Load

BRADSHAW: (Imperial College, London, England)

I would like to make a comment on the eddy-chasing techniques. I think it is important to realize that we can never get any quantitative information out of eddy chasing for calculation methods because we simply cannot measure turbulence quantities with the accuracy that we want to predict. We are always going to have to fiddle around with empirical functions or empirical constants in calculation methods. Nevertheless, I am sure that eddy chasing is important. I think its main function is to give us qualitative ideas about turbulent processes, especially turbulent transport; how (qualitatively) should we treat these triple-products which have been agonizing us all? It would be nice if eddy chasing could tell us how to cope with the pressure-velocity gradient scrambling terms as well. But maybe one could get to chase pressure fluctuations as well as eddies.

As far as Mollo-Christensen's remark about 2.7985 wheels is concerned, I feel we are up in some danger of assuming that turbulence does come in two-wheel and four-wheeled varieties. Maybe the stuff does have 2.7985 wheels!

I will just defend myself against Mark Morkovin who accused me of being confused. The reason I was going on about the interface at some length, and in particular the question of how convoluted the interface is (see Bradshaw & Murlis, IC Aero Report 74-04) is because it seems to me that this is a very important question when one comes to deal with reactions, particularly the fast reactions, and I think we do not know, even to an order of magnitude, what is the effective ratio of the surface area of the interface to plan area below. Paizis' recent measurements (J. Fluid Mechanics) suggest that the ratio is 7. This probably ignores the small-scale indentations which contribute greatly. We need to look into this more carefully.

MORKOVIN:

I would like to ask the reaction people: could the extra indentedness of the interface--which is much less at the high Reynolds number than what you showed--could it truly make a difference with respect to the combustion problems as such?

LIEPMANN:

Let us make sure what you mean by "interface." You mean the edge of the boundary layer? If so, that is not a surface like in the mixing layer, because it is not a layer which is sharp and which separates species. You have to be careful about what you mean by an interface.

If you are talking about Corrsin's super-layer, I do not know if this plays such a definite role in the chemical processes compared to the interface Roshko refers to.

MORKOVIN:

Because, you mean, this is a free field ...

LIEPMANN:

Corrsin's superlayer can at best be considered an "interface" separating the outer flow from "particles" originating from the wall.

BRADSHAW:

It is like the separation, between, say, a boundary layer which has been filled with fuel and a free stream of oxidant.

LIEPMANN:

Right, and in this case it is properly an "interface."

BRADSHAW:

Yes. This is the point I had in mind. All my remarks in general, apply to the internal interface if that really does the job of bringing reactants into contact.

LIEPMANN:

If it is the internal interfaces, I completely agree with you, that the contorsions are essential.

BRODKEY:

In England, I saw some of the smoke experiments run at high Reynolds numbers at Cambridge by Robert Falco. Although the surface is not as convoluted as the free shear layer, there is a lot going on. On the surface, one can see vortex motions that are like a large number of secondary eddies, which might actually constitute far more contact with the irrotational flow than the basic surface would.

MORKOVIN:

That is what I was referring to when I was talking about Falco's work (see the references to my initial presentation).

LIEPMANN:

That is very important because it could be ascertained in any case.

BRODKEY:

Well, it is easier in the case of low Reynolds numbers.

LIEPMANN:

I am supposed to make a few concluding remarks:

The present workshop is a perfect model for its own subject of inquiry: The participants are species drawn from different fields such as chemistry, chemical engineering, fluid mechanics and physics; they are mixed here, and it is hoped that they will react with each other to produce something useful. To assess the success of this endeavor, we will have to wait for the results which should be noticeable in the work produced by the various research groups within a year or so, the usual relaxation time for such a process. In this model--as in the real problem of chemical reactions in turbulent flow--it is necessary to bring the species together to be able to react, i.e. within a few mean free paths or so, only then does the reaction kinetics take over and determines rates and relaxation times. (I will leave out for the time being the case of a premixed reacting gas.) Consequently, it makes sense to con- sider at first the process of mixing uncoupled from the chemical kinetics as a problem of turbulent diffusion.

My own impression of the state of the art in turbulent shear flows, developed over the last few years or so and very much rein- forced by the discussion and papers at the present workshop, is as follows:

The success in the description and analysis of homogeneous, isotropic turbulence resulted for quite some time in a tendency to extend these concepts and techniques to shear turbulence. Experi- ments on turbulent shear flows, in particular the Brown-Roshko experiments, have demonstrated conclusively, in my opinion, the crucial importance of an understanding of the motion and inter- action of the large scale coherent structures in the flow. These structures do not lend themselves naturally to a description in terms of local, Eulerian, power-spectra and space correlation functions familiar from homogeneous turbulence. These vortex structures are more naturally described in terms of a Lagrangian approach with its inherent difficulty both in experimentation and in theory.

This "new look" in turbulent shear flows is the first important ingredient in the present content of the workshop. The second is the advent of the large computer and its use for the prediction of turbulent shear flows. The third is the order of magnitude improvement in experimental technique in recent years.

Unfortunately, there has developed a divergence in direction between experimentation and computation, and I do hope that this workshop, where this divergence has been demonstrated once more, will help in reorienting the directions of both theory and experiment. In my opinion, the former needs it more, but I am clearly prejudiced: the experimenters have adopted the "new look" without delay, and a set of what Prof. Coles called "eddy chasers" has developed. Techniques, in particular, conditional sampling, have been used extensively to approximate to Lagrangian measuring techniques. The study of interfaces, i.e., vortex sheets, separating species in turbulent mixing has come to the forefront. Work on Reynolds stresses and turbulent energy balance has become less important. In the computing schemes, almost without exception, no provision for the large scale coherent motion and for the interface topography exist. To be more precise one should distinguish between different kinds of model equations: Reynolds type equations, i.e., equations obtained from the Navier-Stokes equation by forming averages over velocity or similar moments; model equations, e.g., Saffman's equations in which an attempt is made to model the physically significant interactions directly. Both of these approaches are familiar from kinetic theory and correspond roughly to the use of moment equations or the BGK method ("crooked Boltzmann equation") respectively. In the former the hierarchy of Reynolds equations is broken off using empirical assumptions more or less based on physical intuition. In my opinion, the computing schemes are very useful indeed for interpolation between experimentally explored cases. Extrapolating to new flows is risky. I completely agree with Profs. Spalding and Swithenbank in stressing the need in our technology for the computational methods. In industrial application one does encounter flows for which one desperately needs some information and some results. The attempts to supply this information by various computing schemes is highly recommendable as long as it is kept in mind that success in predicting a specific flow configuration does not imply necessarily the correctness of the assumed model and closure scheme. As long as the physical basis of the computational models differs as much as it does today from experimental facts, we must be prepared for surprises.

I believe in a year or two, another round of the computational olympics would be quite in order and I propose the inclusion of one completely wrong set of experimental data to make certain that the codes are capable of discovering them.

KLINE:

I have already tried that.

LIEPMANN:

I do hope that the increasing emphasis on model computation does not lead us to forget the primary reason for experimentation: the discovery of new phenomena. This is certainly much needed in turbulent shear flows and, in particular, in flows with chemical reaction today. The interrelation between turbulent spots in transition and large eddies, stressed by Prof. Coles, is a case in point. The two-dimensional nature of the large eddies in the mixing zone contrasted with the three-dimensional structures in the boundary layer is another. There is much room and much need for analytical experimentation on basic flows.

In addition, I completely underwrite Prof. Bradshaw's remark about the need of work on very complex but technically important flows, e.g. swirl flows. Experiments here probably require less sophistication in technique, but a great deal of imagination coupled with a thorough understanding of the particular technical applications. Close coupling with industry here is indispensible; the research is very challenging and unfortunately often underrated.

I do not believe that experimental research in turbulence in the U. S. is in particularly good shape today. In the past decades a trend toward theory and computation developed in the student population. The need for more sophisticated instrumentation, longer thesis times, etc. reinforces the trend and finally the decreasing financial support in conjunction with steeply rising expenses for facilities and electronic devices puts an added incentive to choose the "cheaper" and "safer" computational research. In spite of the overall shortages of academic positions, it is today quite difficult to fill a position with a really first-rate, highly trained experimental fluid dynamicist. This direction is not only deplorable but in the long range disastrous since academia has a built-in feedback loop: a purely theoretically inclined faculty is unlikely to bring up first rate experimenters. There is little danger today for the opposite! In view of the crucial importance of the turbulence problem to modern technology, the situation is very bad indeed.

I believe it is now evident that the primary quantity to determine chemical reaction is the mixing on the molecular level, the temperature fluctuations are the <u>result</u> rather than the <u>cause</u> of the reactions. Consequently, techniques to determine the degree of molecular "mixedness" have to be developed. The only general technique seems to me Raman Spectroscopy which in principle is capable of detecting any molecular species. Even with the advent of the high intensity U. V. Lasers, the technique is difficult to apply because of the small scattering crossections and the correspondingly small signal to noise ratio. Measuring methods for specific reactions exist of course, e.g. UV absorption of O_3 is being used to trace the O_3-NO reaction. I.R. spectroscopy is applicable to a whole set of reactions, etc.

There is room here for a continuous development and improvement of optical techniques.

Hot wire anemometry still has some way to go. The importance of large amplitude fluctuations coupled with bimodal distribution functions is now well established and the processing of hot wire signals has to be adapted to these conditions. It is not any more possible to work within the linear approximations to the hot wire response.

I find it difficult to add anything intelligent concerning the chemical aspects of the turbulent reactive flows. Evidently one can distinguish a few limiting cases. Thus reacting flows with very large rates, very fast reactions, are of course in the limit determined by the turbulent mixing alone. The opposite case of very slow reaction becomes in the limit a purely chemical problem. In addition, it is important to note that the turbulent mixing and the chemical reactions are thermally uncoupled only if the heat release in the reaction is small compared to the energy content of the flowing gas. For large heat releases the flow itself, i.e. the structure of the turbulence must be altered and the clean separation between fluid mechanics and chemistry ceases to be valid. I do not know of an observation where this has been clearly demonstrated. It is bound to happen. Finally another different regime is reached when the light emission energy from the reaction becomes significant. In this case the photon gas has to be considered as one species of the reacting gases and the photon mean free path becomes an important length scale. An important and fashionable special case here is evidently the chemical laser.

The regime of strong interaction between turbulence and chemical reaction strikes me as particularly intriguing for future work. It may even be possible to model the effect of heat release on turbulence by a programmed focussed laser spot to separate again chemistry and fluid mechanics.

Finally, it is my hope that since we are now mixed the reaction will continue and that we have developed a respect for each other and an understanding of each other's problems.

LIST OF ATTENDEES

1. ABBOTT, D. E.
    Purdue University
    West Lafayette, Indiana 47907

2. ALBER, Irwin E.
    TRW Systems
    One Space Park, R1/1008
    Redondo Beach, California 90278

3. BALWANZ, W. W.
    Naval Research Lab, Code 4004B
    Washington, D. C. 20375

4. BENHAM, Charles
    Naval Weapons Center, Code 45701
    China Lake, California 93555

5. BEVILAQUA, Paul M.
    Aerospace Research Labs, ARL/LE, WPAFB
    Dayton, Ohio 45433

6. BORGHI, Roland
    ONERA
    29 Avenue de la Division Leclerc
    Chatillon Jous Bagnenx, France 92290

7. BRADSHAW, P.
    Aero. Department
    Imperial College
    London SW7 2B7 England

8. BRANDT, Alan
    Applied Physics Lab
    Johns Hopkins University
    8621 Georgia Avenue
    Silver Spring, Maryland 20910

9. BRAY, K. N. C.
    Southampton University
    2 Holly Hill
    Southampton, England

10. BROADWELL, James E.
    TRW Systems, Room 1044/R1
    One Space Park
    Redondo Beach, California 90278

11. BRODKEY, Robert S.
    Ohio State University
    140 West 19th Avenue
    Columbus, Ohio 43210

12. BRONFIN, Barry R.
    Research Laboratories
    United Aircraft Corporation
    East Hartford, Connecticut 06108

13. BROWAND, F. K.
    Department of Aerospace Engineering
    University of Southern California
    University Park, California 90007

14. BURGGRAF, Odus R.
    Department of Aeronautical Engineering
    Ohio State University
    2036 Neil Avenue
    Columbus, Ohio 43210

15. BUSHNELL, Dennis M.
    Fluid Mechanics Branch
    NASA-LaRC, M/S 163
    Langley AFB, Virginia 23369

16. BYWATER, Ronald J.
    MTS, Aerospace Corporation
    P. O. Box 92957
    Los Angeles, California 90009

17. CERVANTES, Jaime G.
    Ray W. Herrick Labs
    Purdue University
    West Lafayette, Indiana 47907

18. CHAMBERS, Frank W.
    Ray W. Herrick Labs
    Purdue University
    West Lafayette, Indiana 47907

19. CHEVRAY, Rene
    Department of Mechanics
    State University of New York
    Stony Brook, New York 11794

20. COLES, Donald
    California Institute of Technology
    Pasadena, California 91109

21. CRAIG, Roger R.
    Air Force Aerospace Propulsion Lab
    Wright Patterson Air Force Base, Ohio 45433

22. DONALDSON, Coleman
    Aeronautical Research Associates of Princeton
    Box 2229
    Princeton, New Jersey 08540

23. DOPAZO, Cesar
    Department of Mechanics
    State University of New York
    Stony Brook, New York 11790

24. DREWRY, James E.
    Aerospace Research Lab
    United States Air Force, ARL/LF
    Wright-Patterson Air Force Base, Ohio 45440

25. EDELMAN, Raymond
    R & D Associates
    P. O. Box 3580
    Santa Monica, California 90403

26. EGGERS, James M.
    Langley Research Center
    NASA
    Hampton, Virginia 23665

27. ERIAN, Fadel F.
    Department of Mechanical and Materials Science
    The Johns Hopkins University
    Baltimore, Maryland 21218

28. FEJER, Andrew A.
    Technology Center
    Illinois Institute of Technology
    Chicago, Illinois 60616

29. FENDELL, Francis E.
    TRW Systems, Building R1, Room 1016e
    One Space Park
    Redondo Beach, California 90278

30. FIEDLER, H.
    Technical University of Berlin
    Berlin - 12
    Strasse Des 17. Juni, 134/West Germany

31. FOSS, John F.
    Department of Mechanical Engineering
    Michigan State University
    East Lansing, Michigan 48824

32. GERSTEIN, Melvin
    University of Southern California
    Los Angeles, California 90007

33. GOLDSCHMIDT, Victor
    Purdue University
    West Lafayette, Indiana 47907

34. GOULARD, R.
    TSPC - Chaffee Hall
    Purdue University
    West Lafayette, Indiana 47907

35. GRANT, Alan J.
    Shell Research Ltd.
    P. O. Box 1
    Chester, England

36. HARSHA, Philip T.
    Engine Test Facility
    ARO, Inc.
    Arnold Air Force Station, Tennessee 37307

37. HILL, James C.
    Chemical Engineering
    Iowa State University
    261 Sweeney Hall
    Ames, Iowa 50010

38. HUDSON, Dale A.
    United States Air Force, Aeropropulsion Lab
    Wright-Patterson Air Force Base, Ohio 45433

39. JONES, Barclay G.
     223 Nuclear Engineering Lab
     University of Illinois
     Urbana, Illinois 61801

40. KAPLAN, R. E.
     Department of Aerospace Engineering
     University of Southern California
     Los Angeles, California 90007

41. KENTZER, C. P.
     Aeronautical and Astronautical Engineering
     Grissom Hall
     Purdue University
     West Lafayette, Indiana 47907

42. KLINE, S. J.
     Stanford University
     Palo Alto, California 94305

43. KOVASZNAY, Leslie S. G.
     Johns Hopkins University
     Baltimore, Maryland 21218

44. L'ECUYER, Mel
     Department of Mechanical Engineering
     Purdue University
     West Lafayette, Indiana 47907

45. LEE, Jon
     ARL
     Wright-Patterson Air Force Base, Ohio 45433

46. LIBBY, Paul A.
     University of California
     La Jolla, California 92037

47. LIEPMANN, Hans W.
     California Institute of Technology
     1201 California Street
     Pasadena, California 91109

48. LUCKEY, David W.
     Graduate House West, Room 1132
     Purdue University
     West Lafayette, Indiana 47907

49. LYON, Craig A.
     Aero Propulsion Laboratory/TBC
     Wright-Patterson Air Force Base, Ohio 45433

50. MIKATARIAN, Ronald R.
     Lockheed Research and Engineering Center
     P. O. Box 1103, West Station
     Huntsville, Alabama 35807

51. MILLSAPS, Knox
     University of Florida
     Box 13857
     Gainesville, Florida 32604

52. MJOLSNESS, R. C.
     Los Alamos Scientific Lab
     P. O. Box 1663
     Los Alamos, New Mexico 87544

53. MORGENTHALER, John
     Bell Aerospace Company
     Box 1
     Buffalo, New York 14240

54. MORKOVIN, Mark V.
     Illinois Institute of Technology
     Chicago, Illinois 60616

55. MURTHY, S. N. B.
     Purdue University
     West Lafayette, Indiana 47907

56. NAGIB, Hassan M.
     MMAE Department
     Illinois Institute of Technology
     Chicago, Illinois 60616

57. NETZER, David W.
     Department of Aeronautics
     Naval Postgraduate School, Code 57NT
     Monterey, California 93940

58. NEWMAN, Barry G.
     McGill University
     P. O. Box 6070
     Montreal, P. Q., Canada

59. NOVICK, Allen S.
     Detroit Diesel Allison
     Tibbs Avenue
     Indianapolis, Indiana

60. O'BRIEN, Edward E.
     Department of Mechanics
     State University of New York
     Stony Brook, New York 11794

61. OH, Youn H.
     Langley Research Center
     NASA, LaRC, MS-164
     Hampton, Virginia 23665

62. PATTON, James R., Jr.
     Power Program, United States Navy
     Office of Naval Research
     800 North Quincy Street
     Arlington, Virginia 22217

63. RHODES, Robert P.
     ARO, Inc.
     Arnold Air Force Station, Tennessee 37389

64. ROBERTS, Ralph
     Power Program
     Office of Naval Research
     800 North Quincy
     Arlington, Virginia 22217

65. ROGERS, Milton
     Air Force Office of Scientific Research
     1400 Wilson Boulevard
     Arlington, Virginia 22209

66. ROSE, William C.
     NASA - Ames
     19782 Braemar Drive
     Saratoga, California 95070

67. ROSHKO, Anatol
     California Institute of Technology
     1201 East California Boulevard
     Pasadena, California 91109

68. SAMUELSEN, G. S.
     School of Engineering
     University of California
     Irvine, California

69. SHAMROTH, Stephen
     Gas Dynamics
     United Aircraft Research Laboratories
     East Hartford, Connecticut

70. SKIFSTAD, James G.
     Mechanical Engineering
     Purdue University
     West Lafayette, Indiana 47907

71. SMITH, C. R.
     Department of Mechanical Engineering
     Purdue University
     West Lafayette, Indiana 47907

72. SPALDING, D. B.
     Imperial College
     London, England

73. STREHLOW, Roger
     University of Illinois
     105 Transportation Building
     Urbana, Illinois 61801

74. SUMMERFIELD, Martin
     AMS Department
     Princeton University
     Princeton, New Jersey 08540

75. SWITHENBANK, J.
     Department of Chemical Engineering and Fuel Technology
     Sheffield University
     Sheffield, England S1 3JD

76. THOMPSON, H. Doyle
     Mechanical Engineering
     Purdue University
     West Lafayette, Indiana 47907

77. TORDA, T. P.
     Illinois Institute of Technology
     3110 South State Street
     Chicago, Illinois 60616

78. WALKER, J. D. A.
     Mechanical Engineering
     Purdue University
     West Lafayette, Indiana 47907

79. WALSH, Dennis E.
     Aero Propulsion Lab, AFAPL/TBC
     Wright-Patterson Air Force Base
     Dayton, Ohio 45433

80. WILLIAMS, F. A.
     University of California, San Diego
     La Jolla, California 92037

81. WINSTANLEY, David K.
     Purdue University
     West Lafayette, Indiana 47907

82. WOLFSON, Bernard T.
     Air Force Office of Scientific Research
     1400 Wilson Boulevard
     Arlington, Virginia 22209

83. WOOD, Albert D.
     Office of Naval Research, Boston Branch
     495 Summer Street
     Boston, Massachusetts 02210

SUBJECT INDEX

boundary layer
   large scale motion in, 267-
      269, 277, 288, 293, 432
   synthetic, 285-292, 302
bursting and eddies, 288, 451

closure schemes, 40-44, 131,
      145-146, 172-173, 217, 327
coherent structure, 9, 14-20,
      263-266, 297, 302, 315-316,
      317, 390-391, 395-396
   and classical eddy, 272
   axisymmetric configuration, 19
   boundary layers, 265, 267-269,
      277-280, 292-293, 304
   Langrangian approach, 436, 449
   pipe flow, 269-270
   three-dimensionality, 16, 265,
      299, 311
combustion
   laminar, 68-69
   turbulent, 70, 94, 180, 203,
      425-427, 443, 446
'complex' turbulent flows,
      243-244
compressibility effects, 29-32,
      64-67, 156, 185-186, 257-
      258, 295, 327-329, 436
computational methods, 92-99,
      129, 146, 169, 174, 228,
      254, 447, 450
conditioned data generation, 4,
      11, 20-22, 75, 221, 245,
      247, 277, 317, 338, 348,
      362, 369, 381, 412-414
curvature parameter, 251

eddy break-up model, 71
entrainment, 5-6, 274, 304,
      323-326
extinction of flame, 196-200,
      202-203

flame models, 70, 108-109, 157-
      158, 161, 440-442
   diffusion, 50, 70-75, 100,
      189, 194-196, 200-201, 214
   pre-mixed, 69, 72, 101, 223

independence principle, 54
interacting shear layers,
      244-248
interface, 21, 22, 37, 61, 74,
      160, 190, 247, 313-314,
      324, 337, 338, 348-352,
      360, 389, 408, 436, 447-
      448, 450
   foldedness, 338, 350, 366
intermittency, 9, 10, 20, 60,
      74, 246, 352, 412, 428,
      434

jet mixing, 94, 96, 101, 164,
      174-182, 231, 266-267, 391,
      392, 404, 411-415

large Reynolds number flows, 96,
      107, 311, 321, 436
length scale transport equation,
      39, 91, 253, 258, 259, 260
low Reynolds number flows, 99,
      106, 115, 127, 302, 434